Energy 4.0

The text provides insight into renewable technologies and their challenges in terms of design, efficiency, implementation, and solutions to mitigate the problem of energy crisis. It will help readers understand the role of the Fourth Industrial Revolution technologies in developing user-friendly, economic, and implementable solutions to mitigate the present and future energy crisis through cleaner and renewable technologies. This Book:

- Discusses electric energy transportation systems, electricity distribution systems, demand response, metering, smart grids, and cyber-security.
- Covers electric vehicles, vehicle-to-grid technologies, charging-while-driving technology for future roads and highways, and autonomous mobile robots.
- Presents data-driven modelling, forecasting, and optimisation techniques of power, and energy systems.
- Illustrates energy storage technologies, energy-efficient systems, power conversion topologies, and related control techniques.
- Explains the design of control algorithms for solar systems and the design of efficient energy management for solar power.

The Book is primarily written for senior undergraduates, graduate students, and academic researchers in the fields of electrical engineering, electronics and communications engineering, energy, environmental engineering, and computer science and engineering.

Energy 4.0
Trends, Challenges, and Applications

Edited by
Gulshan Sharma, Pitshou N. Bokoro,
and Sudeep Tanwar

CRC Press
Taylor & Francis Group
Boca Raton London New York

CRC Press is an imprint of the
Taylor & Francis Group, an **informa** business

Designed cover image: Shutterstock

First edition published 2025
by CRC Press
2385 NW Executive Center Drive, Suite 320, Boca Raton FL 33431

and by CRC Press
4 Park Square, Milton Park, Abingdon, Oxon, OX14 4RN

CRC Press is an imprint of Taylor & Francis Group, LLC

ISBN: 978-1-032-74451-3 (hbk)
ISBN: 978-1-032-99540-3 (pbk)
ISBN: 978-1-003-53782-3 (ebk)

DOI: 10.1201/9781003537823

Typeset in Times
by KnowledgeWorks Global Ltd.

Contents

SECTION V Conclusions

General Introduction

BACKGROUND AND MOTIVATION

Throughout human history, the pursuit of more efficient energy has been central to every stage of our evolution. Economic growth and industrial development, whether in developing or developed nations, are unsustainable without sufficient energy capacity. The uneven distribution of energy resources and disparities in energy conversion technologies have created significant opportunities for global trade, generating substantial revenue. However, sustaining this global trade of energy requires efficient resource management, investment in reliable infrastructure, and the adoption of modern, efficient technologies.

As the challenges of global warming and climate change intensify, the energy value chain is becoming increasingly complex and technology-driven. This evolution sets the stage for the conceptualisation of energy within the framework of the Fourth Industrial Revolution (4IR). The concept of energy 4.0 emerges from the integration of intelligent systems for sustainable resource management with data-driven algorithms that power 4IR technologies, including artificial intelligence, machine learning, blockchain, big data, deep learning, internet of things, digital twin technology, virtual reality, and pattern recognition. These technologies enable effective decision-making in energy conversion, transactions, and utilisation.

Implementing energy 4.0 is essential for accelerating industrialisation and economic growth in today's technology-driven world. To achieve this, it is crucial to broaden access to knowledge and promote skill development to enhance human capacity. The topics covered in this book address a wide range of issues from the fundamental question of alternative sources of energy and grid integration, converter topologies, and the performance of electric vehicle charging systems to distributed energy systems and sustainable infrastructure.

This book offers readers the opportunity to engage with cutting-edge technologies in the energy sector. It consists of a collection of 14 chapters, grouped into five key themes:

- Alternative Energy Sources and Grid Integration
- Converters and Electric Vehicle Charging Systems
- Intelligent Technologies in Energy Systems
- Distribution Systems and Sustainable Energy Infrastructure
- Conclusions

This structure provides a comprehensive overview of the current state and future directions of energy technologies in the context of 4IR.

SECTION I: ALTERNATIVE ENERGY SOURCES AND GRID INTEGRATION

This section deals with alternative sources of energy and issues related to grid integration of these renewable energy systems. The section consists of three fundamental chapters: Chapter 1 explores the potential of harnessing wind generated by vehicles on highways using vertical-axis wind turbines. This approach presents a viable and cost-effective solution for alleviating traffic congestion, while also integrating solar systems to produce electrical energy. This stored energy can be used for applications such as toll gates and streetlights. Chapter 2 introduces a novel high-step DC to AC converter with controlled functionality pulse modulation technology. The converter delivers three-phase AC power output, which is connected to the grid through an LC filter. A low-voltage renewable power source, such as batteries or photovoltaic arrays, is linked to the converter. The proposed converter increases the low input voltage to high voltage by utilising an inductor integrated with a resonant switched capacitor. Chapter 3 investigates the development of an optimal cooperative transmission strategy in a basic setup involving two tags and a reader. The system is modelled within a stochastic framework, and the optimisation problem is structured as a Markov Decision Process. This strategy considers the current system state and available energy resources to devise an optimal transmission policy aimed at maximising long-term average throughput. Numerical simulations are conducted to assess the performance of the proposed cooperative strategy in comparison to a noncooperative strategy.

SECTION II: CONVERTERS AND ELECTRIC VEHICLE CHARGING SYSTEMS

This section includes topics related to the analysis and modelling of static converters and electric vehicle charging stations. Three chapters are dedicated to this section: chapters 4, 5, and 6. Chapter 4 aims to develop a solar photovoltaic (PV) converter specifically designed for electric vehicles. This involves creating a power electronic converter that can be integrated into any charging station. The focus is on the batteries of plug-in electric vehicles, a type of plug-in hybrid vehicle, exclusively using a solar PV system. The system is connected through a DC-to-DC converter equipped with a maximum power point tracking controller to optimise energy transfer. In chapter 5, an in-depth analysis is presented on the various types of electric vehicles planned for different locations. It also delves into the key subsystems involved, such as energy source availability (both single and multi-source), the use of converters, the necessity of battery storage, and the management of energy at the charging stations. Chapter 6 presents an investigation into the stability of a DC microgrid, focussing on the voltage control of the DC bus within a self-sufficient power production and consumption system. The DC microgrid comprises a solar PV system, a wind turbine, and an energy storage device based on lithium batteries. The energy storage system remains continuously connected to the DC bus to mitigate fluctuations in DC voltage and ensure stable operation.

SECTION III: INTELLIGENT TECHNOLOGIES IN ENERGY SYSTEMS

In **this section**, the concepts and novel applications of intelligent technologies in energy systems are explored. Chapter 7 investigates the applications of the internet of things (IoT) for enhancing the management of charging station networks. The implementation of IoT in these networks involves the use of various IoT sensors and devices. A solid understanding of IoT communication protocols and data analytics tools is essential for successful integration, and an overview of the relevant tools and protocols is provided. This chapter concludes with an analysis of a satisfaction survey conducted among users of IoT-enhanced charging stations, offering insights into their experiences and feedback. In chapter 8, the real-time Large-scale Energy Anomaly Detection data is employed for forecasting and optimising power and energy systems. The energy forecasting includes analysing daily consumption patterns across different areas, energy usage under varying weather conditions, peak demand periods during festivals, and instances where power consumption exceeds the specified capacity during different seasons. In chapter 9, a comprehensive analysis of the implementation of cyber-physical systems (CPS) in the power and energy industry is presented, with a focus on key issues and the role of digital twins. This chapter also includes applications and case studies that underscore the critical importance of CPS in enhancing system performance. Chapter 10 concludes this section by introducing an innovative solution tailored for smart cities: a secure electricity distribution method that achieves phase modulation for signal encryption. The primary objective is to develop a theft-proof system that enhances sustainability and efficient supply of electricity in urban environments. Additionally, this work integrates blockchain technology into the proposed approach, providing transparent and tamper-proof certification of renewable energy sources.

SECTION IV: DISTRIBUTION SYSTEMS AND SUSTAINABLE ENERGY INFRASTRUCTURE

In this **section**, system reliability analysis and power trade models such as applicable to electricity distribution systems are explored. Three chapters are presented in this section. In chapter 11, the primary focus is on managing the increasing load, particularly across various consumer categories. As technology advances, the integration of renewable energy sources into existing power systems is accelerating using Distribution STATCOMs (DSs). At the same time, it is crucial to maintain grid operation within permissible limits, ensuring that prescribed loading conditions are not exceeded. Chapter 12 delves into the mechanisms that have achieved significant progress in addressing the supply–demand imbalance within the power trading industry. The challenges arising from these solutions are critically reviewed. This chapter also highlights how distributed generation and power sharing between systems offer the potential to reduce carbon emissions by promoting the use of renewable clean energy, while also alleviating the electricity supply–demand mismatch problem. Chapter 13 presents simulations based on the mathematical modelling of a three-phase induction motor (IM) under various fault conditions. IMs are essential in modern manufacturing industries, with 90% of machines utilising them as a prime mover. These motors face various stresses during their operation, which can lead to faults and unexpected breakdowns. The mathematical model developed for IMs is designed to predict the motor's performance characteristics under different operating conditions.

SECTION V: CONCLUSIONS

This book concludes in **section V** wherein a comprehensive summary of the contributions of this book is discussed in chapter 14.

About the Editors

Gulshan Sharma, Ph.D., earned his Bachelor of Technology (B. Tech) degree in Electrical Engineering from Punjab Technical University in 2007, Master of Technology (M. Tech) in Electrical Power System Management from Jamia Millia Islamia University, New Delhi, in 2009, and Doctor of Philosophy (Ph.D.) in Electrical Engineering from Malaviya National Institute of Technology, Jaipur, India, in 2015. He was a post-doctoral research fellow at the University of Pretoria, South Africa from 2015 to 2016. He has more than 17 years of teaching and research experience at various institutes at the territory level. Presently, he is working as an Associate Professor in the Department of Electrical Engineering Technology, University of Johannesburg, South Africa. Earlier, he was with Durban University of Technology, South Africa, as a Senior Lecturer in the Department of Electrical Power Engineering and he acts as Departmental Research Committee Chair and Faculty Research Committee Representative in the Department of Electrical Power Engineering, Durban University of Technology. He is registered with the Engineering Council of South Africa. He received Y-rated research award from NRF South Africa in 2021 and the best researcher award from the Department of Electrical Power Engineering, Durban University of Technology in 2021. He is the Associate Editor of the *International Journal of Modelling and Simulation* (Taylor & Francis). Additionally, he serves as an Academic Editor for the *International Transactions on Electrical Energy Systems* (Wiley) and the *Journal of Electrical and Computer Engineering* (Wiley). He has received research funding from various research agencies and has published several book chapters & research papers in international journals and conferences and has been continuously engaged in guiding research activities at graduate/post-graduate and Ph.D. levels. His areas of interests include power system operation and control, renewable power generation, FACTS, and the application of AI techniques to power systems.

Pitshou N. Bokoro, Ph.D., is an accomplished academic and researcher with a distinguished career in the field of Electrical Engineering. He completed his Bachelor's degree in Electrical Engineering from the Durban University of Technology in 2006, followed by a Master's degree in Electrical Engineering from the University of Johannesburg in 2011, and a Ph.D. from the University of the Witwatersrand, South Africa, in 2016. He joined the University of Johannesburg in 2007 as a Lecturer in the Department of Electrical Engineering Technology. He has since advanced to the position of Senior Lecturer, Head of Department, Associate Professor, and currently Head of School. He is professionally registered with the Engineering Council of South Africa and is a senior member of both the South African Institute of Electrical Engineers (SAIEE) and the Institute of Electrical and Electronics Engineers (IEEE). In addition, he serves as a specialist editor for the SAIEE *Africa Research Journal.* As a prolific researcher, he has authored and co-authored over 72 journal articles, including publications in high-impact, quartile-1 journals. He has presented 74 papers at both local and internationally recognised conferences in the field of engineering. His scholarly contributions also extend to authoring two books, and eight book chapters (with three currently in print) published by globally reputed publishers such as Springer Nature, IGI Global, and IET. He has successfully conducted numerous research projects, securing over US$42,611.72 in grants. Notably, he is a recipient of the prestigious Erasmus+ funding in collaboration with the University of Klagenfurt, Austria and the University of Perugia, Italy. He has mentored a significant number of postgraduate students, having supervised to completion of 20 master's students (with 8 currently under supervision) and 6 Ph.D. candidates (with another 7 currently under his guidance). His research interests encompass a wide range of topics, including power quality, distributed generation, virtual sensors, condition monitoring of dielectrics and insulation materials, the reliability of surge protective devices, and the economics of energy.

Sudeep Tanwar, Ph.D., is a full professor at the Nirma University, India. He is also a Visiting Professor with WSG University, Bydgoszcz, Poland; Jan Wyzykowski University, Poland; and the University of Pitesti, Romania. He earned his B.Tech in 2002 from Kurukshetra University, India; M. Tech (Honors) in 2009 from Guru Gobind Singh Indraprastha University, Delhi, India; and Ph.D. in 2016 with a specialisation in Wireless Sensor Network. He has authored 5 books, edited 25 books, and published more than 350 technical articles in top-cited journals and conferences, including *IEEE TNSE, IEEE TVT, IEEE TII, IEEE TGCN, IEEE TCSC, IEEE IoTJ, IEEE NETWORKS, IEEE WCM, ICC, IWCMC, GLOBECOM, CITS,* and *INFOCOM.* He initiated the research field of blockchain technology adoption in various verticals in the year 2017. His H-index as per Google Scholar and Scopus is 68 and 55, respectively. His research interests include blockchain technology, deep learning/machine learning, wireless sensor networks, fog computing, smart grid, and the IoT. He is a member of the Technical Committee on Tactile Internet of the IEEE Communication Society. Recently, he was awarded a cash prize of Rs. 50,000 for publishing papers with an impact factor of 5+ and publication of books with Springer, IET, and CRC under the scheme of "Faculty Awards and Incentives" of Nirma University for the year 2019–2020. He received Best Research Paper Awards from IEEE SCIoT 2022, *IEEE IWCMC-2021, IEEE ICCCA-2021, IEEE GLOBECOM 2018, IEEE ICC 2019,* and *Springer ICRIC-2019.* He also won Dr. KW Wong Annual Best Paper Prize (750 USD) in2021, sponsored by Elsevier (publishers of *Journal of Information Security and Applications*). He has served many international conferences as a member of the organizing committees, such as Publication Chair for *FTNCT-2020, ICCIC 2020,* and *WiMob2019*; and a General Chair for *IC4S 2019, 2020, 2021, 2022*; *ICCSDF 2020*; and *FTNCT 2021.* He is on the editorial boards of *COMCOM-Elsevier, IJCS-Wiley, Cyber Security and Applications-Elsevier, Frontiers of Blockchain, SPY-Wiley, IJMIS-Inderscience, JCCE,* and *JSSS.* He also leads the ST Research Laboratory, where group members works on the latest cutting-edge technologies.

Contributors

Abhishek
Department of Electronics and Telecommunication Engineering
Dayananda Sagar College of Engineering
Bengaluru, India

Sureshkumar Alagarsamy
Department of Electrical and Electronics Engineering
SRM Institute of Science and Technology
Chennai, India

Shekh Kulsum Almas
Department of Electronics and Communication Engineering
Sage University
Indore, India

Neeraja B.
Department of Electronics and Communications Engineering
Chaitanya Bharathi Institute of technology
Hyderabad, India

Mohit Bajaj
Department of Electrical Engineering
Graphic Era Hill University
Dehradun, India

Pitshou N. Bokoro
Department of Electrical Engineering Technology
University of Johannesburg
Johannesburg, South Africa

Biswajeet Champaty
Ajeenkya D.Y. Patil University
Pune, India

Li-Chung Chang
Department of Electrical Engineering
National Taiwan University of Science and Technology
Taiwan, China

Kiran Dasari
Department of Mechatronics Engineering
Manipal Institute of Technology
Bengaluru, India

Ravi Gandhi
Ajeenkya D.Y. Patil University
Pune, India

Sivasankari G.S.
School of Electrical & Electronics Engineering
SASTRA Deemed University
Thanjavur, India

Yunghsiang Sam Han
Department of Electrical Engineering
National Taiwan University of Science and Technology
Taiwan, China

Narayanan K.
School of Electrical & Electronics Engineering
SASTRA Deemed University
Thanjavur, India

Ashutosh Kashiv
Electrical Engineering Department
Prestige Institute of Engineering Management & Research Indore
Indore, India

Preeti Khanwalkar
Department of Electronics and Telecommunication Engineering
Dayananda Sagar College of Engineering
Bangalore, India

Zipho Khumalo
Department of Electrical Engineering Technology
University of Johannesburg
Johannesburg, South Africa

Mengistu Abera Mulatu
Department of Electrical and Electronic Engineering
University of Eswatini
Eswatini

Sneha Nagar
Electronic and Telecommunication Engineering Department
IET DAVV
Indore, India

Shekinah P.
Department of Information Science & Engineering
M. S. Ramaiah Institute of Technology
Bangalore, India

Suresh Panchanathan
Department of Electrical and Electronics Engineering
SRM Institute of Science and Technology
Kattankulathur, India

Deepak Pandey
Department of Electrical and Electronics Engineering
Technocrats Institute of Technology
Bhopal, India

Puja Pohakar
Pimpri Chinchwad College of Engineering
Pune, India

Swetha Ravikanti
Department of Electronics and Telecommunication Engineering
Dayananda Sagar College of Engineering
Bangalore, India

S.R. Mani Sekhar
Department of Information Science & Engineering
M. S. Ramaiah Institute of Technology
Bangalore, India

Gulshan Sharma
Department of Electrical Engineering Technology
University of Johannesburg
Johannesburg, South Africa

Deepak Sharma
Jaypee University of Engineering and Technology
Guna, India

Siddharth Shukla
Department of Electrical and Electronics Engineering
Bansal Institute of Science and Technology
Bhopal, India

Arvind R. Singh
Department of Electrical Engineering
School of Physics and Electronics Engineering
Hanjiang Normal University
Shiyan, Hubei, China

Nagendra Singh
Department of Electrical Engineering
Trinity College of Engineering and Technology
Karimnagar, India

Pradeep Vishnuram
Department of Electrical and Electronics Engineering
SRM Institute of Science and Technology
Kattankulathur, Chennai, India

Prateek Pandey
Jaypee University of Engineering and Technology
Guna, India

Section I

Alternative Energy Sources and Grid Integration

1 Power Generation at Highways Using Vertical Windmill, Efficient Solar Systems, and Internet of Things

Swetha Ravikanti, Preeti Khanwalkar, Abhishek, Kiran Dasari, and Neeraja B.

1.1 INTRODUCTION

India's massive population and rapid economic growth led to a huge demand for electricity. The majority of the energy produced during the last 10 years has mostly relied on fossil fuels. There has been a notable transition in the generation of power from conventional to non-conventional renewable energy sources in India due to the country's limitless supply of renewable energy resources. Wind energy is one of the most important forms of renewable energy for the production of electricity despite the fact that India has not utilized wind energy extensively. For the development of wind energy and other renewable resources, the Ministry of Non-Conventional Energy Sources was established. The wind flow will be affected by the speed, size, and volume of traffic in addition to the cars themselves. Wind power accounts for nearly 8.6% of India's total installed power generation capacity and generated 28,604 million kWh, (MU) in the fiscal year 2015–16, which is nearly 2.5% of total electricity generation. An ideal wind turbine design has to be created based on the findings. The main issue in heavily populated nations like India, China, etc., is traffic congestion. This study suggests that as the population increases, so does the number of cars on the road, which impacts traffic flow and increases congestion at crossings. People spend more time in traffic, leading to productivity losses and increasing fuel waste as a result of frequent traffic jams and signal stalling. In this study, the design of traffic signals is focused on to use traffic management to lessen congestion. Traffic lights, one of the best and most adaptable active traffic controls, are widely employed to manage traffic flow in many places across the world. This is a practical and affordable way to lessen traffic. A novel design for the inner Savonius turbine blade is developed for accuracy and reliability using Ansys CFX software. The evaluation includes considerations of the blade profile and overlap ratio [1]. Similarly, the limitations of individual rotors are addressed by integrating an optimal Savonius rotor within a Darrieus rotor [2]. A new method for assessing the power generation

DOI: 10.1201/9781003537823-2

capabilities of vertical-axis wind turbines (VAWTs) mounted with tubercles is discussed in Ref. [3]. Two VAWT layouts under wind conditions of 12 m/s are analyzed using the CFD software Ansys Fluent and Unsteady Reynolds Averaged Navier-Stokes simulations [4]. Another technique, the air-activated hybrid-type flat VAWT, is investigated with torque and power analyzed at various wind speeds [5–8].

This chapter aims at designing and developing vertical turbines to generate power for highway lighting systems. It also use solar plates to generate energy during the day and night, and it also focuses on traffic density control. In this system, we use a microcontroller 8051 and IR sensors, which are used to provide dynamic traffic control, thus increasing the duration of the green light of the lane in which traffic density is high and hence, regulating and clearing the traffic. Wind power generation capacity in India has significantly increased in recent years. As of 30 June 2018, the total installed wind power capacity was 34.293 GW, the fourth largest installed wind power capacity in the world. Wind power capacity is mainly spread across the South, West, and North regions [9–11].

1.2 PROPOSED SOLUTION

This chapter aims to design and develop highway power generation using vertical-axis turbines and sun energy. The design process is divided into three stages:

1. Energy production
2. Energy measurement
3. Energy Application

The first stage entails:

- Designing a vertical axis turbine.
- Optimal location of the solar panel over the VAWT, with the solar panel operating on the MPPT (maximum power point tracker) principle.
- Power generation with vertical windmills and solar panels.
- Produced energy should be used efficiently, for example, by storing it in a battery.

The second stage entails monitoring the electricity generated by the renewable energy source and uploading it to the cloud via ThingSpeak, where the data is saved and analyzed for the specified period [12].

The third step entails ensuring that the generated power is used for two end applications:

1. EV charging stations
2. Control of streetlights

When air impacts the Savonius type of VAWTs, some electricity is generated and stored. We used light-dependent resistor (LDR) sensors to detect maximum sunlight and then rotate the motor in the proper direction [13–16]. It will also produce electrical energy by vehicle headlights at night. The generated power will be utilized to power up the streetlights at night; we have created a sophisticated energy conservation system that will turn on the lights at night only when there is vehicle movement, allowing us to save 50% of the energy. LDRs are used to identify day/night time, and the lights are turned on/off based on the signal from the LDR controller. Power generation and consumption are monitored via sensor network, with data uploaded to the cloud via Internet of Things (IoT) technology. To reduce the overall size of the system, the solar panel is positioned over the vertical wind turbine. The generated energy is also stored in the battery and then inverted to power the load.

1.2.1 Hardware Implementation

The three-bladed Savonius wind turbine is first designed using SOLIDWORKS software with a height of around, a width of meters, 2.5 ft., with a circumference of around 1.6 m, and an angle of inclination of 33.33 degrees. The wind turbine is designed in SOLIDWORKS software considering the parameters such as turbine height, circumference, angle of inclination of blades, and base height in mm which is shown in Figure 1.1(a). Since it is a three-bladed Savonius-type turbine, it is designed by keeping the angle of inclination of at 33.33 degrees for efficient airflow through the blades [17, 18]. Figure 1.1(b) shows the turbine model designed in different views. Out of four, the first one is the front view, the second one is the view from the left, the third one is the top view, and the final one is the isometric view.

The top view, bottom view, and isometric view of the model look as shown in Figure 1.2 with three blades inclined equally.

The proposed model consisting of three blades is been designed using aluminum steel of thickness about 0.6 mm considering the durability which is less exposed to wear and tear. The flow chart of the proposed work is shown in Figure 1.3 where the power generated from both renewable sources will be monitored and conserved.

To generate maximum voltage from the solar panel, we use maximum power point tracker. This mechanism consists of the hardware as well as the software part.

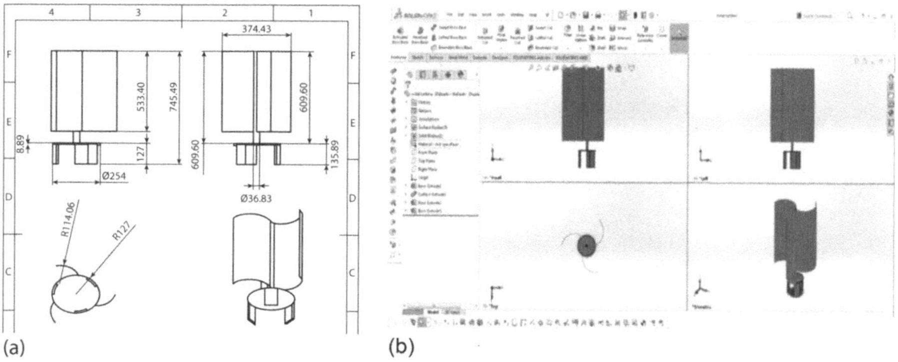

FIGURE 1.1 (a) Layout of turbine design (b) turbine design in different views.

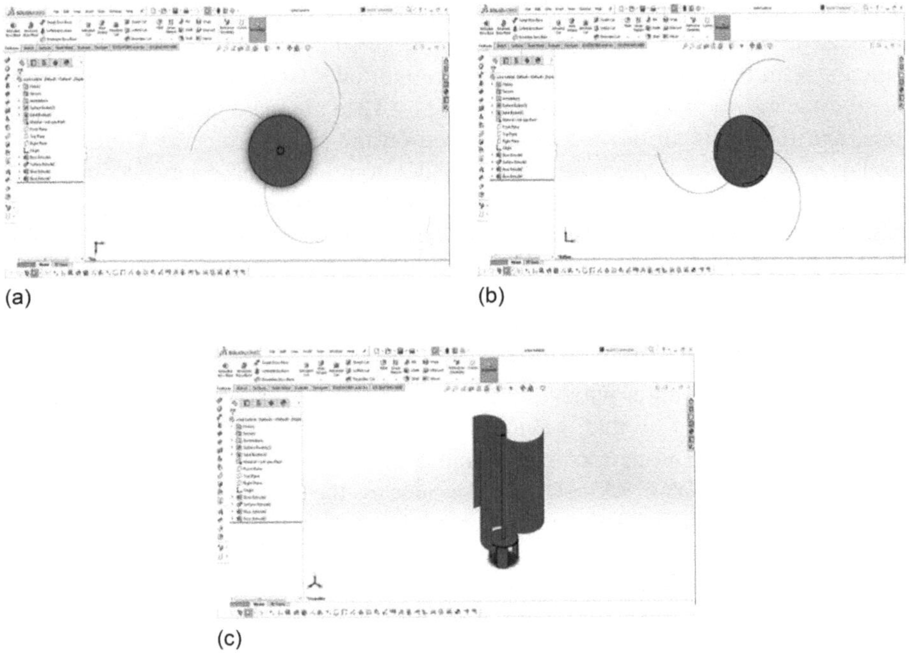

FIGURE 1.2 (a) top view, (b) bottom view, (c) isometric view.

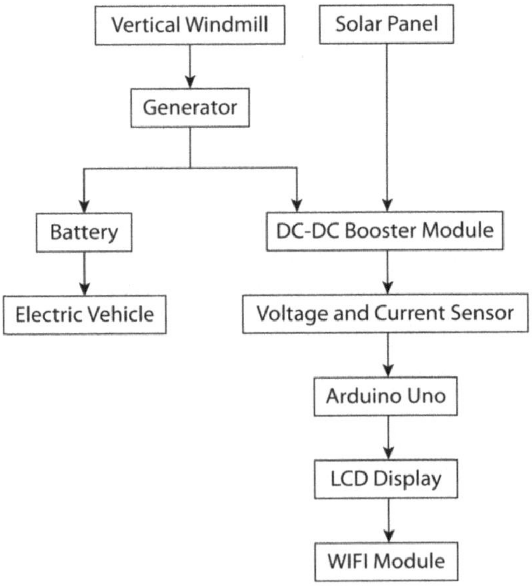

FIGURE 1.3 Monitoring of power generation.

The hardware part consists of the Arduino UNO and two LDR sensors, and the program is dumped through the Arduino IDE software on the computer. The vertical wind turbine's dynamo, which is mounted at the bottom, transforms mechanical energy into electrical energy while the turbine rotates. To transform light energy into electrical energy, a sun-tracking solar panel is also utilized. Figure 1.4(a) shows the complete setup of devices. The voltage generated from both the vertical wind turbine and the solar panels is stored in the battery as depicted in Figure 1.4(b).

(a)

(b)

FIGURE 1.4 (a) Complete setup of devices, (b) storage battery.

FIGURE 1.5 Sun tracking solar panel.

The sun tracking solar panel, depicted in Figure 1.5, is made up of two LDRs, a solar panel, and an 8051 microcontroller board. Two LDRs are positioned on either side of the solar panel, which will provide less resistance when exposed to the light. The DC motor, which is attached to the panel, rotates it in the direction of the maximum light incident on it. The panel is set up so that when two LDRs' light outputs are compared, the panel is rotated toward the LDR that has the highest intensity and hence, the lowest resistance. The panel is rotated at three angles by the DC motor. When the intensity of the light falling on the right LDR is greater, the panel moves to the right, and when the intensity on the left LDR is greater, the panel moves to the left. At noon time when the Sun is ahead, the panel takes the middle position accordingly. In such cases, the panel is constant and there is no rotation [19, 20].

The windmill and solar panel-generated voltage is used to power the intelligent highway as shown in Figure 1.6. To determine whether it is day or night, it uses an LDR sensor. It also uses infrared sensors to turn on the streetlights automatically when a vehicle passes by the highway.

A battery is used to store the power produced by the vertical wind turbine and solar panel, which will later be utilized to charge an electric vehicle.

1.2.1.1 Software Implementation

The voltage and current generated by both wind and solar energy are displayed on an LCD. The wind and solar power are calculated, and the corresponding data is

FIGURE 1.6 Automatic streetlight.

displayed and uploaded to the cloud using an ESP8266 Wi-Fi module via ThingSpeak. To achieve this, a ThingSpeak account is created, followed by the creation of a channel with fields corresponding to the number of data points to be uploaded. The generated API key for the created channel is used in the code to upload the live data, i.e., the voltage, current, and power outputs of both sources. [21–24]. Coding is done in Arduino IDE software. The code is compiled and uploaded to the Arduino Uno board at the baud rate of 115200, which is compatible with Wi-Fi, by selecting the right port to which the board is connected to the personal computer.

1.3 RESULTS AND DISCUSSION

This section deals with the results obtained from the implementation of the paper and discusses their implications. The proposed paper was designed and implemented and the mentioned objectives were met the wind and solar powers can be taken on daily or further analysis. The 2-D plot of the voltage, current, and power of wind and the voltage, current, power of solar, and the total power can be viewed in thing speak by creating respective fields in a channel created, and the previous data can be exported in Excel format as shown in fig.(11-17). The voltage and the current from the respective sources get multiplied to get a power, and the total power is obtained by summing those two. It is obtained that around 25–30 watts of total power can be obtained from this single model depending on the wind speed. From a wind turbine, maximum of up to 10–12 volts can be generated and from solar up to 8–10 volts can be obtained so a total of around 25–30 watts can be obtained at its peak. The 2-D plot of wind voltage is shown in Figure 1.7(a) where with appropriate pressure of wind on the turbine can give the output of a maximum of 18–20 volts.

The 2-D plot of solar voltage in the solar panel as shown in Figure 1.7(b) can give an output of 35 volts maximum. It varies with the intensity of sunlight falling on it. The wind current is shown in Figure 1.7(c), where it is been related to the voltage generated; i.e., it varies with the winding voltage say around 2.5–4mA of current can be obtained. The solar current is shown in Figure 1.7(d), where it is been related to the generated, i.e., it varies with the winding voltage say around 2–2.5 mA of current can be obtained. The wind power is shown in Figure 1.7(e), where the obtained wind voltage and wind current are multiplied to get the power output of the wind turbine. The solar power and total power are shown in Figure(f and g), where the obtained solar voltage and solar current are multiplied to get the power output of the solar panel, where the collected solar and wind power is combined to produce the complete amount of electricity. A maximum of 25–30 watts of total power can be obtained at the maximum wind pressure.

Although these vertical wind turbines do not generate as much energy as horizontal wind turbines, they are ideal for small-scale electricity generation. The higher power level that can be extracted from the airflow is defined by power output. There will be no complete energy extraction from the wind due to mass and momentum conservation. Betz's law demonstrated that the higher the amount of kinetic energy received from the wind should not exceed 60%. As a result, according to it, none of the turbines extract all of the energy from the wind. The Betz's threshold does not immediately apply to vertical turbines since it is dependent on the wind turbine's

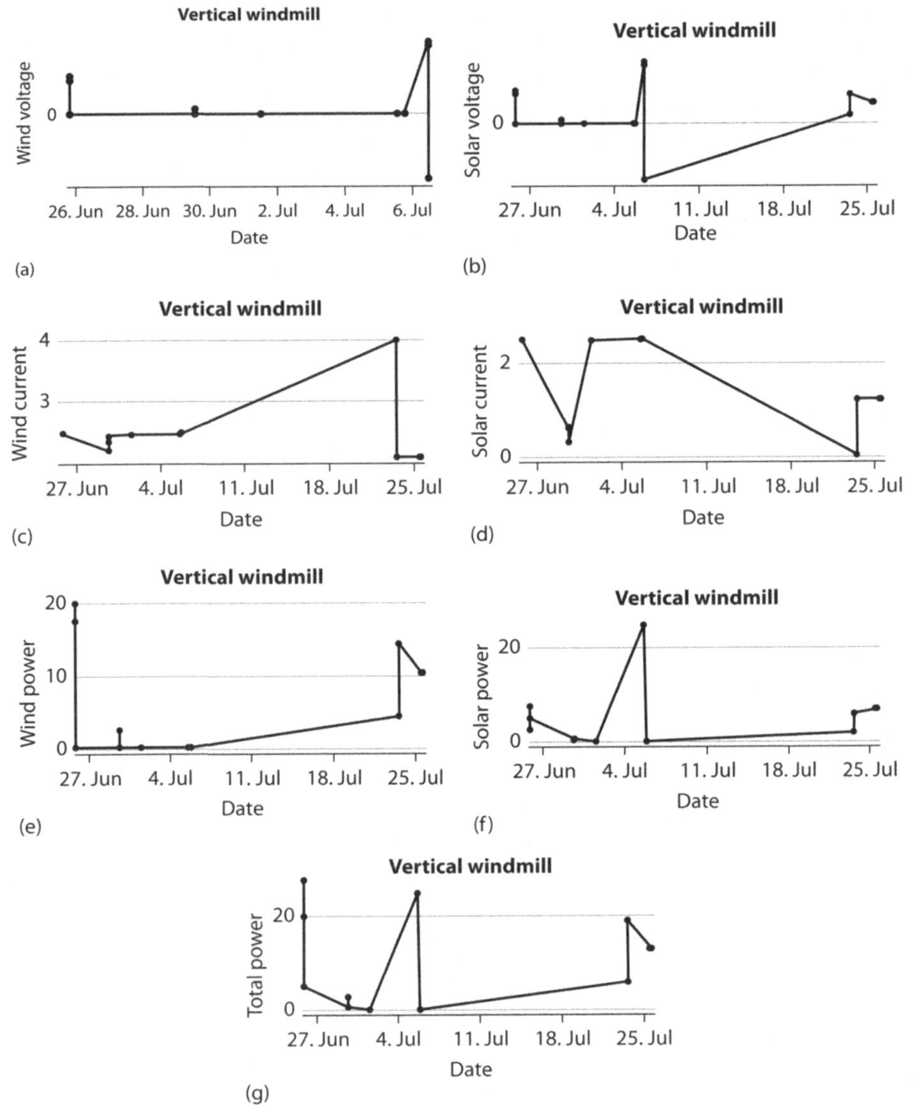

FIGURE 1.7 (a) Wind voltage, (b) solar voltage, (c) wind current, (d) solar current, (e) wind power, (f) solar power, (g) total power.

horizontal axis. A similar statement contends that using a standard VAWT method may improve the theoretical threshold.

There are two types of VAWTs: Savonius and Darrieus type. The basic difference between the two is that the Savonius operates on the traction forces, which means that the turbine rotates because of the force of the wind, whereas the Darrieus operates on the forces from the bottom that is lift forces which are created by pressure differentials created by the wind. The Savonius type turbine model generates more torque but it rotates at a slower rate even with less air pressure. Darrieus achieves significant rotational speed with significantly less torque [18, 19]. Darrieus cannot start on its own, whereas Savonius can if it has three or more blades. Depending on the application, both play a unique role, so for the application on highways, Savonius suits best as it can self-start with high torque.

1.4 CONCLUSION

In today's technological landscape, most human-made machines are designed to limit the greenhouse effect, which contributes to climate change. Renewable energy sources are increasingly favored over non-renewable ones due to their potential to provide clean energy, reduce costs, and minimize environmental damage. VAWTs are particularly advantageous due to their low-cost operation, as they do not require fuel, and their ability to address environmental pollution challenges. VAWTs are suitable for small-scale maintenance and operations investments.

On highways, hybrid models of vertical turbines and solar panels offer a robust source of green power. The combination of wind and solar energy can generate significant amounts of power almost continuously throughout the day, providing an alternative to depleting energy sources. This system can supply power to highways and small villages without relying on traditional energy sources. By combining renewable energy sources, more consistent power generation is achieved.

Single turbines may not produce sufficient power alone, but a series of turbines along highways can generate substantial energy, which can power streetlights, nearby rural areas, and other public places. Additionally, selling excess power to the grid can be profitable. In this chapter, we propose placing vertical turbines along highways to harness wind energy from passing vehicles and convert it into electrical energy. Solar panels will also be installed on highways to capture solar energy during the day and vehicle headlights at night, converting it into electrical energy.

The generated power will be used to illuminate streetlights at night. We have developed a smart energy conservation system that activates the lights only when vehicle movement is detected, saving up to 50% of energy. LDRs will detect day and night conditions, and the controller will turn the lights on or off accordingly. Power generation and consumption will be monitored using a sensor network, with data uploaded to the cloud using IoT technology for efficient management.

ACKNOWLEDGMENT

I would like to thank the Department of Electronics and Telecommunication, Dayanandasagar College of Engineering, Bangalore, India for providing the support and facilities.

Project Grant no: 46S_BE_0058, granted by Karnataka State Council for Science and Technology supported by the Department of Science and Technology, Government of Karnataka, and Government of India.

REFERENCES

1. K.A.H. Al-Gburi, F.B.I. Alnaimi, B.A.J. Al-quraishi, E.S. Tan, A.K. Kareem, Enhancing savonius vertical axis wind turbine performance: a comprehensive approach with numerical analysis and experimental investigations, Energies, 16, 2023, p. 4204. https://doi.org/10.3390/en16104204
2. A. Tripathi, P. Das, T. Aggarwal, Sahil, Efficiency enhancement of a hybrid vertical axis wind turbine by utilizing optimum parameters, Materials Today: Proceedings, 62(pt 6), 2022, 3582–3588, ISSN 2214-7853, https://doi.org/10.1016/j.matpr.2022.04.406.
3. M. Ahmad, M.H. Zafar, Enhancing vertical axis wind turbine efficiency through leading edge tubercles: a multifaceted analysis, Ocean Engineering, 288(pt 1), 2023, p. 116026, ISSN 0029-8018, https://doi.org/10.1016/j.oceaneng.2023.116026.
4. D. Stratila, I. Glasberg, Mălăel, Performance analysis of a new vertical axis turbine design for household usage, Engineering, Technology & Applied Science Research, 14(1), 2024, 12536–12542.
5. D. Rudrapal, S. Acharya, Study of an air-activated flat vertical axis wind turbine, in Proceedings of the 2023 IEEE Guwahati Subsection Conference (GCON), Guwahati, India, 2023, pp. 1–6, https://doi.org/10.1109/GCON58516.2023.10183452.
6. J. Liu, H. Lin, J. Zhang, Review on the technical perspectives and commercial viability of vertical axis wind turbines, Ocean Engineering, 182, 2019, 608–626.
7. R. Kumar, K. Raahemifar, A.S. Fung, A critical review of vertical axis wind turbines for urban applications, Renewable and Sustainable Energy Reviews, 89, 2018, 281–291.
8. Z. Tasneem, A. Al Noman, S.K. Das, D.K. Saha, M.R. Islam, M.F. Ali, M.F.R. Badal, M.H. Ahamed, S.I. Moyeen, F. Alam, An analytical review on the evaluation of wind resource and wind turbine for urban application: prospect and challenges, Developments in the Built Environment, 4, 2020, 100033.
9. S.R. Shah, R. Kumar, K. Raahemifar, A.S. Fung, Design, modelling and economic performance of a vertical axis wind turbine, Energy Reports, 4, 2018, 619–623.
10. H. Belmili, R. Cheikh, T. Smail, N. Seddaoui, R.W. Biara, Study, design and manufacturing of hybrid vertical axis Savonius wind turbine for urban architecture, Energy Procedia, 36, 2017, 330–335.
11. A. Azam, A. Ahmed, H. Wang, Y. Wang, Z. Zhang, Knowledge structure and research progress in wind power generation (WPG) from 2005 to 2020 using CiteSpace based scientometric analysis, Journal of Cleaner Production, 295, 2021, 126496.
12. A. Bavchakar, P.K.N. Chougale, S.S. Belanekar, S.P. Rane, N.B. Sawant, A hybrid model of vertical axis wind turbine-solar power generation for highway and domestic application, in Proceedings of the IEEE 2018 International Conference on Computation of Power, Energy, Information and Communication (ICCPEIC) (pp. 135–139).
13. J. Vignesh, A.S. Christopher, T. Albert, C. Pravin Tamil Selvan, J. Sunil, Design and fabrication of vertical axis wind mill with solar system, Materials Today: Proceedings, 21, 2020, 10–14.
14. M.S. Bhaskar, S. Padmanaban, P. Siano, V. Fedák, H. Vaidya, A. Taur, On the structural implementation of magnetic levitation windmill, in Proceedings of the 2017 IEEE International Conference on Environment and Electrical Engineering and 2017 IEEE Industrial and Commercial Power Systems Europe (EEEIC/I&CPS Europe).
15. A.R. Prabowo, D.M. Prabowoputra, Investigation on Savonius turbine technology as a harvesting instrument of non-fossil energy: Technical development and potential implementation, Theoretical and Applied Mechanics Letters, 10(4), 2020, 262–269.
16. B. Hand, A. Cashman, A review on the historical development of the lift-type vertical axis wind turbine: from onshore to offshore floating application, Sustainable Energy Technologies and Assessments, 38, 2020, 100646.

17. K. P.M.Y.V.Dathu, R. Hariharan, Design of wind turbine blade material for higher efficiency, Materials Today: Proceedings, 33, 2020.
18. N.R. Maldar, C.Y. Ng, E. Oguz, A review of the optimization studies for the savonius turbine considering hydrokinetic applications, Energy Conversion and Management, 226, 2020, 113495.
19. D. Mitra, D. Das, D. Brahma, D. Maiti, T. Mondal, Sun tracking solar panel without microcontroller, in Proceedings of the 2019 International Conference on Opto-Electronics and Applied Optics (Optronix).
20. H. Allamehzadeh, An update on solar energy and sun tracker technology with a dual axis sun tracker application, in Proceedings of the 2019 IEEE 46th Photovoltaic Specialists Conference (PVSC).
21. B. Kim, Design method of a direct-drive permanent magnet Vernier generator for a wind turbine system, IEEE Transactions on Industry Applications, 55(5), 4665–4675.
22. D. Dennis, P.S. Ganesh, J. Joy, L.R. Amjith, B. Bavanish, Computational design analysis of vertical axis wind turbine, Materials Today: Proceedings, 66, 2022.
23. P. Meena, V. Prema, M.S. Bhaskar, D. Almakhles, Bi-furcated stator winding configuration in three-phase induction generators for wind power generation, IEEE Access, 9, 2021, 153188–153198.
24. M. Rambabu, J.S.S. Kalyan, T.R. Vijaya Raju, C.H. Sai Ram, V.M. Shyam, Power generation on high way by using vertical axis wind turbine with solar system International Journal of Advanced Research in Electrical, Electronics and Instrumentation Engineering, 2020.

2 Grid Interconnection of High Step-Up DC to AC Converter for Renewable Source

Siddharth Shukla, Deepak Pandey, and Nagendra Singh

2.1 INTRODUCTION

In industrial applications, it is frequently necessary to convert low DC voltage to the highest voltage levels. Using the step-up multilevel inverter, it is possible to convert DC voltage into high AC voltage. Hiring high-power DC-DC converters to raise them to the DC level and link them with a complete bridge is an additional choice. Because of their straightforward design, power-to-switch switched-mode switches, including fly-back and forward converters, etc., are frequently utilized with high-density DC-DC conversion [1].

The majority of DC to DC converters produced recently have been created with one of the following techniques: The terms voltage-frequency, pump, connected inductor, switched inductor, modified inductor, and switched-capacitor (SC) are interchangeable. For example, the top zero-current measure converters integrated into the resonant SC process are shown. The step-down version and high-step converter are presented, incorporating buck/boost and SC methods, respectively [2].

It suggests a switched-inductor construction with a high-step conversion. The SC-related inductor combination greatly increases the high-voltage power gain. One of their shared characteristics is the ability to regulate the power conversion rate using pulse frequency conversion mode (PWM). Magnetic switches and low-speed switches are used [3].

For converters with a high rate of power conversion instance, only one active switch and one integrated inductor are used. We will discuss the two applications of the DC high converter in this chapter: (1) the display of PV-compliant frames; and (2) a potent import model constructed with MATLAB and Simulink that may be utilized to take into account transient motor-drive behavior.

The objective of this work is to influence and look at the constantly changing actions and exhibition of the structures related to the PV grid.

2.2 PROBLEM DESCRIPTION

Other transformer-based converters, such as front, push-dragging, or fly-back converters, can achieve the maximum power gain by converting the transformer

 DOI: 10.1201/9781003537823-3

conversion component. However, the spillage inductor transformer will cause significant issues, for example, with the initial voltage spike switch and the distribution of high power.

In order to improve the magnitude of the change and achieve higher power outages, a number of convertible structures have been demonstrated. Capacitor switching and welding switching techniques have been widely used to achieve the highest power output. In any case, in these frameworks, the charge flow will pass through the change of policy and increase the existing malfunction. Converting the back-boost fly-back converter is one of the most effective ways to achieve maximum boost; this increase is approved by the integrated inductor. The power of spillage is restored to the final fruit [4].

A variable converter with a capacitor that raises the voltage is very similar to the traditional set type. Due to the malfunction of the diode and the switching current, the switching power and the voltage surges are limited. Therefore, more advanced translators are needed to address the problems mentioned earlier [5].

These converters increase the volume of low-level information to a higher-level information. Some conduction defects are brought about by the protection of the semiconductor segments and the integrated inductors.

With the possible exception of all capacitors, none of the parts in this recommended converter are thought to be accurate. The features of spillage inductors are eliminated from the power consumption.

The huge conversion, enormous size, and tiny size of the high conversion converter are its key characteristics. In most cases, these applications call for an increase in high receiving voltage and high efficiency, employing an advanced pre-up converter [6].

2.3 MODELLING OF DC-AC CONVERTER

Grid-connected power grids are typically utilized in conjunction with photovoltaic boards. Increasing the DC power of the bottom boards to supply the inverter is crucial for transferring power from standard PV boards to typical AC voltages. The user interface is the functional device for this.

To provide a significant increase in energy efficiency, such converters may need to operate at higher bond cycles. Different requirements include high productivity, reduced switching pressure fluctuations, minimal electromagnetic interference, reduced noise, and a small number of objects [7].

2.3.1 DC-DC Boost Converter

This amplified voltage is directly connected to a multi-level inverter for AC voltage levels. Another option is to add the low voltage first and then connect it directly to the bridge inverter. Large DC-DC converters use fly-back and advanced converters due to their standard design [8]. Recently, switched capacitors, switched inductors, tapped inductors, and integrated inductors have all been utilized in the development of numerous high-frequency DC-DC conversions [9].

DC-DC converter has the following advantages:

- The high-speed DC switching converter provides ultra-high voltage.
- Low-voltage transistors are used to improve efficiency.
- High-quality DC input is provided by a resonant converter.

2.3.2 Boost-Up Voltage

When the resonant converter circuit operates in two modes, the first mode is MOSFET S_1 ON and S_2 OFF, and the second mode is MOSFET S_1 OFF and S_2 ON. In the first mode, when S_1 is ON and S_2 is OFF, the two capacitors, C_1 and C_2, connected near the PVA, come in series. Thus, the first capacitor may have voltage V_{c1} and the second capacitor may have voltage Vc_2 [10].

Thus, the total voltage.

$$V = V_{c1} + V_{c2} \tag{2.1}$$

Figure 2.1 shows the DC transformation circuit of the designated high converter is the recommended stage to create a DC converter. It has several integrated inductor-SC (CLSC) units and a power amplification unit for synchronization with reinforcing materials. The suggested converter is coupled to a single integrated inductor, two capacitors (C_1 and C_2), two diodes (D_{11} and D_{22}), and one winding with an integrated inductor switch comprising each CLSC unit.

In order to provide the load with the full voltage, filter capacitors C_1 through C_m are connected in series with the input supply. Two assumptions are used in order to

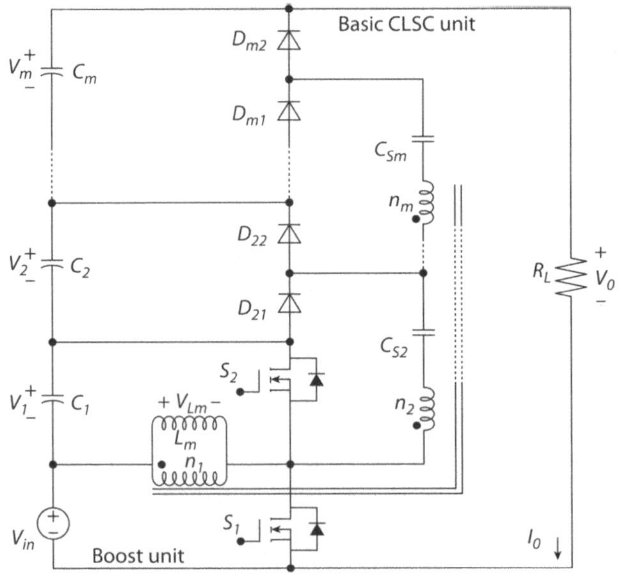

FIGURE 2.1 Step-up DC to AC converter.

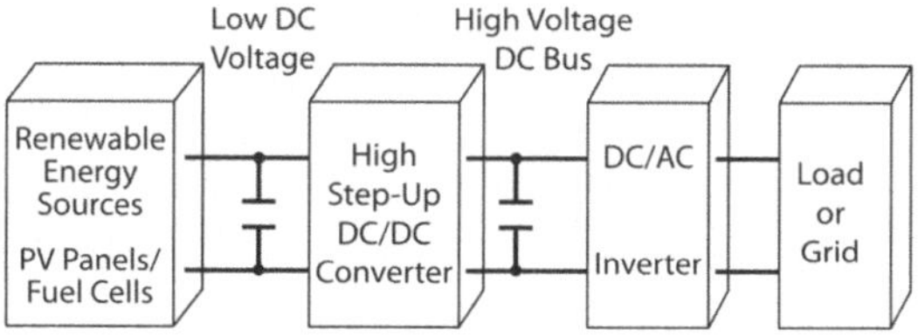

FIGURE 2.2 Solar-based DC-AC converter.

simplify the dynamics of the emission of the electric field: Every component, including the input V_{in} source, is appropriate, and because every capacitance is huge, its voltages can all be thought of as constant.

2.3.3 High Step DC to AC Conversion Using Renewable Resources

The recommended converter's block diagram appears in Figure 2.2. The converter is equipped with a grid, an inverter, a solar or renewable energy source, a high DC-DC stage, and a low-energy DC input power [11].

2.4 DESIGN OF PHOTOVOLTAIC ARRAY

Solar energy is a good solution for the release of solar energy. The solar radiation is converted directly into electricity through a solar photovoltaic module. The solar module contains two solar cells. Photovoltaic modules are made of silicon cells. Silicon solar cell units provide an output of about 120 volts under rotation conditions. When so many cells are connected in series, we get a solar PV module. There are typically 36 cells in the module mode that show an open circuit volume of approximately 20V [12].

The separate cells' locations determine the modules' current rating. Solar PV modules are connected in series in such a way as to get high power at the highest possible current output. According to the irradiation phenomena, photons with higher-than-semiconductor band-gap capacities are caught and produce additional electron holes when exposed to sunshine [13].

2.4.1 Components of Photovoltaic Array

Solar energy is a good solution for the emission of solar energy and electromagnetic solar radiation directly converted to electricity using a solar photovoltaic solar module. The solar module contains a solar cell. The solar cell is basically a p-n junction diode that is designed into a thin layer of semiconductor. Figure 2.3 shows the equivalent circuit, and Figure 2.4 shows the V-I characteristic of a PV system [14]. The current generated by the PV system is given in eqn. (2.2).

$$I = I_{ph} - I_d = I_{ph} - I_o\left(\exp\left(\frac{qV}{KT}\right) - 1\right) \tag{2.2}$$

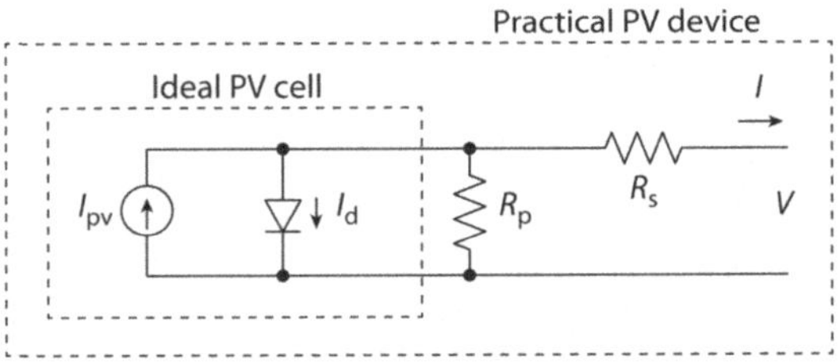

FIGURE 2.3 Equivalent electrical circuit photovoltaic device.

where I_0 is the diode current, q is charges, K is a constant (1.38×10^{-23}J/K), and T is the temperature of the cell in Kelvin.

V is the limited cell volume that can be produced (quadrant of power) or the gas used. Extending the equation provides a simplified cycle and a related equation, in which the n diode ideality factor (between 1 & 2), R_S, and R_{sh} represent a series of low resistance.

$$I = I_{ph} - I_o \exp\left(\frac{q(V + IRs)}{nkT} - 1\right) - (V + IRs) / R_{sh} \tag{2.3}$$

The temperature of the PV cell depends on the following things:

- Current PV (IPV) is directly related to irradiance
- It provides precise alignment between Maximum Power point and open circuit volume

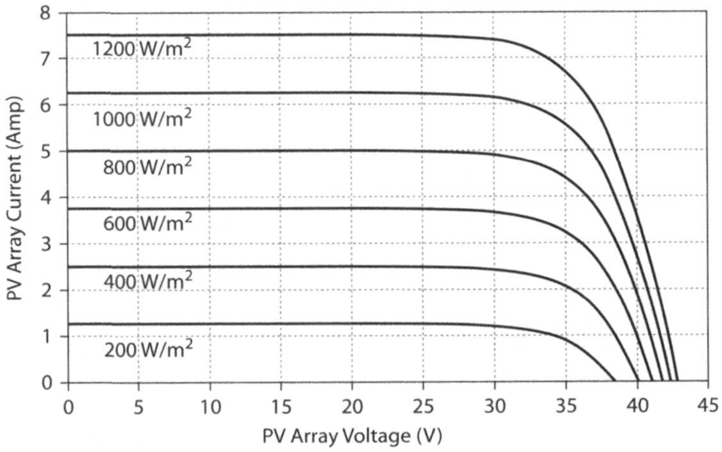

FIGURE 2.4 V-I characteristics of PV system.

2.4.2 Maximum Power to Learn Algorithms

A typical solar panel works beyond the high power point, converting only 30%–40% of solar light into electrical energy. The solar panel's efficiency is increased with the help of an excellent energy-tracking system. The modern power transfer theorem states that when the source impedance of a circuit is similar to the load disturbance, the likelihood of power outages in the circuit is increased. Thus, the issue of inaccurate modeling arises from our concern with high energy tracking. The solar panel is connected to a boost converter on the source side in order to increase the output voltage, which can then be utilized for a variety of purposes, including car charging. By adjusting the converter's duty cycle to grow proportionately, we may compare the source and load bars [15].

2.5 DEVELOPMENT OF GRID INTERCONNECTION

In this chapter, we present a modelling simulation of a grid that is compatible with the PV system using MATLAB/Simulink. The photovoltaic model is in the middle of a grid connection using full electronic control gadgets. Recreation refers to the PV power framework at normal temperatures and solid loads using MATLAB [16].

The first goal of this phase is to develop new learning, illuminating MATLAB/Simulink identified by a framework that comprises dynamic energy management frameworks through Repetitive Power Reserves (R_{es}), such as spotless and productive areas to collect first earth and special grid requirements associated with grid control. Therefore, it is important to commit to the development of specific functions that allow the installation of RES in a high-power inverter and safety. By using these good frameworks, the client is not only a consumer but also a power maker.

Age-rounded average units (distributed generation, DG), including continuous and non-renewable resources, are used in small grid-related networks and small power supplies (up to 1000 kW) associated with an improved operating system. There has been an increase in the number of standard DC sources, e.g., power devices and photovoltaic arrays, or AC frequency intensity that is either less concentrated or greater than the grid density, such as smaller amounts of electricity. These generators require a grid-connected DC/AC converter. Although some generators can be connected directly to the electrical grid, for example those powered by non-current manufacturer, there is a pattern to accept the installation of a used hardware building that converts the power from Battery to DC and then uses an inverter to transfer 50Hz AC mains power [17].

2.5.1 Grid Development

Photovoltaic frames began to be used as isolated frames to power areas of the world where no other energy sources existed. Advances in new applications and concerns about unnatural climate change empower both services and customers to maximize the use of grid-related PV frames. The PV power connected to the grid will reduce energy costs, as it is unthinkable to sell the remaining power supplied to the nearest power supplier. A grid-based PV frame is easy to appreciate as it does not require a battery frame. Grid connectivity for age-old photovoltaic (PV) regulatory frameworks has the advantage of successfully using production management because there are no fixed malfunctions included.

2.5.2 Triple Phase Inverter

The inverter's primary purpose is to increase the amount of DC that is converted to AC. With the aid of Voltage Source Inverter, the boost converter's DC output is transformed into AC. Six PWM variations with a 120-day performance are available for VSI. If we receive three-phase AC voltage, the amplified six inverters add six more thirsty charges in a typical time of 600 by 3600 full. Using IGBT to install thirsty charge has evolved into a contemporary technique.

2.5.3 High Step-up DC to AC Converter Grid Interconnection

The inverter is connected to a DC. DC upgrades booster with gadget power inverters and can convert the power of a DC structure to AC. Inverters can add a variety of an assortment of various parameters such as value, control rate, efficiency, and applications.

DC to AC control ink traditionally takes the DC power provided by the solar panel with a storage battery. Power inverter rectifiers are used in modern life-saving machines in home applications. At the point where the inverter output has no sinusoidal restriction and is connected to the grid.

2.5.4 Concept Theory of Synchronous Reference Frame

This current generation view is developed in the time zone. This concept is widely used because of its simplicity of mathematics and uses only algebraic calculations. Three-quarters of the current load (i_{La}, i_{Lb}, i_{Lc}) are converted into two active components (I_{Ld}) and active (I_{Lq}) by a rotating syllabus in the correct sequence of electrical energy. The basic SRF operating system uses direct (d-q) and inverse (d-q) p as shown in Figure 2.5.

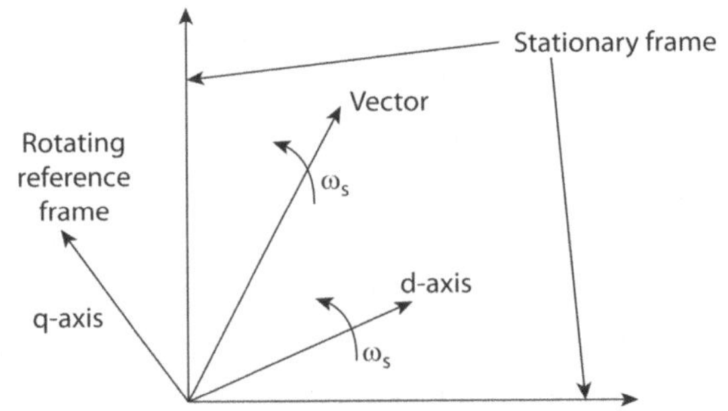

FIGURE 2.5 Vector diagram of reference frame.

Vessel modification method allows for the testing of a specific harmonic component of input signals. The modification of the reference framework is assessed by converting the three-dimensional reference system of abc stationery with a dq-0-linked reference system consisting of two (d) and quadratic axq axis, which are the angular velocity of the third phase rotation given by $\omega e = 2\pi f$ angular frequency of transmission.

We look at two axes, that is, the d axis and the q axis. The q-axis leads the d-axis by 90 degrees. Therefore, this becomes our viewing frame (dq axis). Also, we consider that this reference framework revolves around the same as the system density. Looking down on the image, the phasor rotates at a speed of "w" and frame of the reference rotates at the same speed "w." Therefore, the relative speed between the rotating and independent phasor is zero (0). Therefore, we look at the value of DC and it can be resolved on the d and the q axis. The help we get here using SRF is talking about DC prices. With control applications, DC values are readily available without detecting solid-state errors.

2.5.5 Phase Looked Loop

Grid topologies coupled to a synchronized grid voltage employ closed loops or Phase Locked Loop. To guarantee that power flows directly from the power source to the grid, PLLs are necessary. Power conversion from renewable energy sources makes use of PLLs. PLLs can be applied to monitoring and control tasks.

2.5.6 Synchronous Reference Frame Controlling for Inverter

In this converter synchronous reference from the synchronous reference provided good controlling result for operation this concept provided the synchronized to frequency.

Use of SRF controller benefit

1. It controls frequency.
2. It controls grid voltage.
3. It reduces grid voltage ripple.

The main component of the SRF controller is the PLL, which is a feedback system that removes the phase error of the output. PLL contains the park transform and this Clark transform and is used where the grid is connected with three-phase inverter, controlling the park and the Clark transforms.

PLLs for grid-connected power converters frequently employ synchronous reference frame phase-locked loops or SRF-PLLs. The electrical grid's frequency and phase are measured using SRF-PLL. The cosine signal generated by the PLL utilizing nominal phase makes up the unit amplitude. Grid-connected power converters are controlled closed-loop using reference signals produced by these unit tests. DC offset, harmonics, and input voltage imbalance all impact PLL performance.

Unit vitamins are not affected by synergy or imbalances. There are simultaneous bursts from the default frequency in the measured frequency and phase when the

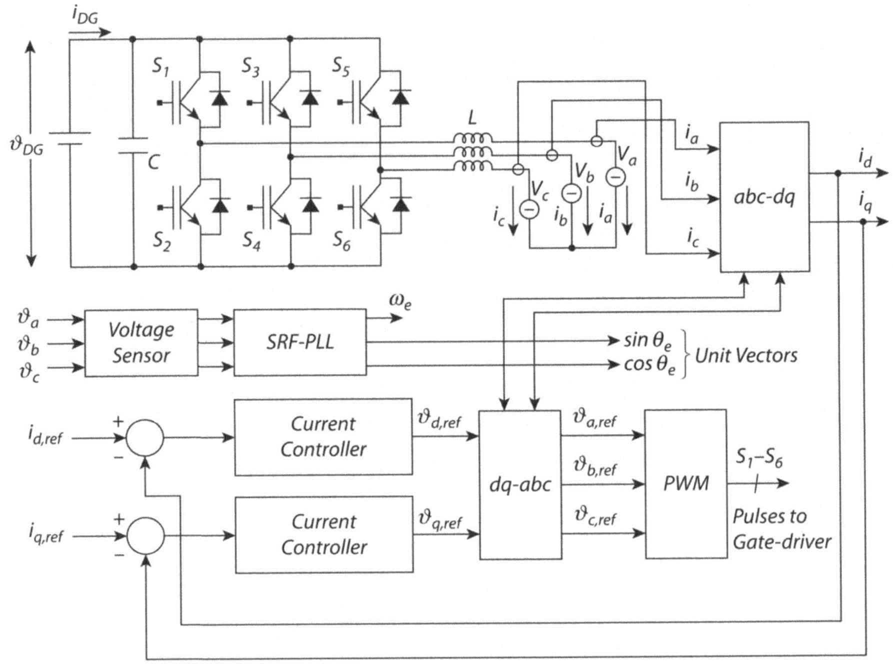

FIGURE 2.6 Three-phase connected grid and SRF-PLL inverter control.

input contains a DC decrease. The current setup of the inverter controller is poor throughput. Hence, DC insertion into the grid will occur for conventional current controller designs where unit technicians employed as references contain DC output, which is less desirable. This is clarified by referring to Figure 2.6, which depicts the grid connected to the inverter and the generation source that is divided up into photovoltaic panels.

When unit vectors containing DC offsets are utilized to construct the voltage indication from v_d, R_{ef}, and v_q, the resulting inverter voltage will also have DC offsets. DC injection into the grid is the outcome of this. As a result, the issue caused by DC overheating in the PLL installation gets more challenging since grid connection rates allow the SRF-PLL's design objectives to be expressed.

1. By installing the worst-case DC input, SRF-PLL must produce a network unit that satisfies the grid connection.
2. The SRF-PLL response time should be very short for the worst DC supply.

2.6 CONCUSION

It is possible to extend the suggested converter for extremely high voltage gain. The suggested converter employs a linked inductor with a drop inductance to achieve smooth switching of the diode. This work offers a no-isolated high step-up DC-AC power converter with tiny on-state resistance for grid application, using low-voltage rated MOSFETs. The converter is built using coupled-inductor and switch capacitor technology.

REFERENCES

1. W. Li, X. He said, "The review of DC / DC converters is not the only one that does not place itself in positions connected to the photovoltaic grid," IEEE Trans. Ind. Electron, vol. 58, no. 4, pp. 1239–1250, 2011.
2. Y. Khan, D. Apparao, S. Gawande, N. Singh, Y. S. Bist, "Performance assessment and working fluid selection of the novel combined helium Brayton cycle and organic Rankine cycle based on solar power tower for sustainable generation", Iran. J. Sci. Technol. Trans. Mech. Eng., vol. 48, pp.1–18, 2024. https://link.springer.com/article/10.1007/s40997-023-00745-8.
3. K. K. Law, K. W. E. Cheng, no-Y. P. B. Yeung, "Design and analysis of transformational tools using step-by-step flexibility," IEEE Trans. Sci. Syst. Mina, Reg. Papers, vol. 52, no. 5, pp. 943–948, 2005.
4. B.B. Macy, Y. Lei, R.C.N. Pilawa-Podgurski, "A resonant based in Ga V 25, 25 V to 100 V Gain-based Dickson switched-capacitor converter with 1011 W/in3 (61.7 kW/L) power density," Proc. IEEE Appl. Electron Power Conf. Expo., 2015, p. 1472–1478.
5. N. Singh et al. "Analysis of heuristic optimization technique solutions for combined Heat–Power economic load dispatch", Appl. Sci.2023, 13(18), 10380; https://doi.org/10.3390/app131810380
6. V. A. K. Prabhala, P. Fajri, V. S. P. Gouribhatla, B. P. Baddipadiga, M. Ferdowsi, "A DC–DC Converter With High Voltage Gain and Two Input Boost Stages," IEEE Transactions on Power Electronics, vol. 31, no. 6, 2016.
7. U. M. Ferdowsi, "A DC-DC converter with a high level of electrical power and two input stages," IEEE Trans. Electron Energy, vol. 31, no. 6, pp. 4206–4215, 2016.
8. M. Uno, A low-power converter that converts a converted capacitor converter and a compatible PWM converter PWN, in: Proceedings of the 35th Int. Telecommun. Power Conf. 'Smart Power Efficiency', Hamburg, Germany, 2013, pp. 1–6.
9. A. Nighoskar, N. Singh, "MPPT techniques used to enhance the efficiency of the solar plant," IJRAR, vol. 6, no. 1, pp. 996–999, 2019.
10. Y. Ye, K. W. E. Cheng, "A family of PWM converters converting single phase," IEEE Trans. Electron Energy, vol. 28, no. 11, pp. 5196–5205, 2013.
11. Y. P. Hsieh, J. F. Chen, T. J. Liang, L. S. Yang, "Converting novel-high-up DC-DC," Ind. Electron, vol. 60, no. 4, p. 1473, 2013.
12. N. Singh, Y. Kumar, "Multi objective economic load dispatch problem solved by new PSO," (Hindawi publication) Adv. Electr. Eng., vol. 2015, pp. 2356–6655, 2015.
13. N. Singh "Analysis of optimum cost and size of the hybrid power generation system using optimization technique", in Proceedings of the 12th IEEE International Conference CSNT-2023 held at Technocrats Institute of Technology (Excellence) Bhopal, Madhya Pradesh (India). 2023. https://ieeexplore.ieee.org/document/10134720
14. N. Singh, "Artificial Intelligence Techniques in Power System Operation and Analysis", Book CRC Press, vol. 1, pp.1–185, 2023.
15. A. Abramowitz, B. Zhao, K. M. Smedley, "High level strengthening one platform for the development of photovoltaic applications," IEEE Trans. Electron Power, vol. 31, no. 5, pp. 3550–3558, 2016.
16. N. Singh, "Renewable energy for sustainable rural development and challenges," IJRAR, vol. 6, no. 1, pp. 185–187, 2019.
17. N. Singh, "Artificial Intelligence Techniques in Power System Operation and Analysis", Book CRC Press, vol. 1, pp. 1–185, 2023.

3 Amplify and Forward-Based Cooperative Communication of Energy Harvesting Tags

Mengistu Abera Mulatu, Yunghsiang Sam Han, and Li-Chung Chang

3.1 INTRODUCTION

Advances in wireless communication technologies, highly integrated microelectronics technologies, wireless sensor networks (WSNs), and energy-harvesting techniques have gained significant attention from both industry and academia. As electronic hardware circuits become cheaper and smaller, a large number of wireless applications are emerging, especially as these tiny wireless devices provide the opportunity for electronic systems to be attached to commonplace objects.

Wireless devices have found wide applications in a variety of fields such as environmental monitoring, military and emergency applications, and medical applications through the formation of autonomous wireless networks in order to collect information. To collect and transmit information, the devices consume energy from their batteries. Once the devices run out of their battery energy, it is often difficult to replace their batteries. If more number of devices stop operating due to the energy shortages, they cannot maintain operation to perform their designated tasks altogether. Energy harvesting (EH) techniques have recently emerged as a promising solution to overcome the battery problem and realize green communications. EH devices collect energy from solar, vibration, thermoelectric effects and RF energy [1, 2].

The recent advances in ultra-low-power, ultra-wideband (UWB) devices and EH techniques [3, 4] will enable networks of self-powered ultra-low-power EH active networked tags [5–9]. Such networks will serve as building blocks for the emerging internet-of-things applications. These small, flexible, and self-reliant (regarding energy) devices can be attached to objects that are not traditionally networked (e.g., books, furniture, walls, doors, keys, bags, clothing, and produce). They can be applied to tracking and monitoring, locating or identifying lost objects and disaster victims, and collecting information regarding the relevant surroundings that can be gathered to a central computer or server. These tags can be located at fixed locations, such as buildings, chemical plants, and forests, or can be attached to objects in motion such as vehicles, animals, or products. EHT design is characterized by its

 DOI: 10.1201/9781003537823-4

ability to network, harvest energy, operate at ultra-low power, communicate with each other, and adapt its energy. EHTs communicate by transmitting and receiving extremely short pulses. EHTs are devices that communicate with one another and with EHT-friendly devices to cooperate and forward information to the intended destination. This communication can be deactivated between pulses, saving considerable amounts of energy.

One paradigm shift in these devices' communication is that the energy required to receive or listen to information is more compared with the energy required to transmit the same information. Regarding throughput, system complexity, size, and energy consumption, energy harvesting tags (EHTs) fit between radio frequency identification (RFID) and WSNs [5]. EHTs are different from RFIDs in that they harvest energy and can be networked. RFIDs are traditionally not active and not networked. EHTs are close to WSNs. However, they are different from WSNs in that they consume much less energy, operate at lower data rates, and transmit mostly ID information. They do not have the capability of WSNs to sense and send data at higher rates. Therefore, the existing WSN transmission techniques and algorithms are too complex for EHTs to apply and necessitate re-designing new transmission techniques and simple algorithms. Efficient use of energy is important to increase the lifetime of wireless devices. Using excessive energy causes battery outages, limiting the lifetime of the device. By contrast, conservative energy consumption reduces the system throughput. In fact, due to the nature of the source of energy, the energy cannot be generated at all times. Moreover, such wireless device applications are typically limited by the size of the battery and the energy it can store. The wireless channel which may significantly reduce communication capacity and range due to signal fading can also make the devices drain their energy to transmit data reliably based on the applications. Therefore, to ensure reliable long-term communication, tags should optimize their energy consumption based on the available energy resources and the wireless channel condition. This requires appropriate designing of resource control mechanisms to attain sustainable communications. Wang et al. [10] proposed an optimal transmission policy that takes into consideration of both the communication reliability and system activity time. They considered a scenario in which each tag individually communicates with the destination. In a network of EH tag-equipped objects, a tag might lack sufficient energy to respond directly to the reader when the reader is outside the communication range, or the stored energy of the tag is below a certain minimum value. Even if a tag attempts to respond to a reader's inquiry when lacking sufficient energy, it greatly reduces the probability that the reader can correctly receive the information, allowing information to be lost. To overcome the problem of reliability, the use of cooperative techniques such as relay concepts among wireless devices has been proposed [11–14].

Cooperative communication has been demonstrated as an important technique to improve the performance of wireless networks. Exploiting both cooperative communications and EH is a promising new direction for designing energy-efficient wireless nodes. However, the emergence of EH necessitates the redesigning of communication systems and networks. To utilize the EH and cooperative communication capacities of wireless devices, an optimal cooperative communication of EH devices needs to be clearly addressed, and this is the focus of this chapter. Consider EH active networked tags and address the problem of transmission optimization of cooperative communications. Consider a time-slotted source-relay-destination system, where a tag (the source) has the option to have another tag (the relay) help to forward its data to the destination. Also, consider a protocol where the relay tag assists the source tag without affecting its own data transmission. All tags under consideration are equipped with EH capability with finite battery size. The source achieves higher throughput if it uses a relay compared to a direct transmission. However, the energy consumed by the relay tag increases, and as a result, the relay tag may not have the energy to transmit its own data in the future. Therefore, the problem of interest is for the source tag to determine whether to transmit data on its own or cooperatively via the assistance of the relay tag in order to maximize the long-term average throughput under the given energy constraints. In order to optimally determine if the neighboring tag should be used or not at a given time, in addition to its own state information (i.e., current battery energy level and the request state), the source tag also has to know the state information at the neighboring tag (relay).

There have been a lot of studies on wireless networks with cooperative communication. Cooperative communication increases system throughput and communication reliability [11, 15]. The most common relaying strategies are decode and forward (DF) and amplify and forward (AF). In a DF strategy, the relay node decodes the received signal, re-encodes it, and transmits it to the destination. In an AF strategy, the relay simply amplifies and forwards the received data to the destination. Cooperative wireless networks with energy-harvesting nodes have also raised significant research interests. In [14], the use of energy-harvesting nodes as AF cooperative relays that assist communication of source and destination is proposed. In [12], the direct transmission with energy gathering and cooperative transmission with energy salvage techniques in wireless networks is studied and outage probability performance is analyzed. In [16], the authors studied energy-efficient scheduling strategies for WSNs with EH. They considered a case where a node may use either direct transmission or a cooperative relay transmission. During cooperation, a relay node drops its own data transmission and is fully engaged in forwarding data from the source node.

Many of these existing techniques and algorithms are too complex to apply for EHTs. In addition, the current WSNs use more energy than the EHTs. EHTs are new architectures that combine the cooperative transmission concept of WSNs for their transmission optimization. There is an increasing interest from the industry on low-energy communications and EH (e.g., [17, 18]), which are key features of EHTs. This is the first step toward future developments of EHTs. In particular, Texas Instrument has commercialized a solar EH development kit [18] to envision the development of sensors. The eZ430-RF2500-SEH is a complete Solar Energy Harvesting development kit to help create a perpetually powered WSN based on the ultra-low-power MSP430 microcontroller. The Solar Energy Harvester module includes a high-efficiency solar (2.25 × 2.25″) panel optimized for operating indoors under low-intensity fluorescent lights, which provides enough power to run a wireless sensor application with no additional batteries.

This chapter considers cooperative energy-harvesting active networked tags (EnHANTs) where data from both tags are equally important in time to the reader. During cooperation, the relay tag assists the neighboring tag without affecting its own data transmission, which can be transmitted during the allocated time slot. It is also assumed that because of the simple EHT circuit design, the relay tag employs simple detection techniques to detect data such as the energy detection (ED) technique [19]. An accurate synchronization timer is employed to detect signals at the reader. Consider a cooperative transmission strategy for EHTs that optimizes the long-term average throughput based on the constraints of available energy resources as well as the channel conditions. For such a system, the transmission optimization problem is formulated as a Markov decision process (MDP).

3.2 MARKOV DECISION PROCESS

The theory of MDPs, also called by several other names including sequential stochastic optimization, discrete-time stochastic control, and stochastic dynamic programming, studies sequential optimization of discrete-time stochastic systems. The basic object is a discrete-time stochastic system whose transition mechanism can be controlled over time. Each control policy defines the stochastic process and values of objective functions associated with this process [20]. MDPs can be solved by linear programming or dynamic programming via value iteration and policy iteration algorithms.

MDPs have attracted many researchers because they are important both from the practical and the intellectual points of view. MDPs provide tools for the solution of important real-life problems. Many business and engineering applications use MDP models. In particular, MDP is widely used for optimization of the communication systems. In an MDP problem, the behavior of the system can be modeled as a set of states and transition probabilities. The decision-making problem is to choose the best possible action in each state in order to maximize/minimize the expected reward/cost over a finite or infinite time horizon. One criterion is that the system should meet the Markov property; that is, the next state of the system should be dependent only on the current state and the current action. A detailed overview of the applications of MDP in (wired and wireless) communication networks is given by Altman [21]. The MDP-based frameworks for wireless networks include transmission scheduling, and routing [22, 23], optimal transmission rate, and power management under the presence of channel state side information [24, 25], where the MDP states are the channel information and buffer size, and action is power and data rate. The objective of the above two studies is to maximize throughput in [24] and the number of received packets per consumed energy [25]. A time-slotted multiple access wireless system with N transmitting nodes, each equipped with an EH device and a rechargeable battery of finite capacity, is studied. The energy arrival process at each node is modeled as a two-state Markov process, such that a node either harvests one unit of energy or none at each time slot. Both MDP and a partially observable MDP (POMDP) models are proposed in [26] to study the scheduling problem in an EH multiple access wireless system, in which the harvested energy at each node is modeled as a Markov process with the goal of maximizing total throughput without knowledge of the states of the EH processes or the nodes' batteries. The authors have also modeled the system as a POMDP and showed the optimality of the Markov process. Other studies utilizing MDP and/or POMDP in wireless communications systems include [16, 27–30]. In a recent study [10], a system of EnHANTs, including the communication model and the EH model is proposed, where the events of identification request and EH are assumed to follow simple Markov processes with the focus on minimizing the communication error probability. However, we note that nearly all of these studies discuss the optimization of the direct-link communication and do not consider the cooperative communication problem. Therefore, we are motivated to apply the MDP to the cooperation communication problem.

A Markov model formulation for the AF cooperation problem is designed by Issariyakul and Krishnamurthy in [31]. They considered the flow and buffer control and automatic repeat request protocols, and by using the Markov model, they showed that increasing the number of relay nodes can significantly reduce the packet error probability. Another MDP framework for the cooperation problem is the use of MDP [32] for finding the optimal cooperation strategies in wireless networks to maximize throughput and provide the highest cooperation gain. In the proposed model, each node uses its local information to decide on the cooperation, while the neighboring nodes are allowed to exchange limited control packets for the purpose of synchronization.

The MDP can be solved efficiently by dynamic programming [33] to provide optimal cooperation behaviors in the wireless communication network, in terms of maximizing a given objective function, such as, total energy consumption, end-to-end delay, bit error rates, or throughput. This chapter uses the MDP version used in [10, 16] for finding the optimal cooperative transmission optimization in EHTs in terms of maximizing the average throughput of the system.

3.3 ENERGY HARVESTING TECHNOLOGIES

EH (also known as power harvesting or energy scavenging) is the process by which energy is collected from external sources (e.g., solar power, thermal energy, wind energy, and kinetic energy), captured, and stored for small, wireless autonomous devices, such as wearable electronics and wireless communication networks. EH from external sources has been used widely in large-scale applications such as EH from windmills and waterwheels.

Some of the fundamental discoveries that are worth mentioning and serve as basis for modern EH technologies include the discovery of the piezoelectric effect by the brothers P. Curie and J. Curie, who realized in 1880 that an electric charge is stored in a solid material, such as a crystal, when the material is subject to mechanical stress, and the discovery of the photoelectric effect by H. Hertz in 1887, who realized that, when a surface is exposed to electromagnetic radiation, the radiation can be absorbed resulting in emission of electrons [34].

EH wireless devices are increasingly being deployed in practice, replacing their traditional, battery-operated counterparts, especially when device inaccessibility or location often makes battery replacement difficult and cost-prohibitive. Potential applications range from machine-to-machine and sensor networks to building automation and monitoring in smart grids [35]. EH technologies are also applicable to tiny impulse radio UWB (IR-UWB) wireless networks. In IR-UWB networks, energy can be harvested to power small, ultra-low-power wireless nodes. The idea of EH was proposed to address the problem of a finite lifetime in a wireless network by enabling the nodes to replenish energy from ambient sources. The commonly available energy sources in the environment, such as solar, mechanical vibration, and thermal energy, can now be efficiently converted into electrical energy through appropriate energy transducers.

3.3.1 Energy Harvesting Methods

- *Solar energy*: Solar photovoltaics is one of the fastest-growing power-generation technologies in the world. Silicon solar cells exploit the photovoltaic effect to convert sunlight into electricity. When the photons of sunlight hit the silicon cell, their energy may be absorbed and transferred to electrons of the silicon, which the electrons escape from their normal positions in the silicon and become part of the current in an electrical circuit. Since solar energy is a convenient harvesting source, lots of implementations of solar EH wireless devices have already been manufactured [2, 36–38]. Solar EH achieves the highest power density [39]. However, one limitation is that the solar energy supply is location dependent and is highly time varying. For example, the amount of solar energy available outdoors is a thousand times more than the amount of energy available indoors [40].
- *Vibrational energy*: Vibrations can generate electric energy. The harvesting device transduces the vibrational energy into electrical energy using three main methods: electromagnetic (inductive) system, electrostatic (capacitive) system, and piezoelectric material. This energy source is currently becoming popular in healthcare applications.
- *Thermal energy*: The thermoelectric effect is the direct conversion of temperature differences to electric voltage using the Seebeck effect. Thomas Johann Seebeck found that a voltage is produced across a circuit made from two dissimilar metals, with junctions at different temperatures. Thermoelectric devices utilize this effect and can generate electricity when there exists a temperature gradient across the device. Compared with vibration-based devices, thermoelectric devices can function for a much longer duration due to the absence of any moving parts.

3.3.2 Energy Harvesting for Ultra-low-power Active Networked Tags

Recently, a new class of ultra-low-power wireless nodes known as EnHANTs prototypes has been designed by researchers at Colombia University (http://enhants.ee.columbia.edu). The existing devices are bulky and require more energy than the energy available in commonplace environments such as indoors. EnHANTs can be attached to objects that are traditionally not networked and provide the infrastructure for various applications. Examples of these applications include locating misplaced items, continuous monitoring of objects (items in a store, boxes in transit), and determining locations of disaster survivors [9]. These devices have special features, such as the ability to communicate with each other, operate at ultra-low-power, achieve energy autonomous via collecting energy from the environment, and adapt communications and networking as per the available energy constraints.

EnHANTs harvest energy from the environment (solar, motion, or thermal) and can be networked similar to sensor nodes. IR-UWB technology is used for short-range communications with low data rates [41–43]. In [3], IR-CMOS circuits with energy consumption in the order of a nanojule per bit have been demonstrated. The authors demonstrated the IR-UWB transmitter and receiver that require 1.65 and 280 nJ/bit, respectively, at a 1 Mbit/s data rate. Similar studies showed that it is possible to arrive at designing IR-UWB transceivers of 3–5 GHz band and 0.1–1 Mbit/s data rates, with energy consumption of 50 pJ/bit and 500 pJ/bit by its transmitter and receiver sections, respectively. Such devices are for EnHANT indoor applications.

3.3.3 Energy Management in Energy Harvesting Active Networked Tags

The energy management of wireless networks is different from non-rechargeable battery-powered wireless devices in many ways. In a battery-powered device, the objectives of energy management design are to minimize energy consumption or to maximize the lifetime achieved under the required performance constraints [44–48]. However, in EH devices, the objective is to maximize the lifetime by taking into account the energy replenishment process. The system continues to operate for a long time with a potentially large amount of energy available to the nodes, which results in an EH device to remain functional for a long period of time. Hence, energy conservation is not the major design issue. For example, an overly conservative energy expenditure may limit the amount of transmitted data by failing to take full advantage of the EH process thus degrading reliability. On the other hand, an overly aggressive use of energy may result in an energy outage, which prevents nodes from functioning properly over a long time. The energy availability constraint, which requires the energy consumption to be less than the energy stored in the battery, must be met at all time. It is known as energy-neutral condition. This constraint complicates the design of an energy management policy, since the current energy consumption decision would affect the outcome in the future.

In general, the harvested energy availability typically varies with time in a non-deterministic manner [49, 50]. Recently, the amounts of energy traces available daily indoors and mobile outdoors in different environments have been published via the CRAWDAD repository [51] by Colombia University. This data available online can be used as an input for designing EH systems and algorithms. Interested readers can refer to http://enhants.ee.columbia.edu for more information.

A number of researches have been conducted on energy management in EH wireless networks. In [50], the complex time-varying nature of energy sources is characterized and modeled analytically. In [52], energy management schemes for a single EH sensor node that achieves the maximum throughput and minimum mean delay are proposed. In [53], Yang and Ulukus proposed a globally optimal off-line scheduling policy in a single-user EH wireless communication system. The authors studied the transmission completion time minimization problem in EH wireless networks and assumed that the EH times and harvested energy amounts were known before the transmission started. In [54], an online algorithm that achieves a sub-optimal utility performance in general energy-harvesting networks equipped with finite-capacity energy storage devices is proposed. They considered Lyapunov optimization techniques with weight perturbation. In [40], utility-optimal energy allocation schemes were considered for systems with predictable or stochastic energy availability. Energy management problems in EH wireless networks have been formulated in some recent works as a MDP [27, 55]. Chen et al. [55] studied the energy allocation problem of a single node using the shortest path approach. A simple distributed heuristic scheme was proposed that solves the joint energy allocation and routing problem in a rechargeable WSN. In [16], energy-efficient scheduling strategies for cooperative communications in EH WSNs are proposed to maximize the long-term utility. The scheduling optimization problems under two different assumptions were formulated and solved using MDP and POMDP. However, some of the EH policies mentioned above are computationally complex to apply to EnHANTs.

3.4 COOPERATIVE COMMUNICATIONS

The history of cooperative communication can be traced back to the work of Cover and El Gamal on the relay channel in 1979 [56]. They modeled a relay channel to include a source node, a relay node, and a destination node. This work analyzed the capacity of the three-node network consisting of a source, a destination, and a relay. However, the concept and idea behind our cooperative communication are quite different from the work on the relay channel. First, recent developments are mainly based on the concept of diversity in a fading channel, while Cover and El Gamal mostly analyze capacity in an additive white Gaussian noise. Second, in the work on the relay channel, the relay's main purpose is to help the main channel (source), whereas in cooperative communication, the total system resources are fixed, and sources act both as information sources as well as relays. Therefore, although [56] is historically importance for recent development on the relay channel, recent work in cooperation has considered a somewhat different view [56].

3.4.1 Relay Strategies

The general relay channel considered in this chapter is the three-terminal communication channel shown in Figure 3.1. The terminals are labeled source one (S_1), source two (S_2), and the destination (D). The information messages originate at the sources, and must travel to D. Each source can also act as a relay for the other terminal when required.

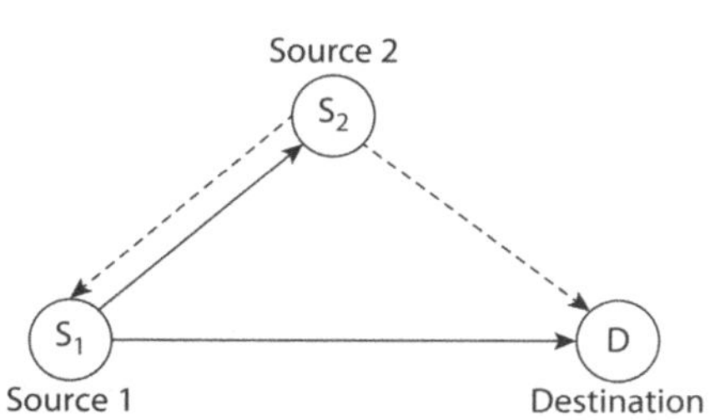

FIGURE 3.1 Three terminal cooperative system.

There are various methods of relaying exist in the literature. Conceptually, data is relayed in two orthogonal phases: first, when S_i transmits its data and (S_j, D) receive, commonly called the broadcast (BC) phase; and second when (S_i, S_j) transmit and D receive, also known as the multiple access phase, where $i, j \in (1,2)$ and $i\ 6 \neq j$. Note that this differentiation is only conceptual since it is possible for communication in both modes to take place simultaneously [57]. There are four different models of relaying that can be classified based on the above two transmission phases.

1. $S_i \rightarrow (S_j, D)$; $(S_i, S_j) \rightarrow D$ (general form of relaying);
2. $S_i \rightarrow S_j$; $(S_i, S_j) \rightarrow D$ (D ignores signal from S_i in first phase);
3. $S_i \rightarrow (S_j, D)$; $S_j \rightarrow D$ (S_i remains silent in second phase);
4. $S_i \rightarrow S_j$; $S_j \rightarrow D$ (multi-hop communication).

From the above classification, most early relaying results were based on the first model. The second and the third are simplified models used basically for analytical tractability such as for the analysis of outage probabilities and the design of space-time codes for fading relay channels, as discussed in [11, 58]. The fourth model of relaying is much older than that of the first. It is also simpler than the other three and is commonly known as multi-hop communication. It does not yield diversity benefits unlike the other three, and it is primarily used to combat signal attenuation in long-range communication.

A relay can also be classified as half-duplex or full-duplex based on the mode of operations. A relay is said to be half-duplex if it cannot simultaneously transmit and receive in the same band. That means the transmission and reception take place on orthogonal channels. On the other hand, if a relay transmits and receives simultaneously at the same time, it is known as full-duplex. In the full-duplex mode, the transmitted signal interferes with the received signal. In theory, it is possible for the relay to cancel out interference from the transmitted signal as it knows the transmitted signal. In this chapter, a half-duplex mode of channel allocation is used [11] as shown in Figure 3.2. Relaying allows sophisticated encoding in the transmit signals of the source node and the relay node and complicates processing and decoding of the received signals at the relay node and the destination. Due to an increasing

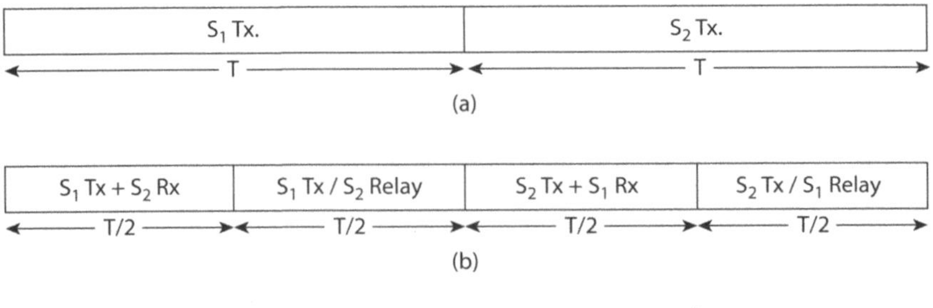

FIGURE 3.2 Example time-division channel allocations for (a) orthogonal direct transmission, and (b) orthogonal cooperative diversity transmission protocols. Let's focus on orthogonal transmissions of the form (b) throughout the chapter with little modification.

number of relaying protocols available in the literature, only the two major types of relaying: AF and DF are summarized.

3.4.1.1 Amplify and Forward (AF)

AF scheme was first proposed by Laneman et al. [11]. In the AF scheme, signal processing can be simplified into three phases: In Phase 1, the source node transmits signals, while the destination and the relay nodes receive the signals due to the broadcast nature. In Phase 2, the relay node amplifies the powers of the signals received from the source node and forwards them to the destination node. In Phase 3, the destination node combines and decodes the signals received from the source node in Phase 1 and the relay node in Phase 2 in order to recover the original information. Compared with other schemes, AF is the simplest. The destination node can receive independent fading signals from the source and relay nodes, thus achieving full diversity gain and good performance with this scheme. The drawback of the AF scheme is that it is prone to noise propagation effects as the relay node amplifies the noise on the source-relay channel.

Suppose, the source transmits $X_s[k]$ for $k = 1,2,\ldots, n$. The relay processes its corresponding received signal $Y_r[k]$ for $k = 1,2,\ldots, n$.. and relays the data it received by transmitting

$$X_r\left[k\right]=\Gamma_r Y_r\left[k-n\right], k=n+1, n+2,\ldots,2n \tag{3.1}$$

where the relay uses the gain $\Gamma_r \leq \sqrt{\dfrac{E_s}{\left|h_{s,r}\right|^2 E_r + N_0}}$ to remain within its energy constraint. Where E_s and E_r are signal energies of the source and relay nodes, respectively and $h_{s,r}$ is the fading coefficient between source and relay nodes.

3.4.1.2 Decode and Forward (DF)

The DF scheme was first presented by Sendonaris et al. [59]. In DF scheme, the relay applies a form of decoding or detection algorithm to the received signals and re-encodes the data into its transmit signals. This decoding and re-encoding process is a non-linear transformation of the received signals. This scheme has the advantage of reducing the effect of noise. However, it has the impact of limiting its performance because of the fading effects. Again, as a simple example, suppose the source transmits $X_s[k]$ for $k = 1,2,\ldots, n$. The relay decodes the signal $Y_r[k]$ it received from the source node and forms an estimate $\hat{X}_s[k]$ for $k = 1,2,\ldots, n$, and relays a re-encoded version of $\hat{X}_s[k]$]. For example, the relay can implement repetition coding by transmitting the signal

$$X_r\left[k\right]=\sqrt{\frac{E_r}{E_s}}\hat{X}_s\left[k-n\right], k=n+1, n+2,\ldots,2n \tag{3.2}$$

Furthermore, the destination processes its received signal by some form of diversity combining techniques. For detailed discussions on DF protocol, interested readers can refer to [60].

3.4.2 Cooperative Communication of Energy Harvesting Tags

IR-UWB is designed for short range (less than 10m) ultra-low-power communications [3]. One challenge in designing these devices is that the received energy per bit is much higher than the transmitted energy per bit. However, in a narrow-band communication system, the transmitted energy is much higher than the received energy. The energy consumption for EHT depends on the energy harvested and stored in its battery. Therefore, the energy spent for communication cannot exceed the energy available in the battery. The throughput performance of EHTs that cooperatively communicate with the destination outperforms the EHTs that independently communicate with the destination. EnHANTs communicate with the destination in two modes: the direct and the cooperative mode based on their energy states.

In the direct mode, also known as the independent state, a tag does not maintain contact with the neighboring tags. It decides the amount of energy for transmitting its data by itself based on the energy stored in the battery. Such an independent transmission mode is cheaper for a tag in terms of energy as it does not spend for the neighboring tag. In the cooperative mode, EHTs jointly decide on the amount of energy spent for the transmission and reception of information. They make joint decisions based on their own battery energies and via exchanges of the battery energy levels of each other. When a large number of tags are involved in cooperative communications, designing control decisions are complex and need large-scale optimization problems and a suitable solution has to be computed. In the next chapters, this problem will be solved by designing an affordable optimal cooperative transmission strategy for EHTs.

3.5 SYSTEM MODEL

3.5.1 Communication Model

Consider a network of EHT-equipped objects, in which each object communicates with a reader by using two modes: the *direct mode*, in which each object independently communicates with the reader, and the *cooperative mode*, in which neighboring objects cooperate to forward information to the reader. Assume that battery energy parameter exchanges are conducted between tags and a reader before any transmission attempt. For simplicity, consider two tags (T_1 and T_2) and a reader (R), as shown in Figure 3.3 (a). T_1 can directly communicate with R if it has sufficient battery energy, or it can communicate with the assistance of T_2 if its energy cannot support direct communication. T_2 cooperates with T_1 only if its stored battery energy is sufficient to receive and relay the data from T_1 and transmit its own data. Similarly, T_2 can also communicate with R either directly when it has sufficient battery energy or with the assistance of T_1 when it lacks energy.

It is assumed that each tag uses UWB pulse position modulation (PPM) [10] to transmit its information to the reader. Let $\mathbf{x}_m^n = \left[x_{1,m}^n, x_{2,m}^n, \ldots, x_{J,m}^n \right]$ be the mth encoded PPM symbol of duration T_s, where $n \in \{T_1, T_2\}$ denotes the transmitted signals from T_1 or T_2, $x_{i,m}^n \in \{0,1\}$ is the ith data bit within the mth symbol of tag n, and J is the number of information bits per encoded PPM symbol.

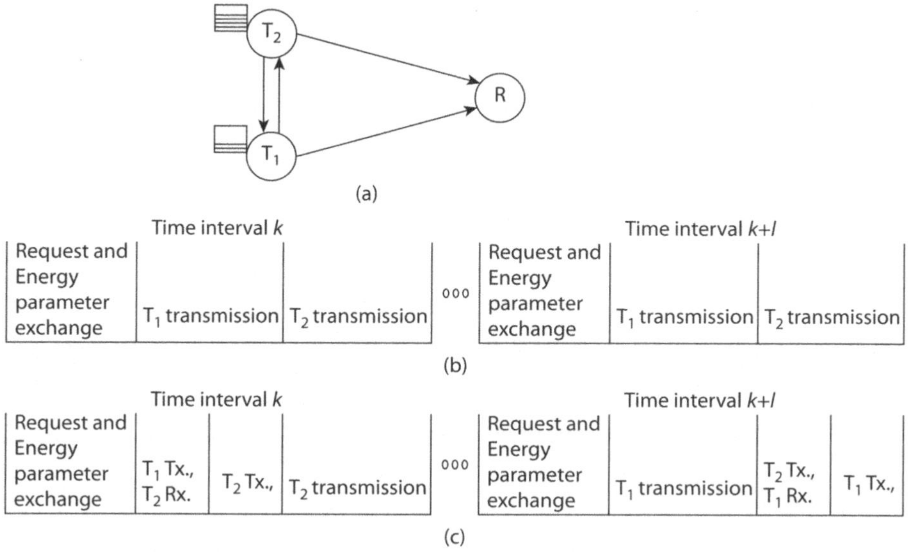

FIGURE 3.3 System Model. (a) Communication model. (b) Direct mode. (c) Cooperative mode: T_2 relays T_1's information at time interval k and T_1 relays T_2's data at time $k + l$, where k and l are positive integers. Request and battery energy parameter exchanges are conducted between tags and a reader at the initial stage.

The transmitted packet from each tag in an analog domain is expressed as follows:

$$s^n(t) = \sum_{m=0}^{L-1} \sum_{i=1}^{J} x_{i,m}^n p\left(t - iT_f - mT_s\right), \tag{3.3}$$

where $p(t)$ is a pulse of duration $T_p < T_f$, L is the number of PPM symbols per transmitted packet, and $T_f = T_s/J$. It is also assumed that the pulses in each symbol are unit pulses and note that a pulse presents only for a bit = '1'. Therefore, the energy of the transmitted packet is equal to the total number of nonzero pulses in the L encoded PPM symbols.

Assume that the communication is organized in equal duration time intervals. Each time interval comprises two equal time slots, allowing data to be transmitted from T_1 and T_2, as shown in Figure 3.3 (b) and (c) [11]. At the beginning of each interval, the reader makes a request to T_1 and T_2. When the tags receive a request, based on their battery energy levels, they can reply to a reader either directly or with the assistance of each other. Figure 3.3 (c) shows that at time interval k, T_1 communicates with a reader via the assistance of T_2, and at time interval $k + l$, T_2 communicates with a reader via the assistance of T_1. A tag helps its neighbor tag without affecting its own data transmission. When both tags get sufficient energy, they independently communicate with the reader as shown in Figure 3.3 (b).

First describe the signal model of the direct mode at any given time interval. When the battery energies of T_1 and T_2 are greater than or equal to the threshold values δ_{T1} and δ_{T2}, respectively, each tag can independently broadcast without assistance. Each tag transmits packets that contain its information symbols. Without loss of generality, each transmitted packet can be represented in terms of symbols. The mth transmitted symbol of a packet from T_1 and T_2 received by R in the direct mode can be modeled as follows:

$$y_{R,m}^{T_1}(t) = \vartheta_{T_1R}\sum_{i=1}^{J} x_{i,m}^{T_1} p\left(t - iT_f\right) + w_R^{T_1}(t) \tag{3.4}$$

$$y_{R,m}^{T_2}(t) = \vartheta_{T_2R}\sum_{i=1}^{J} x_{i,m}^{T_2} p\left(t - iT_f\right) + w_R^{T_2}(t), \tag{3.5}$$

where $w_R^{T_1}(t)$ and $w_R^{T_2}(t)$ are the ambient Gaussian noises with zero mean and equal variance N_0. ϑ_{uv} indicates the effect of path loss which is proportional to $1/d_{u,v}^{\eta/2}$, where $d_{u,v}$ is the distance between tags u and v, and η is the path loss exponent. Since IR-UWB pulses are robust to the effects of small-scale fading, this chapter considers only the channel path loss. Denote $\gamma_{T_1R} = \frac{\vartheta_{T_1R}^2 \varepsilon_{m,x}^{T_1}}{N_0}$ and $\gamma_{T_2R} = \frac{\vartheta_{T_2R}^2 \varepsilon_{m,x}^{T_2}}{N_0}$ to be the received signal-to-noise ratios (SNRs) at R from T_1 and T_2, respectively, where $\varepsilon_{m,\mathbf{x}}^{T_1}$ and $\varepsilon_{m,\mathbf{x}}^{T_2}$ are the transmitted symbol energies (total number of non-zero pulses) of $\mathbf{x}_m^{T_1}$ and $\mathbf{x}_m^{T_2}$ respectively.

Next, the received signal model will be described for the cooperative mode. Consider two tags helping each other when any one tag lacks sufficient energy and a second tag exhibits an energy surplus, indicating that it has sufficient energy to transmit its own data and to receive and relay the data from the neighboring tag. Consider a case in which T_1 needs assistance and T_2 has sufficient energy to relay T_1's packet. By using the AF relay mode at T_2, cooperative transmission occurs during two equal transmission phases of T_1's time slot (Figure 3.3 (c)). During the first phase of T_1's time slot, T_1 broadcasts its data to T_2 and R. During the second phase of T_1's time slot, T_1 becomes silent and T_2 relays the information it received from T_1 to R. The mth transmitted symbol of T_1 received at T_2 and R in the first and second phases of T_1's time slot can be modeled as follows:

$$y_{T_2,m}^{T_1}(t) = \vartheta_{T_1T_2}\sum_{i=1}^{J} x_{i,m}^{T_1} p\left(t - iT_f\right) + w_{T_2}^{T_1}(t) \tag{3.6}$$

$$y_{R,m}^{T_1}(t) = \vartheta_{T_1R}\sum_{i=1}^{J} x_{i,m}^{T_1} p\left(t - iT_f\right) + w_R^{T_1}(t) \tag{3.7}$$

$$y_{R,m}^{T_2}(t) = \Gamma\, \vartheta_{T_2R}\, y_{T_2,m}^{T_1}(t) + w_R^{T_2}(t), \tag{3.8}$$

where $w_{T_2}^{T_1}(t)$ is the ambient Gaussian noise at T_2 with zero mean and equal variance N_0. Γ is the signal amplification factor of T_2, which can be expressed as $\Gamma^2 = \frac{\varepsilon_{m,\mathbf{y}}^{T_2}}{\vartheta_{T_1T_2}^2 \varepsilon_{m,\mathbf{x}}^{T_1} + N_0}$, where $\varepsilon_{m,\mathbf{y}}^{T_2}$ is the transmitted symbol energy of $\mathbf{y}_{T_2,m}^{T_1}$. Another case in this cooperative mode is when T_2 lacks energy and T_1 exhibits an energy surplus. The received signal model for this scenario is determined by replacing T_2 with T_1 and vice versa in (3.6)–(3.8). It is assumed that the reader decodes the information after combining the signals received from T_1 and T_2 by using maximum ratio combining [61]. The end-to-end received SNR at the reader when T_2 is used as an AF relay is expressed as follows [14]:

$$\gamma_{T_1T_2R} = \gamma_{T_1R} + \frac{\gamma_{T_1T_2}\gamma_{T_2R}}{1+\gamma_{T_1T_2}+\gamma_{T_2R}}, \tag{3.9}$$

where γ_{T_1R} is the SNR associated with the direct link T_1R due to the broadcast nature of T_1. $\gamma_{T_2R} = \frac{\vartheta^2_{T_2R}\varepsilon^{T_2}_{m,\mathbf{y}}}{N_0}$, is the SNR associated with the T_2R link. From now on, drop the subscripts m, $\mathbf{x}$, and $\mathbf{y}$ for ease of notation; namely, use ε^n as the transmit symbol energy for tag n. Similarly, the end-to-end SNR at the reader when T_1 is used to relay the signal from T_2 is expressed as follows:

$$\gamma_{T_2T_1R} = \gamma_{T_2R} + \frac{\gamma_{T_2T_1}\gamma_{T_1R}}{1+\gamma_{T_2T_1}+\gamma_{T_1R}}, \tag{3.10}$$

where γ_{T_2R} and γ_{T1R} are the SNRs associated with the direct link T_2R and relay link T_1R respectively. The SNRs at the reader from T_1 and T_2 can be written as follows using either of the modes:

$$\gamma_{T_1} = \begin{cases} \gamma_{T_1R} & ;\ \text{direct mode} \\ \gamma_{T_1T_2R}; & \text{cooperative mode} \end{cases} \tag{3.11}$$

and

$$\gamma_{T_2} = \begin{cases} \gamma_{T_2R} & ;\ \text{direct mode} \\ \gamma_{T_2T_1R}; & \text{cooperative mode.} \end{cases} \tag{3.12}$$

Note that γ_{Ti}, $i = 1,2$ is a function of the transmitted signal energies of the tags. In the absence of the cooperating tag when a tag needs assistance, the tag still broadcasts its data during the first phase of its time slot. There is still a chance for the reader to detect the data directly. However, successful detection with higher probability occurs when a neighboring tag can be in the cooperative mode.

It is assumed that the reader processes the received PPM signal by using a compressive sensing technique and signal detection methods with compressive measurements [10, 62]. This avoids the need to employ high sampling rate A/D converters that are complex and expensive.

To recover the original PPM signal $\mathbf{y}^n \in \mathcal{R}^N$ received at R from tag n by using compressive linear measurements, consider an $M \times N$ measurement matrix $\mathbf{\Phi}$, where $\mathbf{y}^n$ is a sample of $y^n_{R,m}(t)$. $\mathbf{y}^n$ can be recovered from the compressed samples $\tilde{\mathbf{y}}^n = \mathbf{\Phi}\mathbf{y}^n \in \mathcal{R}^M$ by using simple and computationally tractable algorithms, given that $M << N$ at a compression ratio of $\frac{M}{N}$. Assume that there are K PPM symbols. Their corresponding projections can be described as $\tilde{\mathbf{y}}_i^n = \mathbf{\Phi}\mathbf{y}_i^n, i = 1,2,\ldots,K$, where $\mathbf{y}_i^n$ corresponds with the samples of the ith PPM symbol. The reader performs the following decision rule to detect the transmitted PPM signal:

$$\hat{i} = \text{argmin}_{1\leq i\leq K}\left(\tilde{y}^n - \tilde{y}_i^n\right)^T\left(\Phi\Phi^T\right)^{-1}\left(\tilde{y}^n - \tilde{y}_i^n\right), \tag{3.13}$$

When using this detection method, the probability of mis-detecting a symbol is expressed as follows [10]:

$$p_{md}^n\left(\varepsilon^n\right) = 1 - Q\left(-\sqrt{\frac{M}{N}}\gamma_n\left(\varepsilon^n\right)\right)^{K-1}, \tag{3.14}$$

where $Q(x) = \frac{1}{\sqrt{2\pi}}\int_x^{\infty} e^{\frac{-t^2}{2}}dt$, and γ_n is determined based on (3.11) or (3.12).

In a network of EHTs that depends solely on environmental conditions for communications, communication error is caused by two error sources: the *no-response errors* and the *mis-detection errors.* The no-response error (p_{nr}) occurs when the objects do not reply to the reader's requests due to a lack of sufficient energy in their batteries. The mis-detection error (3.14) occurs when the reader fails to successfully decode the data transmitted by an object. Therefore, at any given time interval k, the weighted sum of these two error probabilities, which results in communication error probability in EHTs, is defined as follows [10]:

$$p_e^n\left(\varepsilon_k^n\right) = \left(\beta\cdot p_{nr}\left(\varepsilon_k^n\right) + (1-\beta)\,p_{md}^n\left(\varepsilon_k^n\right)\cdot 1_{\{\varepsilon_k^n > 0\}}\right)\cdot a_k, \tag{3.15}$$

where $p_{nr} = 1_{\{\varepsilon_k^n = 0\}}$ defines the no-response error, ε_k^n is the symbol of energy ε^n at the kth time interval, and $\beta \in [0,1]$ is a weighting factor that constitutes the two errors under one performance metric. β is adjustable depending on the application, that is, whether the no-response or mis-detection error is more important. $a_k = 1$ indicates a reader request and $a_k = 0$ indicates no request. The indicator function $1_{\{X\}}$ is defined as $1_{\{X\}} = 1$ if X is true, and 0 otherwise.

3.5.2 Energy Harvesting Dynamics

The reader is assumed to have no power constraints and each tag is powered by a finite capacity rechargeable energy storage device that can harvest substantial energy from ambient sources such as the sun or vibrations. It is assumed that each tag tracks its current battery energy and that of its neighboring tags through energy parameter exchanges or reader inquiries. Let $B_k^{T_1}$ and $B_k^{T_2}$ be the battery energies of T_1 and T_2, respectively at the beginning of the kth time interval. The tag's energy consumption is constrained by the battery levels; in other words, $0 \leq L\varepsilon_k^{T_1} + \left(\frac{L}{2}\varepsilon_k^{T_1} + C_k^{T_1}\right)\cdot 1_{\{B_k^{T_2} < \delta_{T_2}\}} \leq B_k^{T_1}$ and $0 \leq L\varepsilon_k^{T_2} + \left(\frac{L}{2}\varepsilon_k^{T_2} + C_k^{T_2}\right)\cdot 1_{\{B_k^{T_1} < \delta_{T_1}\}} \leq B_k^{T_2}$.

A tag T_i consumes $L\varepsilon_k^{T_i}$ energy in the direct mode and $\frac{L}{2}\varepsilon_k^{T_i}$ energy in the cooperative mode for the transmission. $C_k^{T_i}$ spent by T_i includes the following energies; energy spent in a passive sleep state, energy to exchange state information before data transmission attempt, and energy to receive data from neighboring tag by T_i during cooperation. Unlike the conventional communication system, in IR-UWB-based EHTs, the energy required to receive information is more than the energy to transmit it. Assume that the energy spent in exchanging information is relatively small and can be ignored. In addition, we do not consider the synchronization issues and accompanying energy consumption due to the simple EHT design and ED techniques employed. Let $h_k^{T_1}$ and $h_k^{T_2}$ denote the amount of energy harvested by T_1 and T_2 during the kth time interval. Energy arrival can be modeled as predictable or random energy models. This study considers only the random energy arrival process and it is assumed that both $h_k^{T_1}$ and $h_k^{T_2}$ take discrete values from the set $\mathcal{H} = \{H_1, H_2,\ldots, H_D\}$. The number of values in this set is critical to the computational complexity of the system. Taking into account the computational complexity, energy arrival here is modeled as a three-state Markov process ($D = 3$) [63] and the state transition probability can be expressed as follows:

$$q^n = \begin{bmatrix} q^n_{H_1H_1} & q^n_{H_1H_2} & q^n_{H_1H_3} \\ q^n_{H_2H_1} & q^n_{H_2H_2} & q^n_{H_2H_3} \\ q^n_{H_3H_1} & q^n_{H_3H_2} & q^n_{H_3H_3} \end{bmatrix}, \tag{3.16}$$

where $q^n_{H_i\,H_j}$ represents the probability of the EH state transferring from the current state, H_i, to the next state, H_j for all $i,j \in \{1,2,3\}$. Let the steady-state probability be expressed as $\left[q^n_{H_1}, q^n_{H_2}, q^n_{H_3}\right]$. The request of the reader at the beginning of the kth time interval is modeled as

$$\alpha_k \sim Bernoulli(r),$$

where $\alpha_k = 1$ indicates a reader request with a probability of r and $\alpha_k = 0$ indicates no request, yielding probability $(1-r)$. The stored energy level at the beginning of the kth time interval linearly increases or decreases, provided that the maximum battery capacity is not exceeded. The energy stored in the batteries of T_1 and T_2 at time interval k for use in the subsequent time interval is determined as follows:

$$B_{k+1}^{T_1} = \min\left(B_k^{T_1} - \alpha_k E_k^{T_1} + h_k^{T_1}, B_{max}^{T_1}\right) \tag{3.17}$$

$$B_{k+1}^{T_2} = \min\left(B_k^{T_2} - \alpha_k E_k^{T_2} + h_k^{T_2}, B_{max}^{T_2}\right), \tag{3.18}$$

where $E_k^{T_1} = L\varepsilon_k^{T_1} + \left(\frac{L}{2}\varepsilon_k^{T_1} + C_k^{T_1}\right)\cdot 1_{\{B_k^{T_2}<\delta_{T_2}\}}$ and $E_k^{T_2} = L\varepsilon_k^{T_2} + \left(\frac{L}{2}\varepsilon_k^{T_2} + C_k^{T_2}\right)\cdot 1_{\{B_k^{T_1}<\delta_{T_1}\}}$. $B_{max}^{T_1}$ and $B_{max}^{T_2}$ are the maximum battery capacities of T_1 and T_2, respectively. A cooperating tag amplifies and forwards the received signal from its neighbor only if at least E_k energy is stored in its battery, otherwise, the cooperating relay remains silent. Denote $A_k = \left(\varepsilon_k^{T_1}, \varepsilon_k^{T_2}\right)$ to be the joint energy per encoded symbols of T_1 and T_2 at the kth time interval expressed in finite discrete values, and $\mathcal{A}$ to be the set of all such possible joint symbol energies by both tags; in other words, $\mathcal{A} = (a_1, a_2, \ldots, a_l)$. $\varepsilon_k^{T_1}, \varepsilon_k^{T_2} C_k^{T_1}$, and $C_k^{T_2}$ are determined based on the hardware design.

3.6 PROBLEM FORMULATION

3.6.1 Performance Measure

In this section, we formulate the problem of maximizing the average throughput for optimal cooperative transmission strategy. This maximization problem is formulated as an infinite horizon MDP problem. We assume that each time slot in the time interval k is sufficient for transmitting a packet of L symbols. The state of the tags at the beginning of the kth time interval is defined as $S_k = \left(B_k^{T_1}, B_k^{T_2}, h_k^{T_1}, h_k^{T_2}, \alpha k\right)$. Let $\mathcal{S}$ denote the set of all possible states. That is, $\mathcal{S} = \left\{s_1, s_2, \ldots, s_{|\mathcal{S}|}\right\}$, where $|\mathcal{S}|$ is the number of possible states. The transmission policy $\pi(s)$ is defined as a mapping from the states $\mathcal{S}$ to the joint energy $\mathcal{A}$. Let $\prod$ denote the set of all possible policies; that is, $\prod = (\pi_1, \pi_2, \ldots, \pi_l)$. Given the current state S_k and the policy π: $\mathcal{S} \rightarrow \mathcal{A}$, the packet throughput of tag n at the kth time interval is determined as

$$U^n\left(\varepsilon_k^n\right) = \begin{cases} R_s\left(1 - p^e\left(\varepsilon_k^n\right)\right)^L \cdot \alpha_k;\ \text{direct mode} \\ R_s \dfrac{1}{2}\left(1 - p_e^n\left(\varepsilon_k^n\right)\right)^{\frac{L}{2}} \cdot \alpha_k;\ \text{cooperative mode} \end{cases}, \tag{3.19}$$

where R_s is the symbol rate. The throughput in cooperative mode is multiplied by 1/2 indicating that the source tag sends half of a packet due to half duplex mode of AF. For a given transmission policy $\pi(S_k) = A_k$, the long-term average throughput of the two tags throughout over all time-intervals is expressed as follows:

$$U(\pi) = \lim_{N \to \infty} \frac{1}{N} \sum_{k=1}^{N} \left(U^{T_1}\left(\varepsilon_k^{T_1}\right) + U^{T_2}\left(\varepsilon_k^{T_2}\right)\right). \tag{3.20}$$

Therefore, the objective is to determine a policy that maximizes the average packet throughput and the optimization problem can be formulated as follows:

$$u: \begin{array}{ll} \underset{\pi \in \Pi}{\text{Maximize}} & U(\pi) \\ \text{subject to} & \text{battery states (3.17) and (3.18)} \end{array} \tag{3.21}$$

It is difficult to compute an optimal policy for (3.21), particularly when numerous state variables are involved. The next section presents how the problem can be cast as an MDP to obtain an optimal policy.

3.6.2 Markov Decision Process Formulation

An MDP provides a framework for modeling sequential decision-making problems, in which the outcomes are partially random and partially controlled by the decision-maker. Figure 3.4 shows the state transition diagram [30].

An MDP is defined using a quadruplet $\left(\mathcal{S}, \mathcal{A}, p_{a_i}\left(s_i, s_j\right), R_{a_i}\left(s_i, s_j\right)\right)$, where $\mathcal{S}$ is the set of states, $\mathcal{A}$ is the set of actions, $p_{a_i}\left(s_i, s_j\right)$ denotes the transition probability from state s_i to state s_j when an action $a_i \in \mathcal{A}$ is taken, and $R_{a_i}\left(s_i, s_j\right)$ is the reward due to transitions from state s_i to state s_j when an action a_i is taken. The goal of an MDP is to choose a policy π that assigns actions to each state and maximizes the average reward. At any time interval k, T_1, and T_2 spend the joint energy a_k per symbol to send their data to the reader either directly or by cooperating. The joint choice a_k causes a state change from S_k to S_{k+1}, yielding the immediate reward of $R_{a_k}\left(S_k, S_{k+1}\right)$. $R_{a_k}\left(S_k, S_{k+1}\right)$ is chosen to equal the throughput and R_s is normalized to one. $\mathcal{V}^n$ denotes such normalized throughput corresponding to tag n. It is expressed as follows:

$$\mathcal{V}^n\left(\varepsilon_k^n\right) = \begin{cases} \left(1 - p_e^n\left(\varepsilon_k^n\right)\right)^L \cdot \alpha_k; \text{ direct mode} \\ \frac{1}{2}\left(1 - p_e^n\left(\varepsilon_k^n\right)\right)^{\frac{L}{2}} \cdot \alpha_k; \text{ cooperative mode} \end{cases} \tag{3.22}$$

and the immediate joint reward is determined as follows:

$$R_{a_k}\left(S_k, S_{k+1}\right) = \mathcal{V}^{T_1}\left(\varepsilon_k^{T_1}\right) + \mathcal{V}^{T_2}\left(\varepsilon_k^{T_2}\right). \tag{3.23}$$

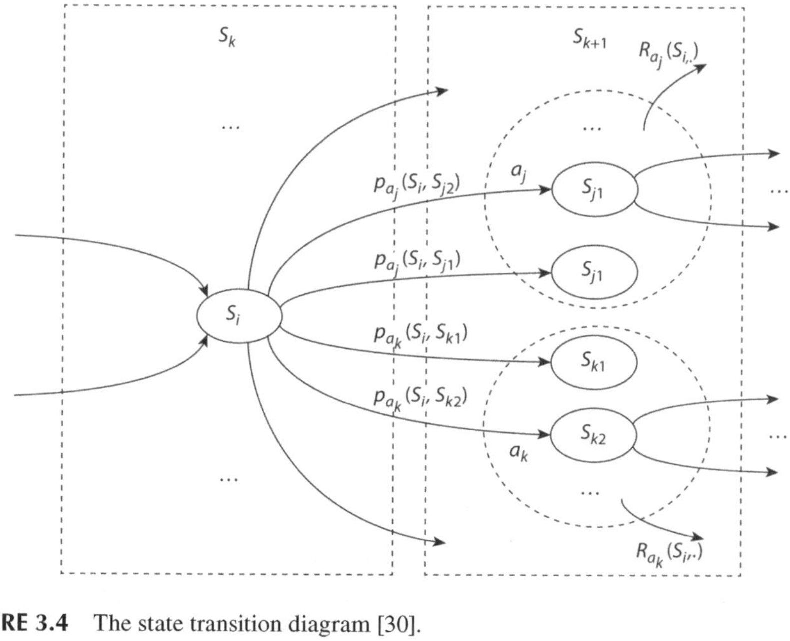

FIGURE 3.4 The state transition diagram [30].

Thus, the infinite horizon average reward of the tags is expressed as follows:

$$R_{ih}(\pi) = \lim_{N\to\infty} \frac{1}{N} \sum_{k=1}^{N} R_{a_k}(S_k, S_{k+1}), \tag{3.24}$$

where $a_k = \pi(S_k)$.

Comparing (3.20) and (3.24) shows that these equations are identical, excepting a scaling factor equal to R_s. This factor does not affect the behavior of the proposed performance metric; therefore, the optimal cooperative transmission policy, π, can be solved as follows by using the MDP problem:

$$\hat{u}: \begin{array}{ll} \underset{\pi\in\Pi}{\text{Maximize}} & R_{ih}(\pi) \\ \text{subject to} & \text{battery state } (3.17) \text{ and } (3.18) \end{array} \tag{3.25}$$

Therefore, the optimization problem in (3.21) can be solved by using the MDP problem in (3.25).

3.7 OPTIMAL TRANSMISSION POLICY

3.7.1 THE VALUE ITERATION

In this section, we solve problem (3.25) by using value iteration MDP. The value iteration MDP is proposed that achieves the maximal average throughput over an infinite horizon. Because all components of S_k are discrete, a finite number of states exists.

When the action a_k is taken, the state S_k yields a transition to any of the next state S_{k+1}. The state transition probability $p_{a_k}(S_k, S_{k+1})$ is the probability that the system will go to state S_{k+1} when action a_k is taken at state S_k during the kth time interval. The state transition probability depends on the state and system parameters r, q^{T1}, and q^{T2} and can be defined as follows:

$$p_{a_k}(S_k, S_{k+1}) = q^{T_1 T_1}_{h_k, h^{T_1}_{k+1}} q^{T_2 T_2}_{h_k, h^{T_2}_{k+1}} p(\alpha_{k+1}) \tag{3.26}$$

The proposed model is a unichain MDP model; according to Section 8.5 of [28], a deterministic, stationary policy exists for unichain models that exhibit average reward criteria, yielding steady-state transition probability distribution. Thus, the optimal policy π^*: $\mathcal{S} \to \mathcal{A}$ can be determined by solving the optimality equation for average expected reward criteria and is expressed as follows [28, 30]:

$$\begin{aligned} &\lambda^* + \upsilon^*(S_k) \\ &= \max_{a\in\mathcal{A}} \sum_{S_{k+1}=g}^{G} p_a(S_k, S_{k+1})\left(R_a(S_k, S_{k+1}) + \upsilon^*(S_{k+1})\right), \end{aligned} \tag{3.27}$$

where λ^* is the optimal average reward and $\upsilon^*(g)$ are the optimal rewards beginning at $g = \left(B^{T_1}_{k+1}, B^{T_2}_{k+1}, H_1, H_1, 0\right), \ldots, \left(B^{T_1}_{k+1}, B^{T_2}_{k+1}, H_3, H_3, 1\right) = G$. For evaluation purposes, the relative value iteration (RVI) algorithm is used to solve (3.27). The solution (λ^*, υ^*) of the optimality equation yields an optimal policy, which is expressed as follows:

$$\pi^*(S_k) = \text{argmax}_{a\in\mathcal{A}} \sum_{S_{k+1}=g}^{G} p_a(S_k, S_{k+1})\left(R_a(S_k, S_{k+1}) + \upsilon^*(S_{k+1})\right). \tag{3.28}$$

RVI is a value iteration algorithm on the relative value function and its basic idea is to iterate the policy search process until the iteration variable converges. Specifically, at the (n+1)-th iteration, we have the iteration variables $\upsilon^n(s)$, $s \in S$, which are the reward iteration values corresponding to different states computed from the previous iteration.

At the beginning, we set $\upsilon^0(s) = 0, s \in S, \epsilon$ to be a very small positive value and select any state (s*) that bears the maximum iteration variable in the system to be the base state. Let $\upsilon(s^*) = \lambda$, which is the average reward. At each iteration, the algorithm normalizes $\upsilon^n(s)$ by subtracting λ^n from each component. The algorithm determines whether the normalized υ^n found in each iteration satisfies the stopping criterion span $(\upsilon^{n+1} - \upsilon^n) < \epsilon$, where the span of a vector υ is defined as span $(\upsilon) = \max_{s\in S} \upsilon(s) - \max_{s\in S} \upsilon(s)$. If not, the algorithm repeats the iteration process by updating the iteration variable. If the stopping criterion is satisfied, we calculate the policy π^n corresponding to $\upsilon^n = \upsilon^*$, and consider π^n as the overall optimal policy π^*. RVI is a value iteration algorithm on the relative value function $\upsilon^{n+1}(s) = \upsilon^n(s) - \lambda^{n+1}$ and is given in Algorithm 1.

Algorithm 1: RVI Algorithm for Solving (13.27)

begin
- Initialize $v^0 = 0$ and specify $\epsilon > 0$.
- Choose $s^* \in \mathcal{S}$.
- $n \leftarrow 0$;
- **repeat**
  - $\lambda^{n+1} = \max_{a\in\mathcal{A}} \left\{ \sum_{s'\in\mathcal{S}} p_a(s^*, s')(R_a(s^*, s') + v^n(s')) \right\}$;
  - **for** $s \in \mathcal{S}$ **do**
    - $v^{n+1}(s) = \max_{a\in\mathcal{A}} \left\{ \sum_{s'\in\mathcal{S}} p_a(s, s')(R_a(s, s') + v^n(s')) \right\} - \lambda^{n+1}$;
    - $n \leftarrow n + 1$;
- **until** $span(v^{n+1} - v^n) < \epsilon$;
- **for** $s \in \mathcal{S}$ **do**
  - $\pi^*(s) \in \text{argmax}_{a\in\mathcal{A}} \left\{ \sum_{s'\in\mathcal{S}} p_a(s, s')(R_a(s, s') + v^n(s')) \right\}$;
- **return** λ^n, v^n, π^*;

3.7.2 Convergence of the RVI Algorithm

The proposed unichain MDP model has finite state space, finite action space, and infinite horizon average reward. According to [28], the RVI algorithm converges to an optimal policy if, for every pair of deterministic policies π_1 and π_2,

$$\min_{\pi_1,\pi_2\in\Pi}\ \min_{(s,u)\in\mathcal{S}\times\mathcal{S}} \sum_{z\in\mathcal{S}} \min\left\{p^{Z}_{\pi_1}(s,z), p^{Z}_{\pi_2}(u,z)\right\} > 0, \tag{3.29}$$

where $p^{Z}_{\pi}(s,z)$ is the transition probability from state s to state z within Z state transitions under the policy π. For our problem, the EH rate $q^{n}_{H_i,H_j}$ by the two tags and the reader request rate r are assumed as $0 < q^{n}_{H_i,H_j} < 1$ and $0 < r < 1$, respectively. In addition, the tags become silent when there is no request from the reader. Therefore, for $\epsilon > 0$, the RVI algorithm converges to an optimal solution (λ^*, υ^*) within a finite number of iterations n.

3.8 COOPERATIVE COMMUNICATION UNDER TIME VARYING CHANNEL ENVIRONMENT

In the previous system model, we assumed that IR-UWB-based EHTs are robust against small-scale channel fading due to small duty cycle and very short pulse widths. However, the presence of strong noise still affects the reception of the pulses having ultra-low-power spectral density. The scattering and reflections with a large number of obstacles in an indoor channel environment also give rise to channel fading. Thus we were motivated to include channel fading parameter in our system model and see its effect on the overall throughput performance of the system.

In this section, we propose a cooperative approach that exploits the received signals coming from both the relay and the source. The transmitted packet from each tag in an analog domain is given by (3.3). θ_l in Figure 3.5 is the channel gain which shows the effects of path loss, and log-normal fading and can be modeled as $\Theta_l = -10\log_{10} d + h' \rightarrow \theta_l = \frac{h}{d^{\eta}}$, where $l \in \{T_1R, T_2R, T_1T_2 \text{ or } T_2T_1\}$ denotes the T_1R, T_2R, T_1T_2 or T_2T_1 link, respectively. Θ_l is in dB, d is the link length, η is the path loss exponent, and h' is the log-normal fading term. Assume $\theta_{T1T2} = \theta_{T2T1}$ for ease of notation.

By using the same communication modes (*direct* and *cooperative*) as in Section 3.2 of this chapter, T_1 and T_2 can communicate with the reader either independently

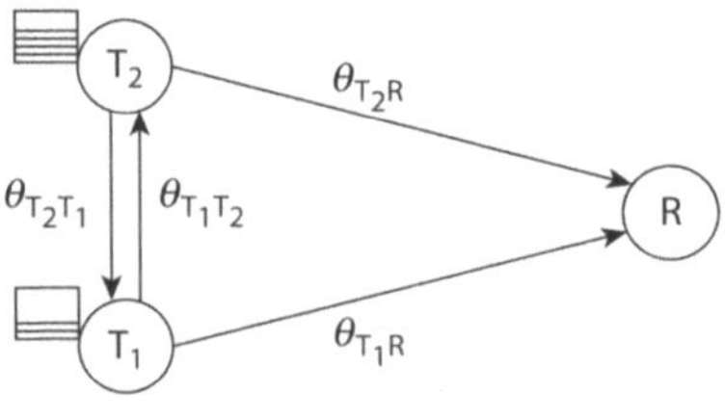

FIGURE 3.5 Communication model under fading channel.

under sufficient battery energy condition or by assisting each other if any one lacks energy.

The channel coefficient is assumed to evolve according to a two-state Markov chain.

Each state corresponds to a degree of channel quality. State 1 corresponds to good channel quality or connectivity, while state 0 corresponds to poor connectivity. Time is slotted and the channel remains fixed within each slot and moves into another state in the next slot following the state transition probability of an underlying Markov chain. The transition probability from state 0 to state 1 is p_{bg} and the transition probability from state 1 to state 1 is p_{gg} as shown in Figure 3.6 [30]. We consider the case that p_{gg} is larger than p_{bg} which is noted as a positive memory channel as described in [64]. The channel state varies according to the two-state Markov chain with the transition matrix of the Markov chain given by

$$P_l = \begin{bmatrix} p_{gg} & p_{gb} \\ p_{bg} & p_{bb} \end{bmatrix} \tag{3.30}$$

for a link l. The SNR of the tags' information at the reader in the direct communication mode is equal to

$$\gamma_{T_iR} = \frac{\theta_{T_iR}\varepsilon^{T_i}}{N_0}, i = 1, 2. \tag{3.31}$$

where, $N_0 > 0$ denotes the variance of the zero-mean Gaussian noise present in the channel. Similarly, denote $\gamma_{T_1T_2} = \frac{\theta_{T_1T_2}\varepsilon^{T_1}}{N_0}$ and $\gamma_{T_2T_1} = \frac{\theta_{T_2T_1}\varepsilon^{T_2}}{N_0}$. We assume that the reader has full access to channel states of both tags. In the cooperative mode of communication, when the two tags help each other, we have two channel state conditions. When the channel state information is a priori known by the tags before transmission, they can dynamically allocate their transmission energies according to the battery states in each transmission interval. Denote this condition as PCSI to indicate the presence of channel state conditions at each tag [65]. The resulting instantaneous SNR at the reader, after MRC, can be expressed as

$$\gamma_{T_iT_jR} = \gamma_{T_iR} + \frac{\gamma_{T_iT_j}\gamma_{T_jR}}{1 + \gamma_{T_iT_j} + \gamma_{T_jR}}, i, j \in \{1, 2\} i \neq j. \tag{3.32}$$

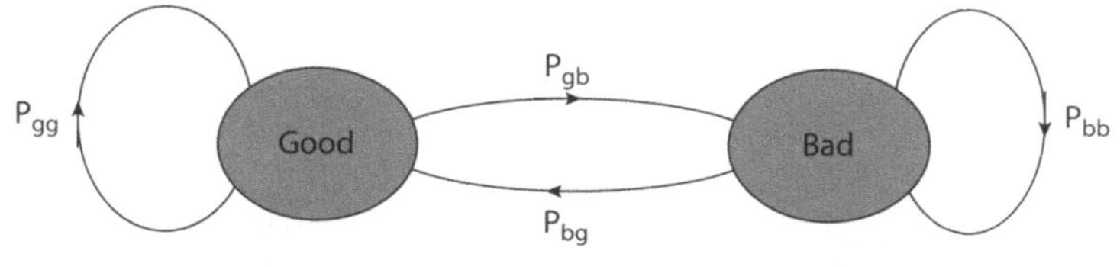

FIGURE 3.6 The two-state Markov chain for the channel coefficient [30].

When the tags do not know the current channel state information, the instantaneous SNR at the reader depends only on the previous knowledge of channel information. Denote this condition as NoCSI to indicate that tags do not know the current channel conditions, and also denote the channel gains θ_{T1R}, θ_{T2R}, and θ_{T1T2} as θ_1,θ_2, and θ_3, respectively for ease of notation. The expected SNR at the reader under direct mode of communication becomes

$$\gamma_{T_iR} = E_{\theta_i}\left[\gamma_{T_iR}\right], i = 1,2. \tag{3.33}$$

and the resulting instantaneous SNR at the reader due to cooperative mode of communication, after MRC, can be expressed as

$$\gamma_{T_iT_jR} = E_{\theta_i}\left[\gamma_{T_iR}\right] + \frac{E_{\theta_3}\left[\gamma_{T_iT_j}\right]E_{\theta_j}\left[\gamma_{T_jR}\right]}{1 + E_{\theta_3}\left[\gamma_{T_iT_j}\right] + E_{\theta_j}\left[\gamma_{T_jR}\right]}, i, j \in \{1,2\} i \neq j. \tag{3.34}$$

where, $E_Z\left[\cdot\right]$ is the expectation operator over the random variable Z. We assume that the channel gains θ_1,θ_2, and θ_3 are random variables that take values from $\{G_b, G_g\}$ according to the two state Markov chains mentioned above and based on the bad or good channel state.

Therefore, the SNR at the reader from T_1 and T_2 using either of the modes can be given by:

$$\gamma_{T_1} = \begin{cases} \gamma_{T_1R}; \text{ direct mode} \\ \gamma_{T_1T_2R}; \text{ cooperative mode} \end{cases} \tag{3.35}$$

and

$$\gamma_{T_2} = \begin{cases} \gamma_{T_2R}; \text{ direct mode} \\ \gamma_{T_2T_1R}; \text{ cooperative mode} \end{cases} \tag{3.36}$$

where, $\gamma_{T_1T_2R}$ or $\gamma_{T_2T_1R}$ for the cooperative mode is obtained from (3.32) or (3.34) based on the channel conditions. Time is slotted in duration of equal time interval and the channel gain remains fixed within each time interval and moves into another state in the next time interval following the state transition probability of an underlying two state Markov chain. At any time interval k, the expected SNR (say $\gamma_{T_iR,k}$) can be computed based on the channel condition in the $k-1$ time interval as:

$$E_{\theta_{k,i}}\left[\gamma_{T_iR,k}\right] = \sum_{\theta_{k,i} \in \{G_b, G_g\}} \gamma_{T_iR,k} \cdot p\left(\theta_{k,i} \middle| \theta_{k-1,i}\right) \tag{3.37}$$

Note that to see the throughput performance under fading channel, we simply replace equations (3.11) and (3.12) by (3.35) and (3.36), respectively, and perform similar MDP formulations as in previous sections (3.2–3.4).

3.9 NUMERICAL RESULTS

In this section, we provide the simulation results to demonstrate the performance of our proposed optimal cooperative transmission scheme under different settings. We compare our results with the direct optimal transmission scheme proposed in [10] for single-link transmission. We consider first the case when the channel is time invariant. We then introduce time-varying channel fading to see its effect on average throughput under various reader request rates. In the simulation, certain parameter sets from the communication model used in [10] are adopted for single-link transmission. It is assumed that each packet transmitted from T_1 and T_2 comprises $L = 4$ PPM symbols, exhibiting a symbol modulation order of $K = 32$. Each symbol can contain 1,2,3, or 4 non-zero pulses; in other words, the symbol energy $\varepsilon_k^n = \{1,2,3,4\}$. It is assumed that each packet contains 4,8,12, or 16 non-zero pulses. The duration of the PPM pulses and symbols are $T_p = 5\ \eta s$ and $T_s = 6.4\ \mu s$. The compression ratio $M/N = 0.1$. The battery thresholds of T_1 and T_2 are assumed to be equal ($\delta_{T_1} = \delta_{T_2} = \delta$) for batteries that exhibit a capacity of $B_{max}^{T_1} = B_{max}^{T_2} = B_{max}$. Furthermore, $\beta = 0.5$ and $C_k^{T_1} = C_k^{T_2} = 10$ are set, assuming that the energy required to receive a pulse is five times more than that required to transmit. Moreover, $d_{T_1R} = d_{T_2R} = 1$, $d_{T_1T_2} = 1/2$, and $\eta = 2$. The number of simulated time slots is $N = 10^6$ and the convergence threshold of the RVI algorithm is $\epsilon = 10^{-4}$. For comparison, the transition probability (q^n) of the EH states $h_k^{T_1}$ and $h_k^{T_2}$ is varied. The probability of mis-detection for the different symbol energies is calculated using (3.14) and is summarized in Table 3.1. Let the direct transmission of a symbol of unit energy yields an SNR of $\frac{\upsilon_{T_1R}^2}{N_0} = 4$ dB at the reader.

3.9.1 Long-Term Average Throughput Performance

We first investigate the long-term average throughput performance of the proposed cooperative transmission strategy under energy deficient, energy balanced, and energy surplus scenarios. We also investigate additional scenarios in which the tags harvest energy according to a two-state Markov process. We set the battery threshold $\delta = 8$. For performance comparison purpose, we consider the optimal, greedy, and conservative direct transmission policies proposed by Wang *et al.* [10]. In the optimal direct transmission policy, the tags can independently communicate with the reader without assistance. The greedy policy chooses the maximum available energy for transmission without accounting for the no-response condition of tags; by contrast, the conservative policy saves as much energy as possible and transmits using

TABLE 3.1
The Probability of Mis-Detection

ε^n	1	2	3	4
$p_{md}^n(\varepsilon^n)$	0.5967	0.5060	0.1966	0.0686

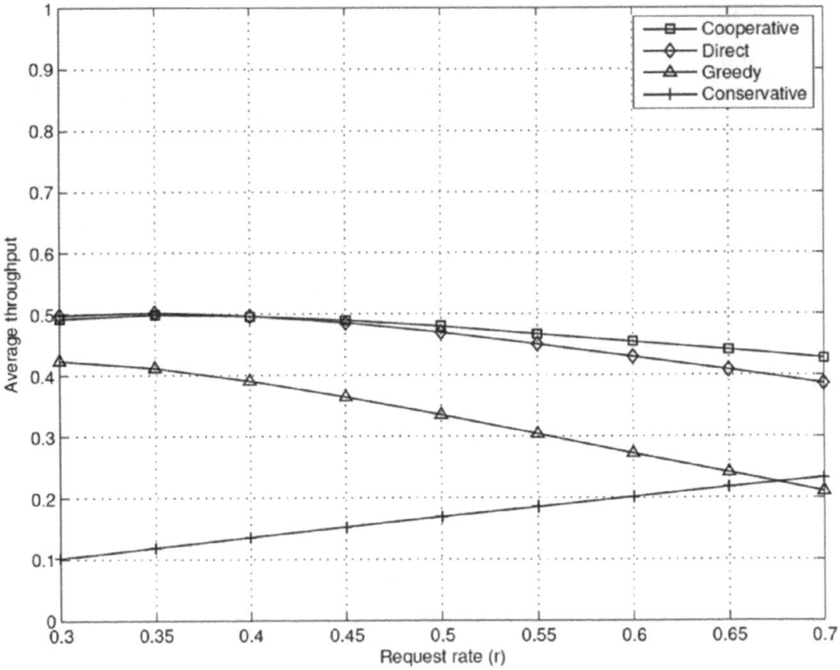

FIGURE 3.7 Performance comparisons for energy-deficient scenario ($\delta = 8$, $B_{max} = 30$).

minimum energy. The optimality of the proposed threshold (δ) is demonstrated by comparing the performances for different δ and B_{max} values. Figure 3.7 provides comparisons of the average throughput performance when the EH parameters of both T_1 and T_2 are $\mathcal{H} = \{0,4,8\}$, and

$$q^n = \begin{bmatrix} 0.7 & 0.2 & 0.1 \\ 0.3 & 0.4 & 0.3 \\ 0.1 & 0.5 & 0.4 \end{bmatrix}.$$

The steady-state probabilities of H_1, H_2, and H_3 are 0.42, 0.34 and 0.24, respectively; this is referred to as an energy-deficient scenario where both tags harvest limited amounts of energy from the environment. Therefore, the tags cannot assist each other in most of the slots. In this case, the performance of our cooperative transmission scheme does not show significant performance difference compared with the direct transmission scheme proposed by [10]. The greedy policy performs worse compared with both the optimal cooperative and optimal direct policies. In addition, the conservative policy yields the worst performance except at $r > 0.65$, where the greedy policy demonstrates an inferior performance.

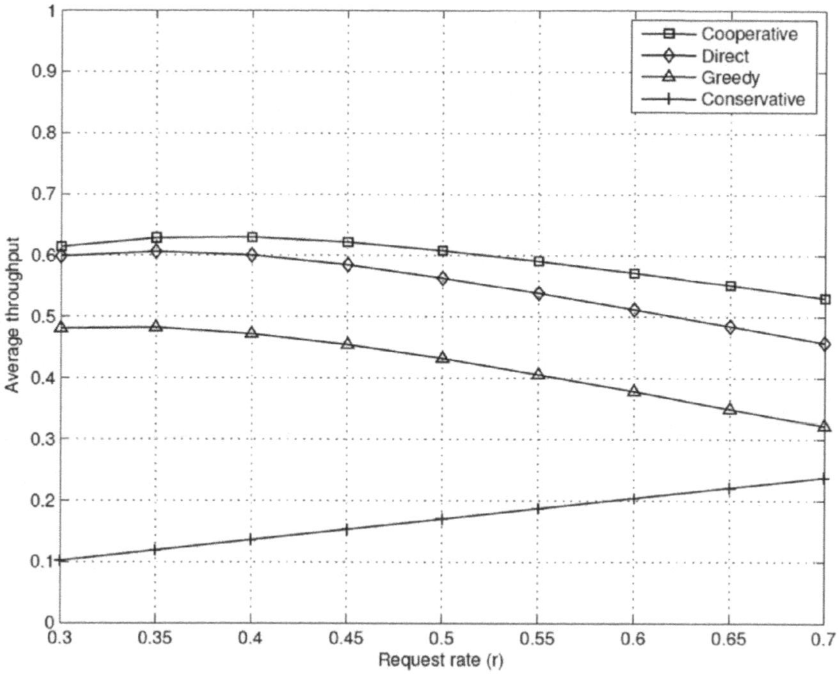

FIGURE 3.8 Performance comparisons for energy-balanced scenario ($\delta = 8$, $B_{max} = 30$).

Figure 3.8 shows an energy-balanced scenario, in which the EH parameters are $\mathcal{H} = \{0,4,8\}$, and

$$q^n = \begin{bmatrix} 0.34 & 0.33 & 0.33 \\ 0.33 & 0.34 & 0.33 \\ 0.33 & 0.33 & 0.34 \end{bmatrix}.$$

The steady-state probabilities of H_1, H_2, and H_3 are 0.33, 0.33 and 0.33, respectively. In this case, both tags harvest a certain amount of energy from the environment and can assist each other in forwarding data in some slots. They can also independently communicate with the reader. In this scenario, the optimal cooperative policy shows superior performance compared with the other policies.

Increasing the request rate increases the average throughput up to $r = 0.4$. Beyond this value, the average throughput decreases because of the energy constraints. Again, the greedy policy performance is inferior to that of both the optimal cooperative and optimal direct policies, and the conservative policy demonstrates the worst level of performance. Figure 3.9 shows an energy surplus scenario in which $\mathcal{H} = \{0,6,12\}$, and

$$q^n = \begin{bmatrix} 0.2 & 0.4 & 0.4 \\ 0.1 & 0.5 & 0.4 \\ 0.1 & 0.3 & 0.6 \end{bmatrix}.$$

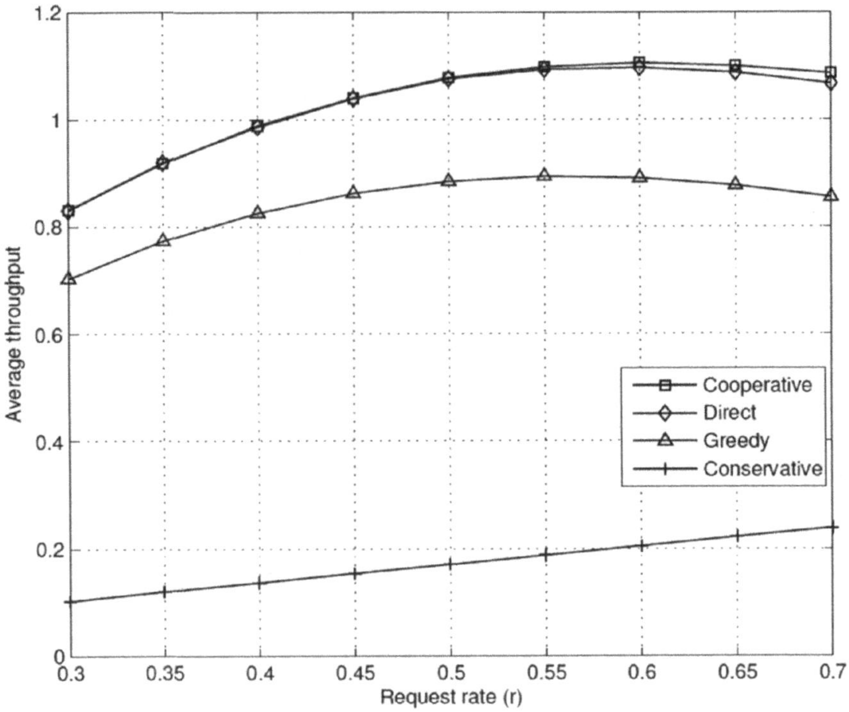

FIGURE 3.9 Performance comparisons for energy surplus scenario ($\delta = 8$, $B_{max} = 30$).

The steady-state probabilities of H_1, H_2, and H_3 are 0.11, 0.39 and 0.50, respectively. Under these conditions, the tag batteries strengthen, becoming able to independently communicate with the reader. When required (particularly at high request rates), the tags can communicate by assisting each other in forwarding packets. The average throughput of optimal cooperative policy nearly fits with the optimal direct transmission policy because of favorable energy-harvesting conditions. This shows that cooperation between tags with favorable energy condition does not guarantee better throughput compared to direct transmission scheme.

The maximum average throughput is achieved at an optimal request rate of $r = 0.6$. Both the greedy and conservative policies yield inferior performance compared with that of the optimal cooperative and optimal direct transmission policies.

We also considered additional scenarios in which the tags harvest energy according to a two-state Markov process when $h_k^{T_1}$ and $h_k^{T_2}$ take values from $\mathcal{H} = \{H_2, H_3\}$, indicating that some energy is always harvested in each slot. An example of this is when both tags are placed outdoors and the environment facilitates harvesting energy. Figure 3.10 summarizes three cases in which the EH parameters are set as: $\mathcal{H} = \{4,8\}, \left[q_{H_2}^n, q_{H_3}^n\right] = \left[0.7, 0.3\right]$ for scenario 1, $\mathcal{H} = \{4,8\}, \left[q_{H_2}^n, q_{H_3}^n\right] = \left[0.5, 0.5\right]$ for scenario 2, and $\mathcal{H} = \{6,12\}, \left[q_{H_2}^n, q_{H_3}^n\right] = \left[0.3, 0.7\right]$ for scenario 3.

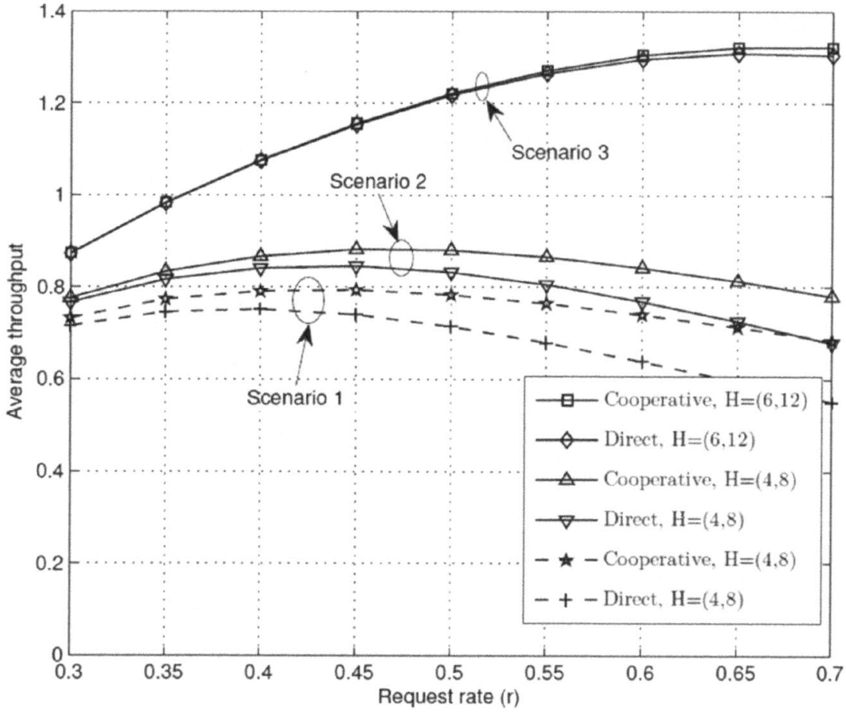

FIGURE 3.10 Performance comparisons for various scenarios of two-state energy harvesting model ($\delta = 8$, $B_{max} = 30$).

At each time interval, the tags harvest either small or moderate amounts of energy to prevent the battery from depleting. This increases the resulting average throughput, and the optimal cooperative policy outperforms the optimal direct transmission policy in scenario 1 and scenario 2. The cooperative policy nearly fits with the direct optimal transmission policy under scenario 3, implying that the tags can independently communicate with the reader without assisting each other when they have sufficient energy.

3.9.2 Effects of Channel Variation

This section presents numerical examples that demonstrates the effects of channel fading. We consider the PCSI and NoCSI on optimum cooperative transmission strategy for the case when the channel gains ($\theta_i, i = 1,2,3$) take values from 0.5 and 1, respectively, under bad and good channel conditions, and channel state transition probability is taken as $P_1 = \begin{bmatrix} 0.8 & 0.2 \\ 0.4 & 0.6 \end{bmatrix}$. The presence of channel fading degrades the overall performance of the system as shown in Figures 3.11 and 3.12. Under both EH conditions, the average throughput of a system with no fading outperforms the a

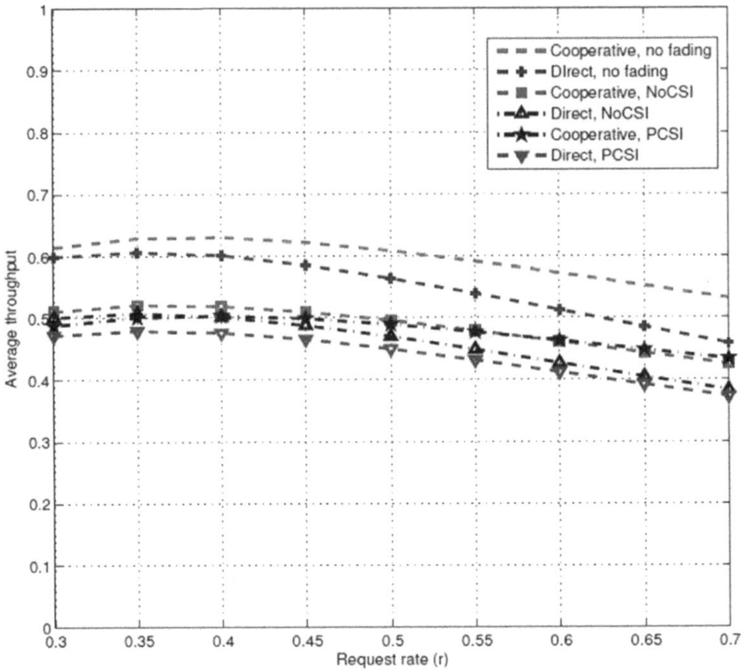

FIGURE 3.11 Effect of channel fading under energy-balanced scenario ($\delta = 8$, $B_{max} = 30$).

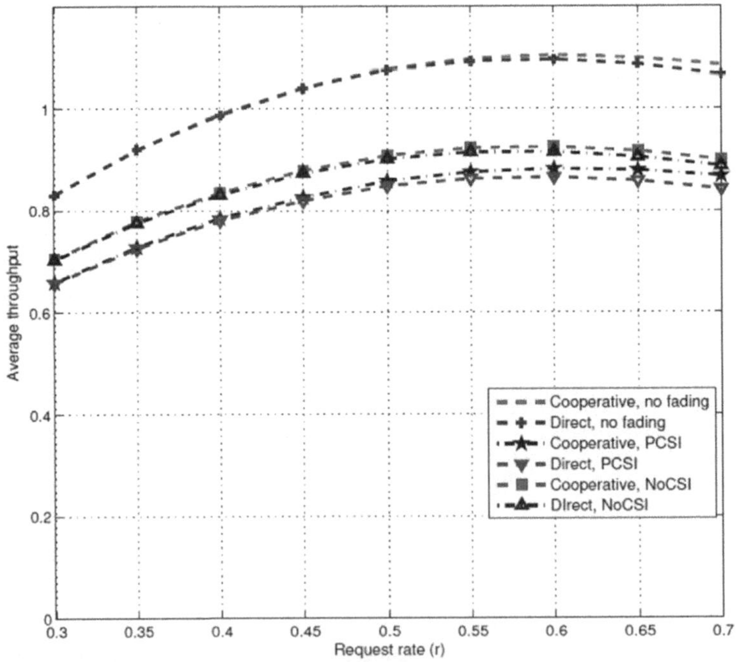

FIGURE 3.12 Effect of channel fading under energy surplus scenario ($\delta = 8$, $B_{max} = 30$).

system when fading is considered. In fact, the overall effect does not prevent the system from transmission. This is because, due to the broadcast nature of the source tag, the reader exploits the signals coming from both the relay tag and the source tag to detect through MRC in the first and second cooperation phases. Thus, the throughput outperforms the noncooperative scheme regardless of the quality of the channel.

The average throughput performance of the system when the channel gain is not a priori known (NoCSI) is slightly better than the PCSI case. This is mainly due to the expectation operation which favors the high channel state transition probability. In our simulation, we considered the case when $p_{gg} > p_{bb}$ that results in better long term expected throughput.

3.9.3 Effects of Battery Size

The battery size of tags has a direct effect on long-term throughput. When the environment is favorable for EH, a significant amount of energy overflows due to small battery size. Increasing the battery size also impacts the overall size of the EHTs. Despite this limitation, in this section, we investigate the long-term average throughput performance under various harvest conditions for increased battery size. Figures 3.13–3.15 show the effect of increasing battery capacity under the three energy-harvesting scenarios. In the three scenarios, when B_{max} becomes 100, the average throughput increases, particularly under favorable energy-harvesting conditions.

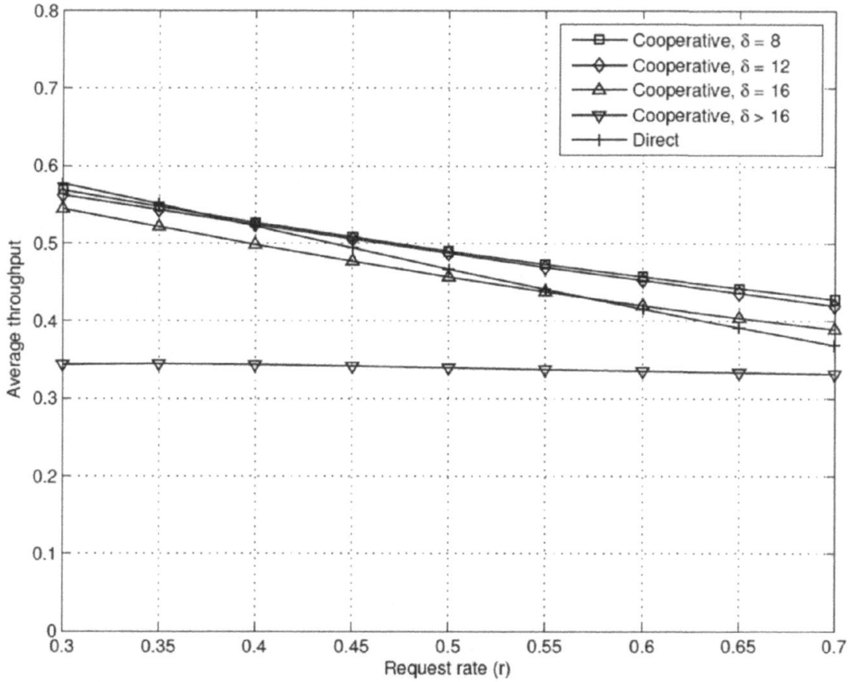

FIGURE 3.13 Performance comparisons in energy-deficient scenario (B_{max} = 100).

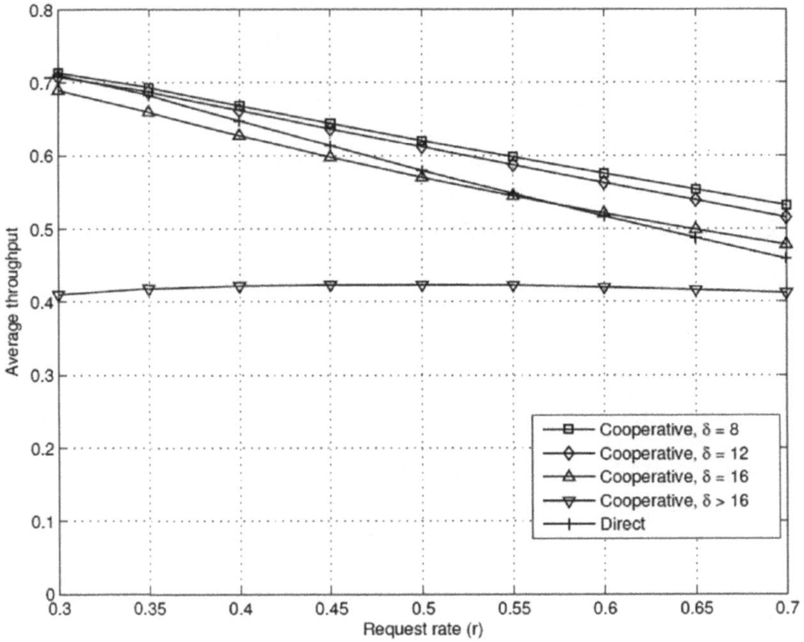

FIGURE 3.14 Performance comparisons in energy balanced scenario ($B_{max} = 100$).

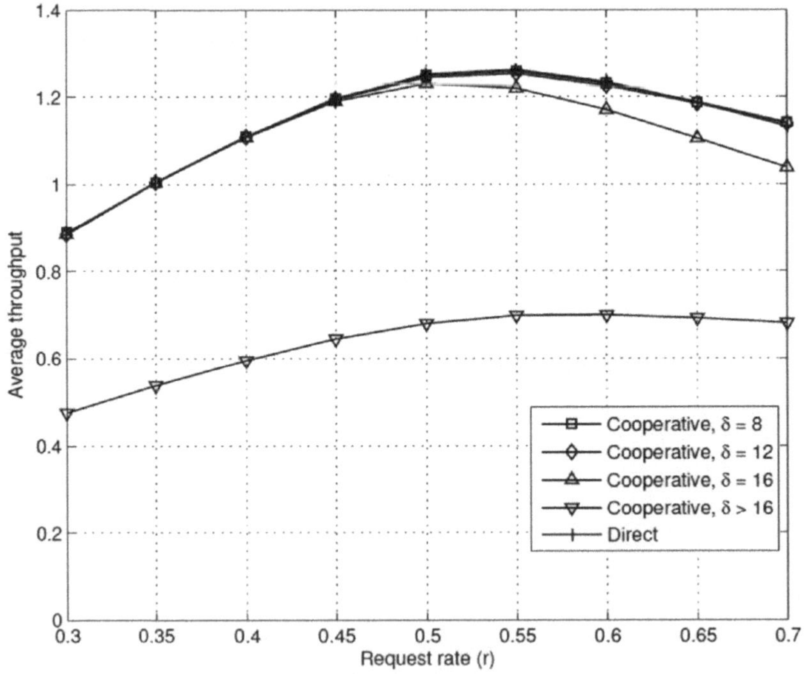

FIGURE 3.15 Performance comparisons in energy surplus scenario ($b_{max} = 100$).

In energy-deficient (Figure 3.13) and energy-balanced scenarios (Figure 3.14), the throughput increase is limited because of unfavorable energy-harvesting conditions. The maximum battery capacity cannot be achieved under energy-deficient and energy-balanced conditions, particularly when the request rate increases. $\delta = 8$ achieves superior performance compared with the other threshold settings.

In an energy-surplus scenario (Figure 3.15), the environment is favorable for harvesting energy and the tags can achieve the maximum battery capacity. The average throughput substantially increases. In this case, the *direct*, $\delta = 8$, and $\delta = 12$ achieve the optimum performance.

3.10 CONCLUSION

This chapter developed an optimal cooperative communication strategy for objects equipped with EHTs. These objects can communicate with a reader either directly or with the assistance of neighboring objects, depending on the varying availability of energy. The effect of channel fading on overall system performance was also examined. The MDP RVI algorithm was employed to obtain numerical results for various EH and request rate conditions, providing insight into the dynamic nature of the system.

The study focused on optimizing the long-term average throughput of two cooperating tags through effective energy resource management techniques. By considering the stochastic properties of energy availability and channel conditions, the proposed strategy adapts to real-time changes, ensuring efficient use of harvested energy. The simulation results demonstrate that the proposed AF-based optimal cooperative policy significantly outperforms the direct transmission policy, especially under certain energy-harvesting conditions where cooperative transmission becomes more advantageous.

Overall, the findings highlight the potential of cooperative communication in enhancing the performance of energy-harvesting wireless networks. By leveraging the cooperative capabilities of EHT-equipped objects, it is possible to achieve higher throughput and better energy efficiency, contributing to the prolonged operational lifetime of wireless networks.

Despite the promising results, this research has also limitations. The study primarily focused on a simple three-terminal scenario involving two tags and one reader, which may not fully represent more complex real-world networks. Additionally, the model assumes perfect knowledge of channel states and energy availability, which might not always be feasible in practical scenarios. The computational complexity of the MDP RVI algorithm could also be a constraint for real-time applications, especially in large-scale networks.

Future work can extend this finding by exploring more complex network topologies with multiple tags and readers. Incorporating imperfect information about channel states and energy levels could also make the model more realistic and applicable to practical scenarios. Developing more efficient algorithms to reduce computational complexity and enhance real-time applicability is another potential area of improvement.

REFERENCES

1. J. A. Paradiso, and T. Starner, "Energy scavenging for mobile and wireless electronics," *IEEE Pervasive Comput.*, vol. 4, no. 1, pp. 18–27, 2005.
2. V. Raghunathan, S. Ganeriwal, and M. Srivastava, "Emerging techniques for long lived wireless sensor networks," *IEEE Commun. Mag.*, vol. 44, no. 4, pp. 108–114, 2006.
3. M. Crepaldi, C. Li, J. R. Fernandes, and P. R. Kinget, "An ultra-wideband impulseradio transceiver chipset using synchronized-ook modulation," *IEEE J. Solid-State Circuits*, vol. 46, no. 10, pp. 2284–2299, 2011.
4. B. Vigraham, and P. Kinget, "A self-duty-cycled and synchronized UWB receiver SoC consuming 375pJ/b for- 76.5 dBM sensitivity at 2Mb/s," in *2013 IEEE International Solid-State Circuits Conference (ISSCC)*, 2013, pp. 444–445.
5. M. Gorlatova, P. Kinget, I. Kymissis, D. Rubenstein, X. Wang, and G. Zussman, "Challenge: ultra-low-power energy-harvesting active networked tags (EnHANTs)," in *Proceedings of the 2009 ACM Mobile Telecommunication Conference*, 2009, pp. 253–260.
6. M. Gorlatova, P. Kinget, I. Kymissis, D. Rubenstein, X. Wang, and G. Zussman, "Energy harvesting active networked tags (EnHANTs) for ubiquitous object networking," *IEEE Trans. Wireless Commun*, vol. 17, no. 6, pp. 18–25, 2010.
7. M. Gorlatova, A. Wallwater, and G. Zussman, "Networking low-power energy harvesting devices: measurements and algorithms," *IEEE Trans. Mob. Comput.*, vol. 3, no. 9, pp. 1853–1865, 2013.
8. M. Gorlatova, R. Margolies, J. Sarik, G. Stanje, J. Zhu, B. Vigraham, M. Szczodrak, L. Carloni, P. Kinget, I. Kymissis, and G. Zussman, "Prototyping energy harvesting active networked tags (EnHANTs)," in *Proc. IEEE INFOCOM'13 mini-conference*, Apr. 2013.
9. M. Gorlatove, "Energy Harvesting Networked Nodes: Measurement, Algorithms, and Prototyping," Ph.D. dissertation, Columbia University, New York, 2013.
10. Z. Wang, A. Tajer, and X. Wang, "Communication of energy harvesting tags," *IEEE Trans. Commun*, vol. 60, no. 4, pp. 1159–1166, 2012.
11. J. N. Laneman, D. N. Tse, and G. W. Wornell, "Cooperative diversity in wireless networks: Efficient protocols and outage behavior," *IEEE Trans. Inf. Theory*, vol. 50, no. 12, pp. 3062–3080, 2004.
12. K. Ishibashi, H. Ochiai, and V. Tarokh, "Energy harvesting cooperative communications," in *2012 IEEE 23rd Int'l Symp. on Personal Indoor and Mobile Radio Commun. (PIMRC)*, pp. 1819–1823, 2012.
13. G. G. De Oliveira Brante, M. T. Kakitani, and R. D. Souza, "Energy efficiency analysis of some cooperative and non-cooperative transmission schemes in wireless sensor networks," *IEEE Trans. Commun*, vol. 59, no. 10, pp. 2671–2677, 2011.
14. B. Medepally, and N. B. Mehta, "Voluntary energy harvesting relays and selection in cooperative wireless networks," *IEEE Trans. Wireless Commun*, vol. 9, no. 11, pp. 3543–3553, 2010.
15. X. Chen, Y. Liu, L. X. Cai, Z. Chen, and D. Zhang, "Resource allocation for wireless cooperative IoT network with energy harvesting," *IEEE Trans. Wireless Commun.*, vol. 19, no. 7, pp. 4879–4893, 2020.
16. H. Li, N. Jaggi, and B. Sikdar, "Relay scheduling for cooperative communications in sensor networks with energy harvesting," *IEEE Trans. Wireless Commun*, vol. 10, no. 9, pp. 2918–2928, 2011.
17. "Greenpeak technologies," http://www.greenpeak.com/.
18. "Texas instruments msp430 solar energy harvesting development tool," http://www.ti.com/tool/ez430-rf2500-seh.

19. Z. Zou, D. S. Mendoza, P. Wang, Q. Zhou, J. Mao, F. Jonsson, H. Tenhunen, and L.-R. Zheng, "A low-power and flexible energy detection IR-UWB receiver for RFID and wireless sensor networks," *IEEE Trans. Circuits Syst. I Regular Papers*, vol. 58, no. 7, pp. 1470–1482, 2011.
20. E. A. Feinberg, A. and Shwartz, *Handbook of Markov Decision Processes: Methods and Applications*. Kluwer Academic Publishers Boston, 2012.
21. E. Altman, "Applications of Markov decision processes in communication networks. In: Feinberg, E.A., Shwartz, A. (eds)," *Handbook of Markov Decision Processes*, Springer, pp. 489–536, 2002.
22. E. Fainberg, Y. A. Kogan, and A. N. Smirnov, "Optimal control by the retransmission probability in slotted aloha systems," *Perform. Eval.*, vol. 5, no. 2, pp. 85–96, 1985.
23. C. E. Perkins, and E. M. Royer, "Ad-hoc on-demand distance vector routing," in *Proceedings of the Second IEEE Workshop on Mobile Computing Systems and Applications, 1999. WMCSA'99*. IEEE, 1999, pp. 90–100.
24. A. T. Hoang, and M. Motani, "Buffer and channel adaptive modulation for transmission over fading channels," in *IEEE International Conference on Communications, 2003. ICC'03*, vol. 4. IEEE, 2003, pp. 2748–2752.
25. C. Pandana, and K. R. Liu, "Near-optimal reinforcement learning framework for energy-aware sensor communications," *IEEE J. Sel. Areas Commun.*, vol. 23, no. 4, pp. 788–797, 2005.
26. P. Blasco, D. Gunduz, and M. Dohler, "Low-complexity scheduling policies for energy harvesting communication networks," in *2013 IEEE International Symposium on Information Theory Proceedings (ISIT)*. IEEE, 2013, pp. 1601–1605.
27. C. K. Ho, and R. Zhang, "Optimal energy allocation for wireless communications with energy harvesting constraints," *IEEE Trans. Sig. Process.*, vol. 60, no. 9, pp. 4808–4818, 2012.
28. M. Puterman, *Markov Decision Processes: Discrete Stochastic Dynamic Programming*. John Wiley and Sons, 2005.
29. D. P. Bertsekas, D. P. Bertsekas, D. P. Bertsekas, and D. P. Bertsekas, *Dynamic Programming and Optimal Control*. Athena Scientific Belmont, MA, 1995, vol. 1, no. 2.
30. M. A. Mulatu, L-C. Chang, Y. S. Han, G. A. Mengistu, and D.-F. Tseng, "Threshold-based cooperative communication of energy harvesting active networked tags," *Int. J. Ad Hoc Ubiquitous Comput.*, vol. 24, no. 4, pp. 225–236, 2017.
31. T. Issariyakul and V. Krishnamurthy, "Amplify-and-forward cooperative diversity wireless networks: model, analysis, and monotonicity properties," *IEEE/ACM Trans. Netw.*, vol. 17, no. 1, pp. 225–238, 2009.
32. G. N. Shirazi, P.-Y. Kong, and C.-K. Tham, "Markov decision process frameworks for cooperative retransmission in wireless networks," in *Wireless Communications and Networking Conference, 2009. WCNC 2009. IEEE*. 2009, pp. 1–6.
33. J. Shen, V. Lesser, and N. Carver, "Minimizing communication cost in a distributed Bayesian network using a decentralized MDP," in *Proceedings of the Second International Joint Conference on Autonomous Agents and Multiagent Systems*. ACM, 2003, pp. 678–685.
34. F. Iannello, *Energy Harvesting-Aware Design of Wireless Networks*. New Jersey Institute of Technology, 2012.
35. D. Gunduz, K. Stamatiou, N. Michelusi, and M. Zorzi, "Designing intelligent energy harvesting communication systems," *IEEE Commun. Mag*, vol. 52, no. 1, pp. 210–216, 2014.
36. J. Taneja, J. Jeong, and D. Culler, "Design, modeling, and capacity planning for micro-solar power sensor networks," in *Proceedings of the 7th International Conference on Information Processing in Sensor Networks*. IEEE Computer Society, 2008, pp. 407–418.

37. F. Simjee, and P. H. Chou, "Everlast: Long-life, supercapacitor-operated wireless sensor node," in *Proceedings of the 2006 International Symposium on Low Power Electronics and Design, 2006. ISLPED'06*. IEEE, 2006, pp. 197–202.
38. X. Jiang, J. Polastre, and D. Culler, "Perpetual environmentally powered sensor networks," in *Fourth International Symposium on Information Processing in Sensor Networks, 2005. IPSN 2005*. IEEE, 2005, pp. 463–468.
39. S. Chen, P. Sinha, N. B. Shroff, and C. Joo, "Finite-horizon energy allocation and routing scheme in rechargeable sensor networks," in *INFOCOM, 2011 Proceedings IEEE*. IEEE, 2011, pp. 2273–2281.
40. M. Gorlatova, A. Wallwater, and G. Zussman, "Networking low-power energy harvesting devices: Measurements and algorithms," *IEEE Trans. Mob. Comput.*, vol. 12, no. 9, pp. 1853–1865, 2013.
41. M. Crepaldi, C. Li, K. Dronson, J. Fernandes, and P. Kinget, "An interference robust s-OOK IR-UWB transceiver chipset," *IEEE J. Solid-State Circuits*, vol. 46, no. 10, pp. 2284–2299, 2011.
42. D. C. Daly, P. P. Mercier, M. Bhardwaj, A. L. Stone, Z. N. Aldworth, T. L. Daniel, J. Voldman, J. G. Hildebrand, and A. P. Chandrakasan, "A pulsed UWB receiver SoC for insect motion control," *IEEE J. Solid-State Circuits*. vol. 45, no. 1, pp. 153–166, 2010.
43. D. D. Wentzloff, F. S. Lee, D. C. Daly, M. Bhardwaj, P. P. Mercier, and A. P. Chandrakasan, "Energy efficient pulsed-UWB CMOS circuits and systems," in *IEEE International Conference on Ultra-Wideband, 2007. ICUWB 2007*. IEEE, 2007, pp. 282–287.
44. A. Sinha, and A. Chandrakasan, "Dynamic power management in wireless sensor networks," *IEEE Des. Test Comput.*, vol. 18, no. 2, pp. 62–74, 2001.
45. R. Min, M. Bhardwaj, S. Cho, A. Sinha, E. Shih, A. Wang, and A. Chandrakasan, "An architecture for a power-aware distributed microsensor node," in *2000 IEEE Workshop on Signal Processing Systems, 2000. SiPS 2000*. IEEE, 2000, pp. 581–590.
46. S. Singh, M. Woo, and C. S. Raghavendra, "Power-aware routing in mobile ad hoc networks," in *Proceedings of the 4th Annual ACM/IEEE International Conference on Mobile Computing and Networking*. ACM, 1998, pp. 181–190.
47. R. C. Shah, and J. M. Rabaey, "Energy aware routing for low energy ad hoc sensor networks," in *2002 IEEE Wireless Communications and Networking Conference, 2002. WCNC2002*. vol. 1. 2002, pp. 350–355.
48. Q. Li, J. Aslam, and D. Rus, "Online power-aware routing in wireless ad-hoc networks," in *Proceedings of the 7th Annual International Conference on Mobile Computing and Networking*. ACM, 2001, pp. 97–107.
49. A. Kansal, and M. B. Srivastava, "An environmental energy harvesting framework for sensor networks," in *Proceedings of the 2003 International Symposium on Low Power Electronics and Design, 2003. ISLPED'03*. IEEE, 2003, pp. 481–486.
50. A. Kansal, J. Hsu, S. Zahedi, and M. B. Srivastava, "Power management in energy harvesting sensor networks," *ACM Trans. Embedded Comput. Syst. (TECS)*, vol. 6, no. 4, p. 32, 2007.
51. M. Gorlatova, M. Zapas, E. Xu, M. Bahlke, I. Kymissis, and G. Zussman, "Crawdad data set Columbia/enhants (v. 2011-04-07)," 2011.
52. V. Sharma, U. Mukherji, V. Joseph, and S. Gupta, "Optimal energy management policies for energy harvesting sensor nodes," *IEEE Trans. Wireless Commun.*, vol. 9, no. 4, pp. 1326–1336, 2010.
53. J. Yang, and S. Ulukus, "Optimal packet scheduling in an energy harvesting communication system," *IEEE Trans. Commun.*, vol. 60, no. 1, pp. 220–230, 2012.
54. L. Huang, and M. J. Neely, "Utility optimal scheduling in energy-harvesting networks," *IEEE/ACM Trans. Netw.*, vol. 21, no. 4, pp. 1117–1130, 2013.

55. J. Chen, L. Clavier, Y. Xi, A. Burr, N. Rolland, and P.-A. Rolland, "Interference modeling and relay selection issue for amplify-and-forward IR-UWB sensor networks," in *2011 4th IFIP International Conference on New Technologies, Mobility and Security (NTMS)*, 2011, pp. 1–5.
56. T. Cover and A. E. Gamal, "Capacity theorems for the relay channel," *IEEE Trans. Inform. Theory*, vol. 25, no. 5, pp. 572–584, 1979.
57. S. A. Chakrabarti, and B. Aazhang, "Cooperative communications: Fundamental limits and practical implementation," in *in Cooperation in Wireless Networks: Principles and Applications*. F. H. P. Fitzek and M. D. Katz, Eds. Springer, 2006, pp. 2–23.
58. R. U. Nabar, H. Bolcskei, and F. W. Kneubuhler, "Fading relay channels: Performance limits and space-time signal design," *IEEE J. Sel. Areas Commun*, vol. 22, no. 6, pp. 1099–1109, 2004.
59. A. Sendonaris, E. Erkip, and B. Aazhang, "User cooperation diversity. Part i. System description," *IEEE Trans. Commun.* vol. 51, no. 11, pp. 1927–1938, 2003.
60. K. Azarian, H. El Gamal, and P. Schniter, "On the achievable diversity-multiplexing tradeoff in half-duplex cooperative channels," *IEEE Trans. Inform. Theory*, vol. 51, no. 12, pp. 4152–4172, 2005.
61. D. Brennan, "Linear diversity combining techniques," *in Proc. IEEE*, vol. 91, no. 2, pp. 331–356. 2003.
62. M. A. Davenport, P. T. Boufounos, M. B. Wakin, and R. G. Baraniuk, "Signal processing with compressive measurements," *IEEE J. Sel. Top. Signal Process.*, vol. 4, no. 2, pp. 445–460, 2010.
63. S. Mao, M. H. Cheung, and V. W. Wong, "An optimal energy allocation algorithm for energy harvesting wireless sensor networks," in *Proc. 2012 IEEE Int'l Conf. Commun.*, 2013, pp. 265–270.
64. D. Zhang, and K. M. Wasserman, "Transmission schemes for time-varying wireless channels with partial state observations," in *Proceedings. Twenty-First Annual Joint Conference of the IEEE Computer and Communications Societies*, vol. 2. IEEE, 2002, pp. 467–476.
65. J. Yang, and D. R. Brown III, "The effect of channel state information on optimum energy allocation and energy efficiency of cooperative wireless transmission systems," in *2006 40th Annual Conference on Information Sciences and Systems*. IEEE, 2006, pp. 1044–1049.

Section II

Converters and Electric Vehicle Charging Systems

4 Solar PV Incorporated with Boost Converter Using Perturbation and Observation Method of Maximum Power Point Tracking Technique for EV Charging Application

Suresh Panchanathan, Pradeep Vishnuram, Mohit Bajaj, and Arvind R. Singh

4.1 INTRODUCTION

The recent era experienced a significant shift away from conventional fossil fuel for energy generation and toward a sustainable and greener generation that utilizes renewable resources such as solar and wind energy [1]. Renewable energy resources have indisputable merits, including minimal operating costs (free fuel) and the ability to expand capacity quickly, and the main point is that these are carbon-free power generation resources [2]. Regarding energy production, climate change and global warming have emerged as major risks in the recent decade. Traditional energy production methods like burning natural gas, carbon, and natural gas have significantly increased global temperature, resulting in events such as the extinction of biodiversity and melting glaciers [3]. Then, a boost converter is created for electric car applications over a typical boost converter using parallel IGBT and MOSFET combinations [4–6]. Researchers who have observed these changes in our world have recommended corporations and governments invest in ecologically friendly green technologies [7–9]. Efficiency is a critical issue to overcome to make sustainable resources appropriate for investment and development as energy sources [10].

Borole et al., in Ref. [11], propose electric vehicles (EVs) to reduce pollution caused by increasingly growing automobiles. To increase the voltage, a DC-DC converter is utilized and separated from a solar photovoltaic (PV) array. To charge, a bidirectional DC-DC converter has been incorporated as discussed in this chapter. Two DC-DC converters are modeled and simulated with the help of MATLAB.

DOI: 10.1201/9781003537823-6

For grid and solar PV electricity, the author states that the objective of this work is to make a power electronic converter that can charge plug-in electric vehicles (PEVs). Charge, propulsion, and regenerative braking were all possible. As soon as the vehicle is at a complete stop, the battery can either be charged from the grid system or the solar PV system. This is called the "RB" procedure, and it uses the movement of the car's wheels to charge its device while moving [12].

The authors in Ref. [13] imply that plug-in hybrid vehicles (PHEVs) have a large battery pack that can only run on electricity for a certain amount of time. The adjacent socket can be used to replenish the battery pack. The converter is used as both a DC and an AC battery adapter, so it can be used for both. A converter, a charger, and a bidirectional DC/DC converter are the main power electrical interfaces that make electric and plug-in hybrid cars bigger and heavier [14].

The article [15] says that for solar energy, an MPPT method with current protection has been proposed. With this plan, some electric cars can be charged with the amount of electricity that can reach the car's rated range. A solar PV array can be installed directly on the rooftop of the car to keep the battery charged even without the use of an external source.

The author of Ref. [16] examines the current situation of PEVs and hybrids from the perspective of battery chargers, the power levels at which they charge, and the infrastructure supporting such vehicles. The study will also look at how they're being used. On-board and off-board charger systems with unidirectional or bidirectional power flow are divided into two categories. Unidirectional charging reduces the amount of hardware required and simplifies connecting concerns. Through the process of bidirectional charging, it is possible to inject energy from the batteries back into the grid. Most on-board chargers have power restrictions because of weight, space, and cost restrictions. The availability of charging infrastructure minimizes the need for and expense of on-board energy storage.

The authors in Ref. [17] propose with growing concern about the environment and increasingly strict outflow limits; EVs have been investigated as a viable mode of transportation. In EVs, battery charger topologies perform a critical performance to increase the number of EVs on the road. To look at electric cars, a plug-in battery charger comes with them. In recent years, resonant inverter, LLC, and SEPIC support have been made. There is a prototype battery that this system can charge. The on-board charger is modeled with the help of MATLAB/Simulink, and the results are checked.

The approach recommends a unique algorithm-based MPPT has been proposed in this research for extracting the highest amount of power from PV systems. The suggested MPPT control algorithm is based on the perturb and observe (P&O) approach, which is widely utilized. Different current methods are examined for tracking the complexity and the speed of operation. The outcome of the investigation enhances the maximum power point tracker (MPPT) for a PV system [18].

Reference [19] states the MPP of the PV generator fluctuates with solar insolation and has nonlinear V-I properties. It is used to fit the PV systems and to run the PV array at full power. An interleaved dual boost converter is used for both of these things. Only load voltage information is used to build the tracking algorithm, which also considers the current in the array. Power output expressions can be determined using interleaved dual boost converters and boost converters.

The proposed system incorporates an affordable and reliable approach for tracking the MPP so that it does not fail to do so when subjected to a sudden shift in the amount of solar insolation. Combining the classic P&O method of MPPT technology eliminates the divergence issue and steady-state power oscillation. The suggested approach reduces distortion and reduces capacitor voltage ripples. This designed system provides a dependable and efficient charging system for electric cars, and it might be a potential option for future electric automobiles.

This work aims to take a fresh method of PV power generation to improve power extraction in solar panels. This highlights the importance of direct current to direct current conversion into solar power generation for both efficiency and cost reduction, as it enables people to utilize the greatest amount of solar power available at any given time, resulting in increased profits and sustainability. The results of this converter are compared to the PV system with a boost converter. It has been demonstrated that the interleaved dual boost converter has the benefit of being more efficient and reducing the amount of ripple content.

4.2 DC-DC CONVERTERS

DC-DC converters are power electronic circuits that convert alternating current to a different voltage level in the field of power electronics. Electronic, linear, switched mode, magnetic, and capacitive are some of the numerous types of conversion methods. Direct current to direct current switched mode converters, the circuits described in this chapter are the electronic devices used to modify the voltage level of DC electrical power. In contrast to alternating current, DC cannot increase or decrease the voltage. Hence, DC-DC converters are necessary. In many respects, a DC-DC converter can be considered the DC version of a transformer.

This converts supplied power to a new impedance level. Irrespective of the voltage output, the output power is entirely derived from the input. This means that the energy is not generated within the converter. Output voltage is usually regulated in most direct current to direct current converters. LED power sources of high efficiency, which are the DC-DC converter that modulates the current through the LED and simple charge pumps that double or triple the voltage output, are the two examples of limitational exceptions. DC-DC converter is meant for improving wind turbine and solar PV energy gathering. Rectification with a power field effect transistor, which has a lower ON resistance, is another major development in DC-DC converters that decreases the switching losses.

DC-DC synchronous, low-powered converters use an electro-mechanical vibrator led by a step-up voltage transformer supplying a semiconductor rectifier or synchronous rectifier and a vacuum tube connection on vibrators before power semiconductors became widely available. Power can often flow in one direction, that is from input to output, with a typical DC-to-DC converter. All switching regulator topologies could be made bidirectional by replacing all diodes with active rectification where each control input is independent of the others.

4.2.1 Types of DC-DC Converters

Following are some of the types of converters available in power electronics applications.

- BOOST converter
- BUCK converter
- BUCK-BOOST converter
- CUK converter

4.2.1.1 Boost Converter

A power converter having a greater DC voltage output than the DC voltage input is referred to as a boost converter. This type of converter is also known as a step-up converter developed in the year 1960. It is a kind of switching mode power supply that has at least one energy storage device in addition to a minimum of two semiconductor switches, such as a diode and a transistor. A capacitor filters the output voltage, generally in conjunction with the inductors that are typically added to the output of the converters. This helps to reduce the ripples in the output voltage. This phenomenon occurs due to the voltage amplification provided by the boost converter, which is commonly referred to as a step-up converter. It is important to note that the output current of the boost converter is typically lower than the input current from the power source. This is due to the principle of power conservation, where power (P = VI) must always remain constant.

Figure 4.1 shows a boost converter circuit with Vs as the supply voltage, L as the inductor, S as the MOSFET switch, a diode, and a capacitor. Semiconductor switching devices are used in the construction of these converters.

Without the boost converter: Linear regulated circuits, such as direct current power-regulated circuits in the semiconductor switching devices, access the voltage from the input of the unregulated supply, i.e., AC power supply, resulting in the huge power wastage. The voltage drop is always proportional to the power loss.

With boost converters: Boost converters convert uncontrolled alternating current or the direct current voltage input to regulated direct current voltage output in switching devices. Mostly boost converters are widely incorporated in switched mode power supply (SMPS) devices, where the input voltage can be converted and be filtered using a capacitor and rectifier in the SMPS with AC mains input supply.

4.2.1.2 Turn ON Mode

Because when the switch is turned on, the connection between the two components is considered to be a short circuit, which offers no resistance to the flow of current. Consequently, upon activation of the switch, the entirety of the electrical current will

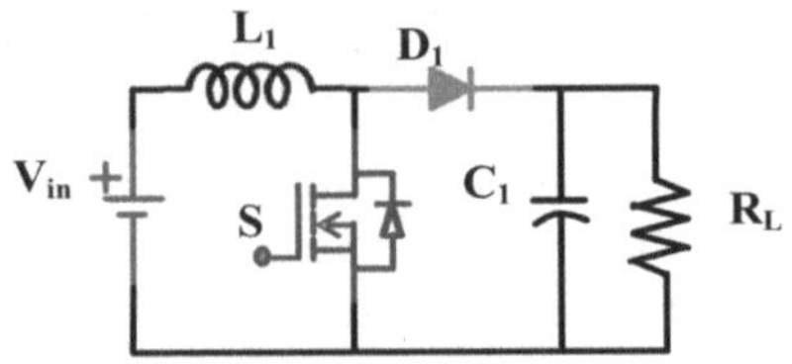

FIGURE 4.1 Boost converter circuit diagram.

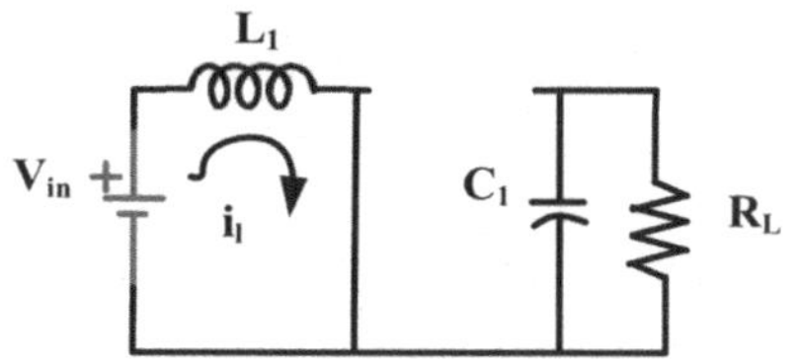

FIGURE 4.2 Turn ON mode of boost converter circuit.

traverse through the switch and afterwards retrace its path directly back to the power source. This mode is illustrated in Figure 4.2.

During this particular state, the inductor gets charged, and the current increases with the presence of inductor (L_1) which is given as equation (4.1):

$$i_L = \frac{1}{L}\int v\,dt \tag{4.1}$$

when the t=0, the switch is toggled ON. The current across the inductor begins to build as inductor L stores energy. Diode D will be disabled. The output capacitor supplies the load. Imagine that the switch is active for T_{on} seconds, equal to D*T_s, where D represents the duty cycle, and T_s indicates the switching period. At the end of the switch-on state, the current through the inductor is given as

$$\Delta L = \frac{1}{L} V_{in} D T_s \tag{4.2}$$

4.2.1.3 Turn OFF Mode

During the turn-off mode, the polarity of the inductor is reversed. During operation, the energy held in the inductor is wasted at the load's resistance as it is gradually released. Since the inductor is still functioning as a source in addition to the input source, it supports maintaining a unidirectional current flow across the load and boosts the voltage output.

Figure 4.3 shows the turn OFF mode of the converter where the boost rectifier diode D begins to conduct to maintain inductor current in the same direction. Inductor current decreases. Capacitor C starts changing. This mode ends and further starts the next cycle, and it goes on. The diode will be short-circuited during turned off condition of the switch and inductor starts discharging through RC circuit.

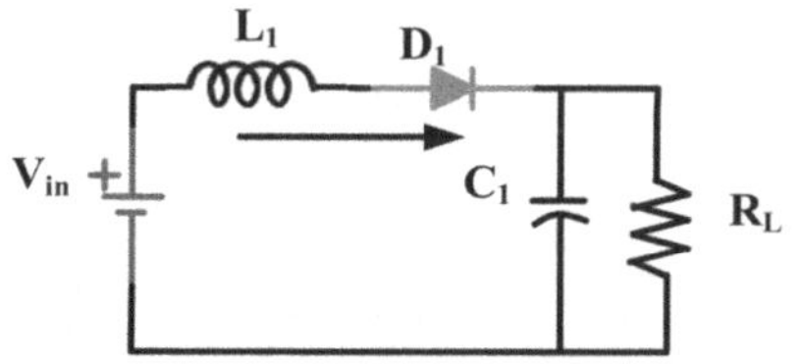

FIGURE 4.3 Turn OFF mode of boost converter circuit.

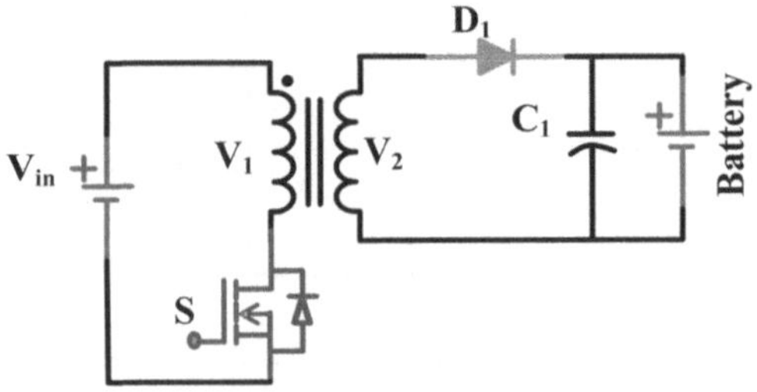

FIGURE 4.4 Flyback converter circuit diagram.

The current flowing through the inductor is now:

$$i_L = -i_L \int \left(V_{in} - V_{out}\right) dt + i_L \tag{4.3}$$

4.2.1.4 Flyback Converter

DC-DC converters having a unique property are known as flyback converters, as shown in Figure 4.4. This is accomplished by placing a transformer in the circuit's midsection, which stores energy before releasing it to the load. A flyback is a converter which performs a conversion indirectly because it does not have a link directly between the load and the input at times of any operational state/face due to the design being unique. Only a few components are used in the flyback converters. Connecting the transformer's secondary windings is a unique feature that distinguishes this converter from other isolated converters. Isolated power converter can be operated in both voltage control mode and the current control mode. Dominant stability is achieved during the current control mode.

The flyback converter stores energy and sends it to the load during the ON state of the switch. When turned OFF, the flyback converter sends the energy to the load. There is no limit to how much voltage the flyback converter can gain. Because this converter doesn't move energy immediately, it is only good for low-power applications (usually below 100 W). This converter can also make a lot of different outputs by tapping into the secondary winding of the inductor it's connected to.

4.2.1.5 Turn ON Mode

The switch is turned off during this time, as shown in Figure 4.5. The switch is not turned on when the transformer's primary is connected directly to the power source.

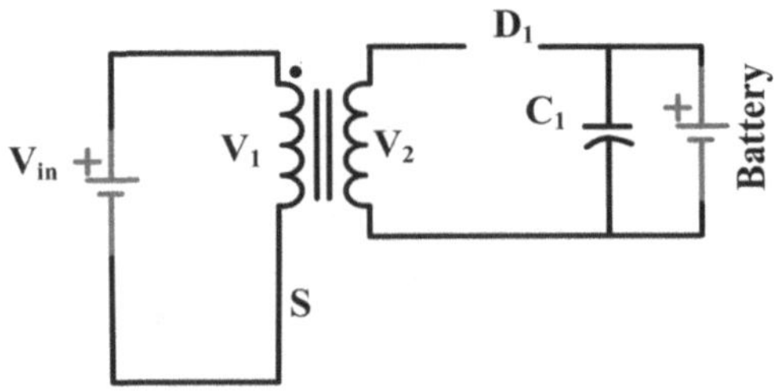

FIGURE 4.5 Turn ON mode flyback converter.

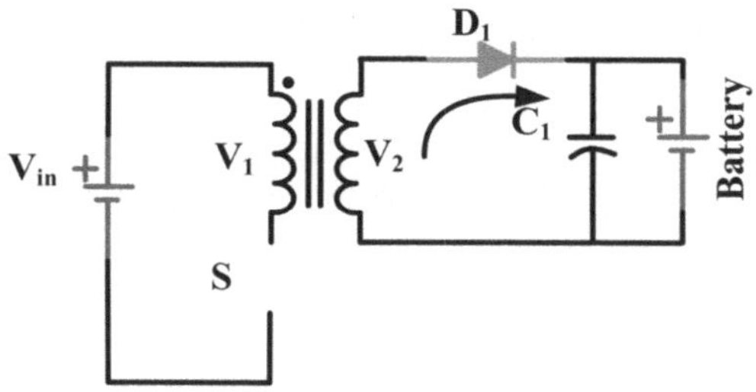

FIGURE 4.6 Turn OFF mode flyback converter.

A transformer's main current and magnetic flux goes up, storing more energy inside the transformer. Because the voltage in the secondary winding of the diode D_1 is negative, the diode is under blocking mode.

4.2.1.6 Turn OFF Mode

The main current and magnetic flux goes down when the switch is pushed. In this case, the secondary voltage is positive, letting current flow from the transformer and pushing the diode toward forward bias, as shown in Figure 4.6. The core of the transformer provides a significant amount of the energy needed to charge the capacitor and supply power to the load.

A lot of power is sent from a transformer to its load side when it is turned off (and the output capacitor).

4.2.2 Electric Vehicle Battery

Battery electric vehicles and hybrid electric vehicles (HEVs) are so named because their electric motors are powered by batteries. For instance, lithium-ion batteries can store a lot of energy (electric charge).

The batteries utilized in EVs differ from those employed for starting, lighting, and ignition (purposes. The batteries in question are designed specifically for deep-cycle applications, with a prolonged lifespan as its primary characteristic. EV batteries have a notable power-to-weight ratio, specific energy, and energy density. The preference for smaller and lighter batteries stems from their ability to decrease the overall weight of the vehicle, enhance its performance, and exhibit more efficiency. There are a lot of new battery technologies, but they don't have as much energy as liquid fuels. This affects how far the cars can go when they're all electric.

Because lithium-ion batteries have a lot of energy per unit weight compared to other types of batteries, they are used in most small consumer gadgets, like cell phones and laptops. In addition to being powerful for their size, these batteries also function well in extreme temperatures and show almost no self-discharge. Although it is possible to recycle many components of lithium-ion batteries, doing so has proven difficult for manufacturers. Lithium-ion batteries are used in most of today's PHEVs and EVs, even though the specific chemistry isn't the same as in consumer electronics batteries. Research and development efforts are focused on lowering their costs, increasing their lifespans, and preventing them from overheating.

4.2.3 Smart Grid

Smart grids continue to evolve with advancements in technology and changes in energy systems. Smart grids will likely see greater integration of renewable energy sources like solar and wind power. Advanced forecasting techniques and real-time monitoring will enable better management of intermittent renewable generation, ensuring grid stability and reliability. The proliferation of distributed energy resources such as rooftop solar panels, energy storage systems, and EVs will continue. Smart grids will need to effectively manage these distributed resources, optimizing their use and facilitating two-way communication between consumers and the grid. With the increasing frequency of extreme weather events and cyber threats, smart grids will focus on enhancing resilience and cybersecurity measures. This may involve deploying advanced sensors, predictive analytics, and AI algorithms to detect and mitigate risks in real-time. Smart grids will leverage demand response programs and demand-side management techniques to balance supply and demand more efficiently. Consumers will have more incentives to adjust their energy usage based on grid conditions, leading to a more flexible and responsive energy system. As EV adoption grows, smart grids will need to accommodate the increased demand for charging infrastructure. Integration with smart charging solutions and vehicle-to-grid (V2G) technology will enable more efficient use of EV batteries and support grid stability. Microgrid deployment will increase, especially in areas prone to grid outages or where energy independence is desired. Smart grid technologies will enable seamless integration and coordination between microgrids and the main grid, allowing for islanding during emergencies and smoother transition back to the main grid.

4.3 SOLAR MPPT CONTROL TECHNIQUES

MPPT controllers in solar applications all use the actual reading process, but slight differences in the algorithm they use to achieve MPP. Sensors on the terminals of the module or array of every DC-DC converter with MPPT read the voltage and current values. They insert those numbers into the algorithm and issue a command to adjust the converter's duty ratio to bring the operation point closer to the MPP. The basics of an MPPT controller for solar power systems are shown in Figure 4.7.

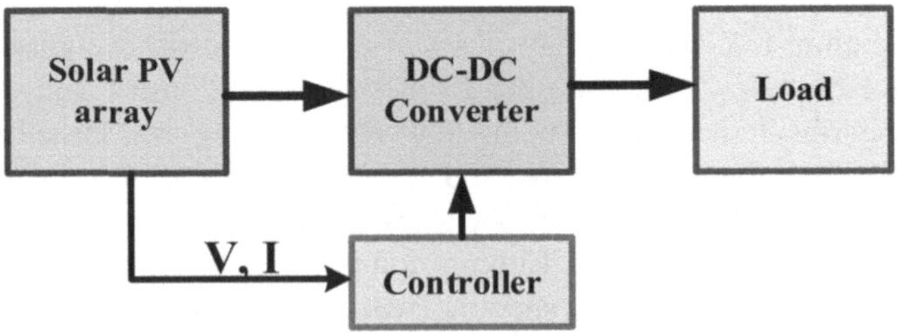

FIGURE 4.7 Basics of an MPPT controller for solar power systems.

First, "Perturb and observe" or "Hill climbing Search" is a very easy and basic strategy. This entails making little modifications to the duty ratio, determining whether the power obtained is higher or lower. If the change results in less power being used, the duty ratio will shift in the opposite direction. This approach is typically employed because of its flexibility in programming and implementation.

4.3.1 P&O Algorithm

This method is often used to keep track of the MPP. In this method, a small change is made to the PV module's power to make it change. It is common practice to compare the current PV output power to its past average regularly before making adjustments. If the output power increases, the same thing happens again; the perturbation is turned around if it doesn't. By using this technique, the voltage of an individual PV module or the entire array can be adjusted. People look at the PV module voltage to see if it has gone up or down to see if the power has changed. More energy can be generated by increasing the voltage. The PV module's MPP is located to the left of its operational point.

Because of this, there needs to be more disturbance on the right to reach MPP. If the voltage goes up, it takes more movement to the left to get the PV module to work at its best point. The maximum power point tracking (MPPT) charge controller is located in this setup between the PV module and the battery. When connected to both the PV module and the batteries, it measures both of their respective voltages. It checks the voltage to see if the battery is fully charged. As soon as the battery is full, the charger stops. This is to keep the battery from getting too hot. The DC/DC converter is turned on to begin charging the battery if it is not fully charged. In the next step, the microcontroller will measure voltage and current to determine how much power comes out of the output and compare it to how much power was measured before. To extract the most energy from the PV cell, the pulse width modulation (PWM) duty cycle is increased when P_{new} leads P_{old}. Similarly, when P_{new} lags P_{old}, then the duty cycle is decreased until P_{old} is reached again. This MPPT algorithm is simple, easy to make, and cheap while still giving good results. For controlling MPPT, the proposed algorithm is a cost-effective and simple choice. It's called the P&O technique used to figure out how much more power the PV system is putting out now than before. This is called the power difference or $P_{new}–P_{old}$ (P). If P is greater than zero, the same action is repeated for more perturbation; if it is less than zero, it changes direction. The P&O method is the most common MPPT algorithm for increasing the efficiency of a PV system. This is because it is easy to set up, has a high level of reliability, and increases the system's efficiency. Figure 4.8 shows the flowchart for P&O algorithm.

4.3.2 PV Module

A solar module is a device designed to generate electricity by conversion of solar energy. PV refers to a photodiode's unbiased functioning state, wherein the current that flows through the system is attributable to the incident light. Solar energy is converted into electrical energy by PV systems. It is a nonpolluting and extremely

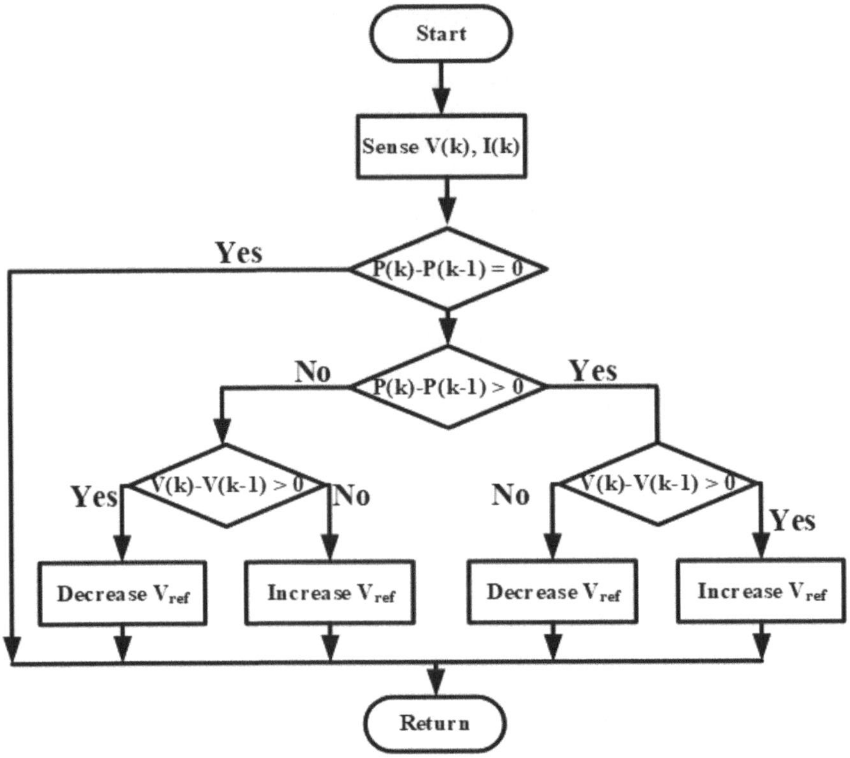

FIGURE 4.8 Flowchart of P&O algorithm.

dependable power source suitable for various applications. The solar cell comprises a Shockley diode, a current source, and a resistor. A current which is proportional to the irradiation of 1000 W/m^2 is produced by the current source.

An alternative model of a solar cell may be used to derive the voltage and current characteristics of a solar cell. Solar PV cells are connected in parallel and series configurations to maximize power output, resulting in solar PV arrays. Solar cells used in series connections should be similar as much as practicable. PV modules linked in series get the same amount of solar irradiance. A series assembly is made up of sub-assemblies that are connected in series, and each of these sub-assemblies has a unique level of irradiance.

4.3.3 Maximum PowerPoint Tracking

Because the efficiency of the device is low which is around 13%, it is advised that the device be operated at its maximum capacity. This will ensure that the maximum amount of total power is accessible to the load, regardless of the temperature or the amount of irradiation. As a consequence of this, raising the power output improves the efficiency of the solar PV module. An MPPT is utilized in order to extract the maximum amount of power from PV modules and then send the energy to where it is required. A few factors to consider while evaluating the MPPT algorithm include monitoring rate, consistency, implementation, easy operation, price, and monitoring effectiveness.

PV cells have a complicated relationship with irradiance, temp, and resistance value, resulting in a I-V curve; by modifying the duty ratio, the load impedance can be changed.

To maximize power, a DC-DC converter is installed between the PV module and the load module. The converter's duty cycle is modified until the maximum power is reached. In existing systems, the output voltage V_{pv} and current I_{pv} of PV cells are tracked to account for a variety of different operating conditions. This data is used to create the I_{pv}-V_{pv} curve. The highest power point is determined by a graphical slope (dI_{pv}/dV_{pv}), corresponding to the curve's knee. Maximum power point algorithms seek peak power points using various control circuits, concepts, or logic, allowing to extract the maximum power from the arrays. Increasing the energy generation of a solar PV array can be accomplished by the application of a variety of methodologies, such as the P&O technique, incremental conductance, parasitic capacitance, stable voltage-based maximum output monitoring, and current controlled-based peak power tracking. MPPTs use any of these techniques, switching between them according to the array's operating circumstances.

4.4 PROPOSED SYSTEM

The boost converter is employed here in the proposed system and Figure 4.9 shows the block diagram of the proposed system. Boost converters are analyzed, and their functionality is evaluated to the existing converter in the proposed system. In power systems, the classic boost converter is used for large applications that require large

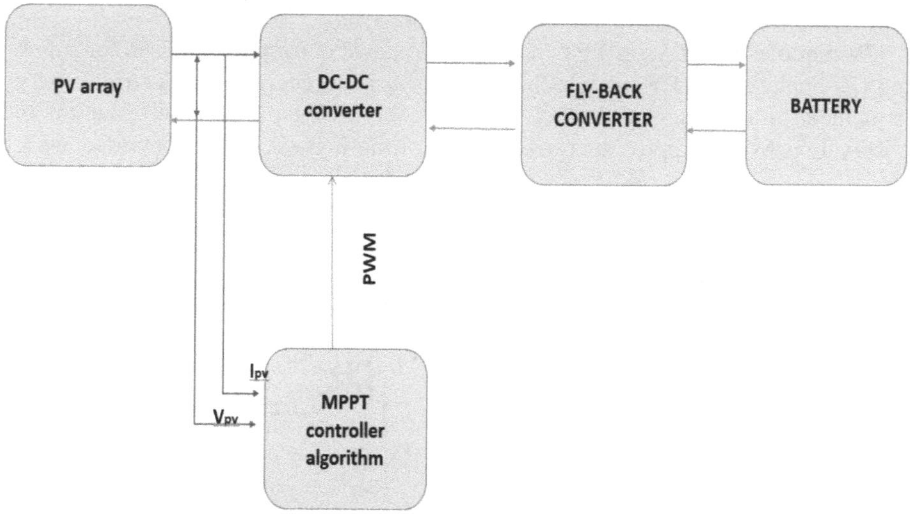

FIGURE 4.9 Block diagram of proposed system.

current fluctuations, which results in conduction losses. Because of the inductance values, most high step-up converters have a high input current. It is essential to boost the switching frequency to decrease the inductance value, minimizing the voltage stress within switches.

Furthermore, soft-switching is required to reduce switching losses while also enhancing productivity. The DC/DC converter's performance in the present system was compared to that of the proposed converter. The proposed converter features a 31.06 V input voltage and a 178 V output voltage with an output power. The converter's efficiency is 92.5%, which is greater than the other converter types studied. In addition, the suggested converter has a low switching voltage stress and uses less input current. According to the time response study, the recommended converter achieves constant current and voltage faster than the existing converter employed in the present system.

The suggested system's converter might be employed in high-static-gain situations. Compared to a conventional boost converter, the modified boost converter employed in our suggested system has twice the static gain value, and the switching voltage is half of the peak value. Based on the boost converter design, the suggested system offers a converter with low switching stress and high static gain in our suggested system. Compared to existing converters, the suggested technique has a lower input inrush current for the inductor, a faster settling time and a shorter output voltage peak duration.

The suggested system converter could be used in high-static-gain applications. While contrasted to a traditional boost converter, the modified boost converter used in our suggested system has double the static gains rate and a changing voltage ½ the magnitude of the output voltage. Based on the design of the boost converter, we offer a converter with high static gain and low switching stress in our suggested system. Compared to existing converters, the suggested technique has a lower input inrush current for the inductor, a faster settling time, and a shorter output voltage peak duration. The circuit diagram is illustrated in Figure 4.10.

PV modules or arrays, MPPT modules, and boost converters are used for supplying DC applications. PV module that transforms solar energy into electrical energy is included in this suggested block diagram. The MPPT uses the perturbation and observation MPPT approach to track the all-time highest energy provided with a

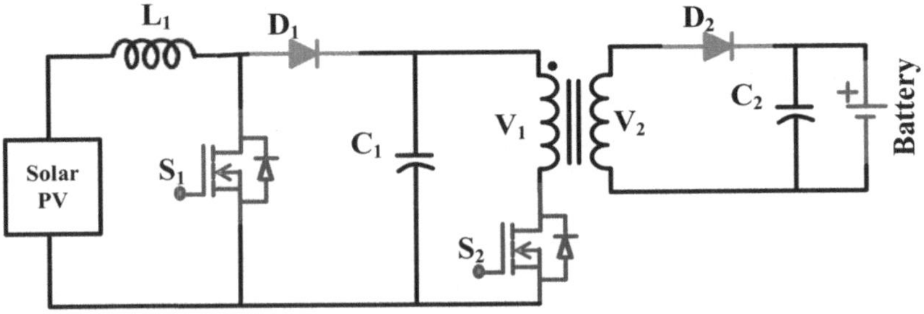

FIGURE 4.10 Proposed system circuit diagram.

solar module, and this information is sent into the DC-DC boost converter. The voltage is then stepped up to the required amount by the boost converter and sent to the flyback converter, with the flyback converter's output voltage delivered to the DC load, which is the battery. Mainly, the DC-DC converter is discussed here.

The converter in our suggested system that converts fluctuating DC value to constant DC supply, which is useful for various DC applications. The main advantage of large step-up DC converters is the generation of significant voltages, i.e., 10 times larger output may be generated from the input voltage. The boost converter is commonly employed in high-voltage applications.

Once the voltage output exceeds the supply voltage, the boost converter operates upon the step-up principle. It is linked with a switched capacitor and coupled inductor to increase the boost converter's efficiency. During large step-up activity, current ripples are higher in power devices, creating conductance inefficiency and switching down the current.

For applications involving high voltage, the switching voltage is very close to the output voltage of the converter. In order to improve the boost converter's overall efficiency, it has been changed into a standard interleaved boost converter. It has two switches, which will help reduce the fluctuation of the current. However, the diode has a reverse recovery difficulty. Because it transmits energy through a capacitor and an inductor, the voltage switching is more than that of a boost converter. A converter's static gain is employed for large input voltage range applications.

The voltage flow through the switch is nearly equivalent to the total of the source and output voltage. Buck converter, boost converter, buck-boost converter, and other converters may be used in MPPT design. Our suggested system has chosen the boost converter since it offers greater flexibility and higher performance. The current does not always flow in a buck converter, but a boost converter can produce a constant current flow. Compared to buck converters, this is still another advantage of the boost converter. The suggested system is simple and small, and it will be particularly beneficial in situations where high-efficiency performance is required 24 hours a day, seven days a week.

4.4.1 System Description

The proposed DC-DC boost converter comprises several essential components, including an inductor denoted as L, a switch represented by a MOSFET, a rectifier diode labelled as D, and an output capacitor. For applications involving high voltage, the switching voltage is very close to a conventional boost converter which is the most common type of DC-DC converter used in PVs because of its uncomplicated circuit design, low voltage stress on its components, and high conversion efficiency.

4.4.1.1 Design of Proposed Converter

In our proposed system design, a high-power boost converter meeting the following criteria was created and simulated in MATLAB. A switching frequency (F_s) of 10 kHz was selected for the overall design because the converter is of modest power. As a result, the stress on the components must be kept to a minimum. The switching frequency for a low-power type can be substantially greater, even into the MHz range. The inductor and output capacitor values are the most important components to be calculated in the boost converter circuit.

Here, the duty cycle d was determined as 0.57 and the value of inductor was estimated to be L = 560 H. Aside from storing and transferring energy, the output capacitor reduces the converter output voltage ripple, V_{out}. Therefore, it is preferable to minimize the voltage ripple caused by the introduction of capacitors.

The converter used in our suggested system comprises various conventional converters connected in such a way that the stress on converter would reduce the damage due to conductivity and switching and thereby increasing the total system performance. Each of the numerous converters runs at 1/n of the sum of switching frequencies and working current because the current and switching frequency strains are evenly distributed among them. Lower source current ripples, reduced switching and conduction loss, higher thermal performance, and greater performance are all advantages of the topology over the standard version. The effective ripple frequency is also larger than the traditional DC-DC boost converter because the power channels of the converter are mixed at the output of the capacitor. Because the currents in each converter linked in parallel which is as half as that of the typical converter, where the inductor value for each of the two converters may be decreased. As a result, the inductor value, which was 560 H for the traditional converter, was lowered to 240 H for each interleaved converter's converters in this design. Proper current sharing between the paralleled converters is essential to ensuring that neither converter is overloaded. MATLAB was used to simulate a Trina Solar PV module, which was utilized to power the suggested boost converters.

The reliability of both the current and emerging systems is compared as the solar supply power changes. The PV working temperature is assumed to remain stable at 25 °C throughout the simulations. PV output power changes in parallel with irradiation. The present and proposed system converters react to these changes in various ways. It can be shown that the present boost converter has a greater efficiency than the suggested converter at low PV output power. However, the suggested converter's efficiency improves when the PV output power grows.

This is to be expected, as the PV generation has high power, converter losses are high, causing a reduction in converter efficiency.

However, the proposed converter's efficiency is better at large PV generating power since converter losses are significantly decreased because of its offering conductance & switching loads among the multiple converters. Since a PV system only uses 30% to 100% of its rated power for 80% of its operation time, it's important to take advantage of the benefits of current and future DC-DC boost converter topologies to make the system more efficient in every scenario. As a result, present converter functions at low irradiation, while the proposed converter runs at high irradiation. The two identical outputs were simulated to illustrate the concept, although the Solar panel has only a single outlet. Even when the PV module is subjected to irradiance below 800 W/m^2, the modern boost converter gets into action but the proposed system is designed to work when the PV module receives an irradiance of 1000 W/m^2.

Flyback converters are specific DC-DC converter that uses a transformer in the middle of the circuit to store energy before releasing it to the load. The flyback converter in our proposed system is indirect since it does not have a direct link between the input and the load at any time during operation due to its unique design. Our proposed system's flyback converter has a small number of components. The primary feature that sets this converter against others is the connection of the secondary windings of the transformer.

A DC supply, a transformer, a diode, a MOSFET switch, a filter capacitor, and a load make up the flyback converter. Because of the operation of the switch at such a high frequency, the uncontrolled input voltage can be assumed to remain constant. It also allows for quick duty cycle control to get the desired output. In a flyback transformer, the primary and secondary conductors do not carry current at the same time. The operation is in continuous mode in this case.

Nonlinear converter systems are used. As a result, it must be linearized at the quiescent point to apply linear control techniques to the converter. The proposed design makes use of a maximum power tracker for tracking the PV array's peak output and a flyback converter to boost the input power. It can perform beyond a large number of input voltage fluctuations. Since the transformer's main and secondary windings are mechanically separated, any voltage ripples in the secondary winding would not transfer to the primary winding, limiting the device to a single or many isolated voltage outputs. When compared to other switching converters, it is both inexpensive and energy efficient.

Solar energy is collected by using the flyback converter technique in our proposed system. This method can also enhance the output voltage of a PV module using an inverter due to the massive bulk and low efficiency of traditional inverters. In order to overcome this drawback and increase efficiency while decreasing the number of switching devices, a flyback converter is designed and coupled in parallel. In this method, the output voltage of the PV source is increased by employing a flyback converter in the role of a boost converter. Higher efficiency and reduced switching loss are accomplished by converting the DC power output of the solar panels to AC power. Flyback switching devices receive gate pulses from MPPT.

As a result, the PV system's low power is enhanced, and the converter's size is reduced. This method is more efficient than the existing system's traditional method. This is analyzed, studied in the suggested system, and then implemented in MATLAB/Simulink.

To get a higher voltage output, a DC-DC converter designated as a boost converter improves the voltage that goes into it. It finds widespread application in battery-powered equipment, LED drivers, and even certain renewable energy systems, among other places where the output voltage must exceed the input voltage. A capacitor, an inductor, a semiconductor switch, a diode, and a boost converter are the conventional components. A series combination of a switch, diode, inductor, and capacitor constitutes the fundamental circuit configuration. A PWM signal regulates the switch, which is often a MOSFET. The output voltage is established by the duty cycle of the PWM signal. The inductor stores energy when the switch is ON and releases it to the load when the switch is opened. The primary component of the circuit is the inductor, which operates like a battery and discharges its charge whenever the switch is opened. The current needs and desirable output voltage ripple determine the inductor value. With a diode, the current can flow in one particular direction which is only from the input to the output but not in the other direction. A Schottky diode is a common type of diode because of how quickly it recovers. The output voltage has a lower ripple and is smoothed out by the capacitor. The value of this parameter is selected according to the power demands at the load end and the desired output voltage ripple. The duty cycle of the switch is controlled by a control circuit that receives feedback from its output voltage. A microcontroller or specialized controller integrated circuit might achieve the same. In order to fulfill the intended performance requirements, a boost converter must be meticulously designed with element choices, control strategy, and optimization of efficiency in the account. The design and analysis process is able to utilize simulation technologies such as MATLAB/Simulink. When designing a boost converter, it is important to take into account the desired output voltage ripple, load current, switching frequency, and efficiency requirements when choosing the values of the inductor (L) and capacitor (C).

4.4.1.2 Design of Inductor (L)

It is common practice to express the output voltage ripple as a percentage. For the majority of uses, it is recommended that the ripple be less than 5% of the final voltage. Equation (4.4) is used to get a rough estimate of the ripple voltage (ΔV_L) across the inductor. In the designed work, when the switch is turned ON, the energy is transferred to the output via the flyback converter. The energy goes to the load when the switch is turned off. Duty cycle (D) and maximum load current are the two variables that decides the peak inductor current

$$\Delta V_L = \frac{V_{out}(1-D)}{f\,s.L} \tag{4.4}$$

where

V_{out} = output voltage
D = calculated duty cycle
fs = switching frequency

Equation (4.5) is used to find a suitable inductance (L) value by estimating the ripple in the output voltage. By rearranging equation (4.4), it is given as

$$L \geq = \frac{V_{out}(1-D)}{fs.\Delta VL} \tag{4.5}$$

Peak inductor current and switching frequency (fs) are the two variables that can be used to determine the inductance value given in equation (4.6).

$$L \geq \frac{V_{in\,min}(1-D)^2}{2fs.\Delta IL} \tag{4.6}$$

where

$V_{in\,min}$ = minimum input voltage

ΔI_L = inductor current ripple (peak to peak)

4.4.1.3 Design of Capacitor (C)

Reducing ripple and smoothing down the output voltage is the task of the output capacitor. In the same way, as with a boost converter, the value of the capacitor (C) is dependent on the output current, ripple voltage, and switching frequency and is calculated by equation (4.7):

$$C \geq = \frac{I_{out}\left(1-D\right)}{fs.\Delta Vc} \tag{4.7}$$

The transformer, which isolates the circuit while altering the voltage, is an essential part of a flyback converter. The desired output voltage can be found by the transformer's turns ratio using equation (4.8):

$$N_P = N_S = \frac{V_{in\,max}\left(1-D\right)}{V_{out}} \tag{4.8}$$

where N_P and N_S are a number of turns in the primary and secondary.

4.5 SIMULATION OF THE PROPOSED SYSTEM

Power is kept at 249.6, voltage is kept at 31.06, current is kept at 8.036, temperature and irradiance are kept at 25 and 1000, respectively, and the output voltage is 42.64. Figure 4.11 shows the simulation diagram of the proposed system using Simulink and MATLAB. The solar PV system was analyzed with the use of a DC-DC boost converter and a flyback converter. The model predictive controller's effect on P&O performance has been studied, and it has proven to be responsive. MPPT with the P&O approach was used to charge the battery, which provided adequate backup. As a result, it provided energy storage and could be used with DC loads. As a result, depending on the requirements, it can be used in low-power applications.

Figure 4.12 is the MATLAB/Simulink graphical result representation of the state of charge (SOC), voltage, and current graph of the proposed system's Simulink circuit design/model. The above result graph shows that the output current and voltage are similar because of the boost converter. A repeating sequence generator sets the operating frequency of 10 kHz. This generator produced the pulse signal that was compared to the MPPT unit's signal before providing the signal to the gate switch. The input of the duty cycle is unnecessary when employing the MPPT technique, as the algorithm autonomously iterates and determines the appropriate duty cycle. In the absence of MPPT implementation, the user would have been required to manually input the duty cycle. The shifting of solar irradiance results in corresponding changes in both the maximum power point and the required duty cycle for the model to operate effectively. Furthermore, it undergoes evolution. Nevertheless,

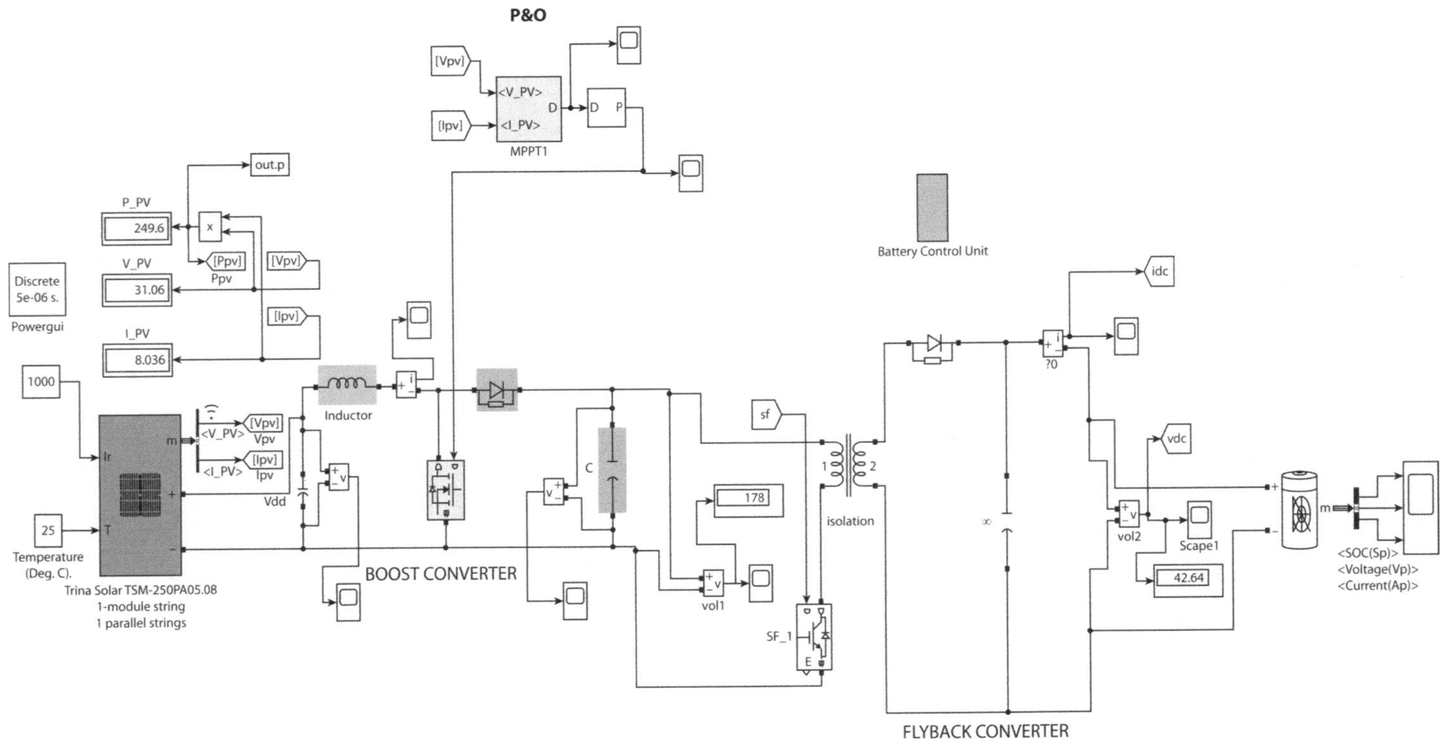

FIGURE 4.11 Proposed system circuit of MATLAB/Simulink.

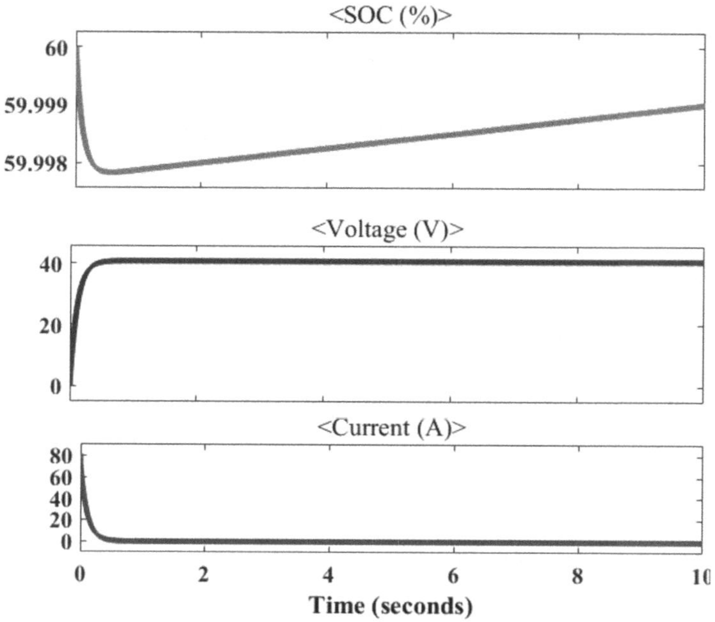

FIGURE 4.12 SOC, voltage, current graph of proposed system.

when employing a continuous duty cycle, it becomes impossible to track the maximum power point, leading to a decrease in system efficiency. The plot technique in MATLAB was employed to construct the aforementioned waveforms. A minor decrease in power occurs between the solar panel input and the boost converter output. The observed phenomenon can be attributed to the losses incurred by the boost converter during switching, as well as losses in the inductor and capacitor.

Figure 4.13 is the MATLAB/Simulink graphical representation of the capacitor voltage for the proposed system's Simulink circuit design/model.

The capacitor voltage of our suggested system is shown in the graph above, and it is 31.06 V. The capacitor voltage ripples are minimized in our proposed system by increasing the capacitor value by calculation while developing the boost converter for our proposed system.

Figure 4.14 shows the MATLAB/Simulink graphical representation of the capacitor voltage for the proposed system. Here the voltage is taken across the capacitor and the same can be available for charging the battery which is considered in our system.

The output graph is drawn between the capacitor voltage and time in seconds. It is seen from the simulation graph that the voltage reaches up to 246 V with a short span of time of 0.8 sec. Also, it can be seen that the overshoot of the voltage is highly reduced by the contribution of the PI controller. The ripples of the output voltage are also highly reduced by the same. This gives rise to a fast response of charging the battery. The capacitor voltage of our suggested system is 246 V and the voltage ripples are highly minimized by increasing the capacitor value. The same

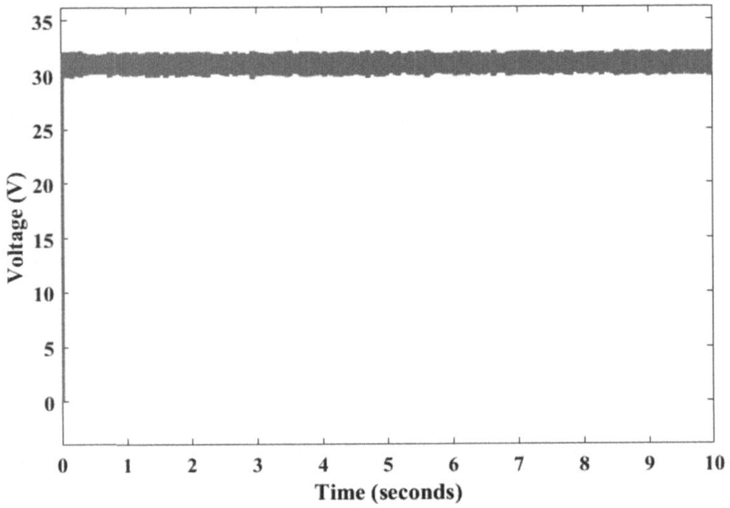

FIGURE 4.13 Capacitor voltage graph (I) of proposed system.

is calculated while developing the design parameters of the boost converter for the proposed system.

Figure 4.15 is the MATLAB/Simulink graphical result representation of the Inductor Current Graph of the Proposed system's Simulink circuit design/model.

The inductor current of our suggested system is shown in the graph above, and it is 42 A. The inductor current is minimized in our proposed system by increasing the inductor value by calculation while developing the boost converter for our proposed system.

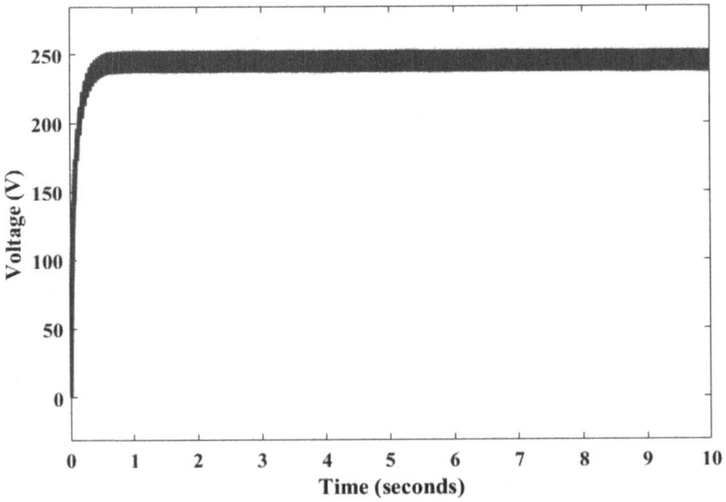

FIGURE 4.14 Capacitor voltage graph (II) of proposed system.

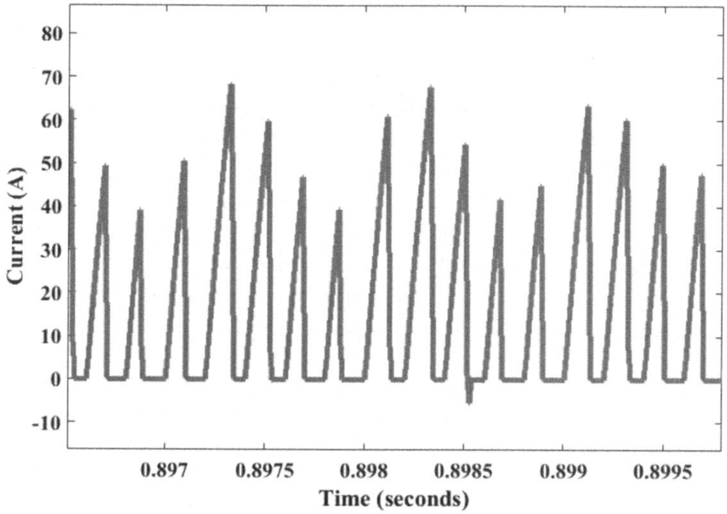

FIGURE 4.15 Inductor current graph of proposed system.

Figure 4.16 is the MATLAB/Simulink graphical result representation of the Output Voltage of the flyback converter of the Proposed system's Simulink circuit design/model.

The initial voltage output in our proposed system was 31.06 V. The above graph depicts the output voltage of the flyback of the proposed system; here, the boosted voltage from the boost converter is 176 V. The output voltage from the flyback converter is 42.64 V, and the load uses this voltage afterward. In our proposed system, the load is the battery.

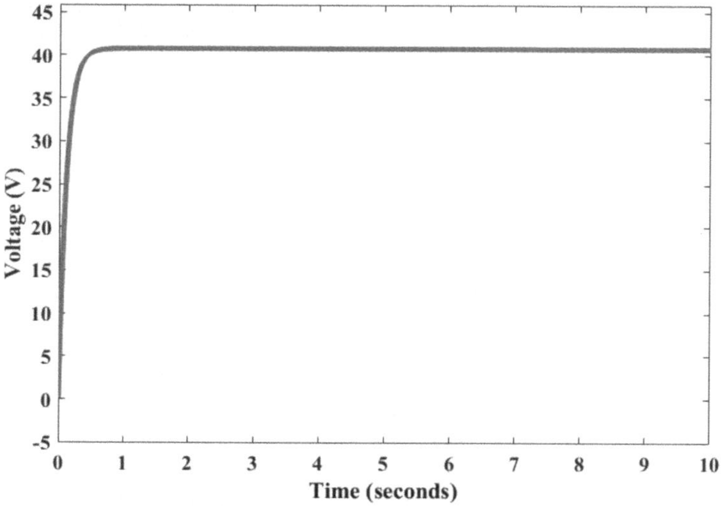

FIGURE 4.16 Output voltage of flyback converter.

The SOC of the battery is stable in the proposed system. In the existing system, however, it is not stable. Capacitor ripples have been decreased when they were previously high. MPPT technique was employed in the design of a solar PV Array. The charging current was managed with the use of a DC-DC Boost converter to attain the EV's rated current, which in this case was achieved using MATLAB/Simulink software, and it was proved that the charging current was kept below the EV's rated limit, yielding satisfactory results. The model performs under varying solar irradiance at nominal circumstances (25 C and 1000 W/m^2). Furthermore, extra charging circuitry is no longer required, and the solar PV array may be directly connected to the EV for charging, making it more cost-effective, dependable, and compact. This method is also simple to adopt and requires.

4.6 CONCLUSION

Renewable energy, often green energy, will play a critical role in energy generation in the next years. The existing system's SOC percent of the battery at load is only 65%. However, our proposed system's SOC percent of the battery at load is 85%. The battery's SOC is an important metric that reflects its performance. Accurate SOC estimation not only protects the battery by preventing overcharging or discharge and increasing its life, but it also enables the application to use reasonable control mechanisms to meet the goal of energy saving.

Furthermore, the boost converter in our suggested system consumes less input current and has low switching voltage stress. In contrast, the boost converter in the existing system consumes a large amount of input current and has a high switching voltage stress. When contrasting the time response evaluation of the suggested converter compared to such boost converter in the existing system, it was observed that the proposed converter resolves to constant state current and voltage faster than the boost converter in the existing system. It was proved that the charging current was kept below the EV's rated limit, yielding satisfactory results. The model performs under varying solar irradiance at nominal circumstances (25 C and 1000W/m^2). Furthermore, extra charging circuitry is no longer required, and the solar PV module may be directly connected to the EV for charging. Because of this, it is more feasible, credible, and stackable.

REFERENCES

1. Owusu P. A. & Asumadu-Sarkodie S. | Shashi Dubey (Reviewing Editor) (2017). A review of renewable energy sources, sustainability issues and climate change mitigation, Cogent Engineering, 3(1), https://doi.org/10.1080/23311916.2016.1167990
2. Strielkowski W., Civín L., Tarkhanova E., Tvaronavičienė M. & Petrenko Y. (2021). Renewable energy in the sustainable development of electrical power sector: a review. Energies, 14, 8240. https://doi.org/10.3390/en14248240
3. Fawzy S., Osman A. I. & Doran J. et al. (2020). Strategies for mitigation of climate change: a review. Environmental Chemistry Letters 18, 2069–2094. https://doi.org/10.1007/s10311-020-01059-w
4. Wang J., Wang B., Zhang L., Wang J., Shchurov N. I. & Malozyomov B. V., (2022). Review of bidirectional DC–DC converter topologies for hybrid energy storage system of new energy vehicles, Green Energy and Intelligent Transportation, 1(2), 100010, https://doi.org/10.1016/j.geits.2022.100010.

5. Kavi A. & Sarkar S. (2022). A review on electric vehicle powertrain system components and the challenges involved in their large-scale introduction. International Research Journal of Engineering and Technology, 9(3), pp.72. https://doi.org/10.13140/RG.2.2.28292.81284.
6. Dimitrov B., Hayatleh K., Barker S., Collier G., Sharkh S. & Cruden A. (2020). A buck-boost transformerless DC–DC converter based on IGBT modules for fast charge of electric vehicles. Electronics, 9, pp. 397. https://doi.org/10.3390/electronics9030397
7. Shahzad M., Qu Y., Rehman S. U. & Zafar A. U., (2022). Adoption of green innovation technology to accelerate sustainable development among manufacturing industry, Journal of Innovation & Knowledge, 7(4), pp. 100231, https://doi.org/10.1016/j.jik.2022.100231.
8. Bradu P., Biswas A. & Nair C. et al. (2022) Recent advances in green technology and Industrial Revolution 4.0 for a sustainable future. Environmental Science and Pollution Research., 30, pp. 124488–124519, https://doi.org/10.1007/s11356-022-20024-4
9. Fallah Shayan N., Mohabbati-Kalejahi N., Alavi S. & Zahed M. A. (2022). Sustainable development goals (SDGs) as a framework for corporate social responsibility (CSR). Sustainability, 14, pp. 1222. https://doi.org/10.3390/su14031222
10. Rasoulinezhad E. & Taghizadeh-Hesary F. (2022) Role of green finance in improving energy efficiency and renewable energy development. Energy Efficiency 15, 14. https://doi.org/10.1007/s12053-022-10021-4
11. Borole M., Bansode A., Kachave A. & Ghuge N. N. Modelling of DC-DC converter for solar based electric vehicle, in: Proceedings of the 3rd International Conference on Communication & Information Processing (ICCIP) 2021, Available at SSRN: https://ssrn.com/abstract=3916372 or http://dx.doi.org/10.2139/ssrn.3916372
12. Singh A. K., Badoni M. & Tatte Y. N. (2020). A multifunctional solar PV and grid based on-board converter for electric vehicles, IEEE Transactions on Vehicular Technology, 69(4), pp. 3717–3727. https://doi.org/10.1109/TVT.2020.2971971.
13. Lee Y.-J., Khaligh A. & Emadi A. (2009). Advanced integrated bidirectional AC/DC and DC/DC converter for plug-in hybrid electric vehicles. IEEE Transactions on Vehicular Technology, 58(8), pp. 3970–3980, https://doi.org/10.1109/TVT.2009.2028070.
14. Annamalai M. C. & Amutha prabha N., (2023). A comprehensive review on isolated and non-isolated converter configuration and fast charging technology: for battery and plug in hybrid electric vehicle, Heliyon, 9(8), p. e18808, ISSN 2405-8440, https://doi.org/10.1016/j.heliyon.2023.e18808.
15. Sharma S. & Swami R. (2022). Solar powered electric vehicle using maximum power point tracking. Material Science & Engineering International Journal. 6, pp. 132–135. https://doi.org/10.15406/mseij.2022.06.00190.
16. Yilmaz M. & Krein P. T. (2013). Review of battery charger topologies, charging power levels, and infrastructure for plug-in electric and hybrid vehicles, IEEE Transactions on Power Electronics, 28(5), pp. 2151–2169. https://doi.org/10.1109/TPEL.2012.2212917.
17. Daniel A. S. & Chandrakala K. R. M. V., "Design of an isolated onboard plug-in electric vehicle charger," in: 2021 7th International Conference on Electrical Energy Systems (ICEES), Chennai, India, 2021, pp. 147–149, https://doi.org/10.1109/ICEES51510.2021.9383717.
18. Ali A. I. M., Sayed M. A. & Mohamed E. E. M., (2018) Modified efficient perturb and observe maximum power point tracking technique for grid-tied PV system, International Journal of Electrical Power & Energy Systems, 99, pp 192–202, https://doi.org/10.1016/j.ijepes.2017.12.029.
19. Marei M. I., El-Helw H. & Al-Hasheem M., "A grid-connected PV interface system based on the DAB-converter," in: 2015 IEEE 15th International Conference on Environment and Electrical Engineering (EEEIC), Rome, Italy, 2015, pp. 161–165, https://doi.org/10.1109/EEEIC.2015.7165534.

5 A Comprehensive Study on Off-Grid and Hybrid Charging System for Electric Vehicles

Shekh Kulsum Almas
and Nagendra Singh

5.1 INTRODUCTION

Electric vehicles (EVs) have become a key component of the automobile industry's recent spectacular shift toward sustainability, reducing carbon emissions and decreasing reliance on fossil fuels [1]. As the world seeks cleaner and more efficient transportation alternatives, the focus has inevitably turned toward optimizing the charging infrastructure for EVs. One of the most promising avenues in this regard is the development of off-grid and hybrid charging systems (HCSs) [2].

Traditional charging infrastructure primarily relies on grid-connected charging stations, which draw power from centralized electrical grids. While effective in urban areas with well-established grid networks, this approach encounters challenges in remote or underdeveloped regions where grid connectivity is limited or unreliable [3]. Off-grid charging systems offer a compelling solution by utilizing renewable energy sources (RES) such as solar, wind, or hydroelectric power to supply electricity to EVs, independent of the grid [4].

However, off-grid solutions come with their own set of complexities, including the intermittency of RES and the need for energy storage systems to ensure round-the-clock availability [5–7]. HCSs bridge this gap by integrating both grid-connected and off-grid elements, thereby leveraging the strengths of each approach while mitigating their respective weaknesses. These hybrid systems not only enhance reliability and flexibility but also contribute to grid stability through smart energy management [8].

The potential benefits of off-grid and HCSs extend beyond addressing infrastructure challenges. They offer opportunities for energy independence, resilience against grid outages, and significant reductions in greenhouse gas emissions. Moreover, they promote decentralized energy production and consumption, empowering communities to take control of their energy needs and fostering local economic development [9–11].

The use of RES like wind, solar, and biomass increases day by day in various sectors. EV charging stations are also designed using solar and wind-connected systems in stand-alone or grid-connected forms [12]. The hybrid power generation system

DOI: 10.1201/9781003537823-7

is a very popular, effective, and economical system used for EV charging stations, especially in remote areas [13].

We explore the complexities of hybrid and off-grid EV charging systems in this extensive study. This chapter explores the technological advancements, regulatory frameworks, economic considerations, and environmental impacts associated with these innovative solutions.

5.2 LITERATURE REVIEW

The shift to environmentally friendly modes of transportation has sparked a great deal of interest in off-grid and hybrid EV charging solutions. This literature review aims to synthesize existing knowledge and identify key research trends within this rapidly evolving field.

Article [1] discusses the different techniques used for stand-alone systems and hybrid EV charging systems. Article [2] shows how to integrate the utility grid (UG) with RES for large EV charging stations. The charging station may be connected to an AC supply system, an AC-DC (grid-connected solar), or a microgrid system; it depends on the size of the charging stations and the availability of the power sources [3–5].

Solar with a grid-connected system is suggested by article [5]. Similarly, the authors of Ref. [6] show a study on fast charging devices, modes of control systems, and management of multiple charging points in a single charging station. A comparative study of the HCS and battery storage system is discussed in articles [7, 8]. They also discuss how to connect the solar system with the UG.

Articles [9, 10] explain EV energy management, the methodologies used for energy management, and the modes of operation of charging stations. Article [11] explores several potential hybrid vehicle charging systems using solar and grid systems and suggests the most suitable AC-DC-based charging station for multiple users.

A DC charging station that uses both AC and DC systems along with battery backup is discussed in Ref. [11]. A comparative analysis of the effects of several EV charging stations linked to renewable energy systems is provided in article [12]. The effects of EV charging stations on the environment are also covered by the writers. In article [13], the topology of the DC-to-DC converter and instructions for connecting the hybrid grid-connected system to numerous DC charging outlets are displayed.

Articles [14, 15] present the advantages and disadvantages of the converter topology. Converters are the most crucial component of an EV charging station because they are responsible for converting power from AC to DC. Additionally, they facilitate communication between EV chargers and charging stations.

5.3 CLASSIFICATION OF CHARGING TECHNIQUES

EV charging stations are designed mainly in three ways: (1) off-grid, (2) integrated with the grid, and (3) hybrid power generation sources. Further, it is classified as dependent on the use of power sources like AC-operated, AC-DC combined-operated, or DC-operated systems using solar power [16].

5.3.1 EV Charging Station Connected with Grid

When a proper grid system is available, like in a home or any commercial place, grid-connected EV charging is best suited. In this charging system, an AC supply is available as an energy source. With the help of a converter, AC power is converted into DC power and utilized for EV charging [17]. The AC supply system may be used as a single-phase or three-phase system; it depends on the design and requirements of the station. Such a charging system is widely used in commercial charging systems [18].

5.3.2 Off-Grid EV Charging Station

Off-grid EV charging systems can be designed with the aid of RES like solar, wind, and biomass, even in remote areas without access to power. Off-grid EV charging systems require energy storage systems for backup. Off-grid EV charging systems require energy storage systems for backup. For energy storage, battery systems, a supercapacitor, and a fuel cell are used [19].

Figure 5.1 displays the parts of off-grid charging stations. The primary part is the energy source, a power electronics converter that transforms electrical power from AC to DC or the other way around. The energy storage center, which primarily uses batteries, is the second major component. The energy management system (EMS) is a crucial component that is necessary to regulate and control the flow of electricity. The EV charger, which is also connected to the EMS used to charge the EV, is the final but crucial part. Such off-grid charging stations supply both AC and DC power, and power is managed by a converter. Such charging stations are used for local load satisfaction.

According to the availability of power, off-grid charging stations are divided into three categories.

- AC-based architecture
- DC-based architecture
- AC- and DC-based architecture

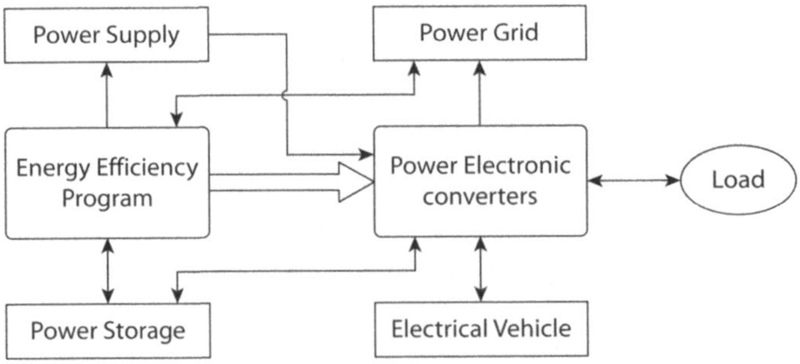

FIGURE 5.1 Basic block diagram of off-grid charging station.

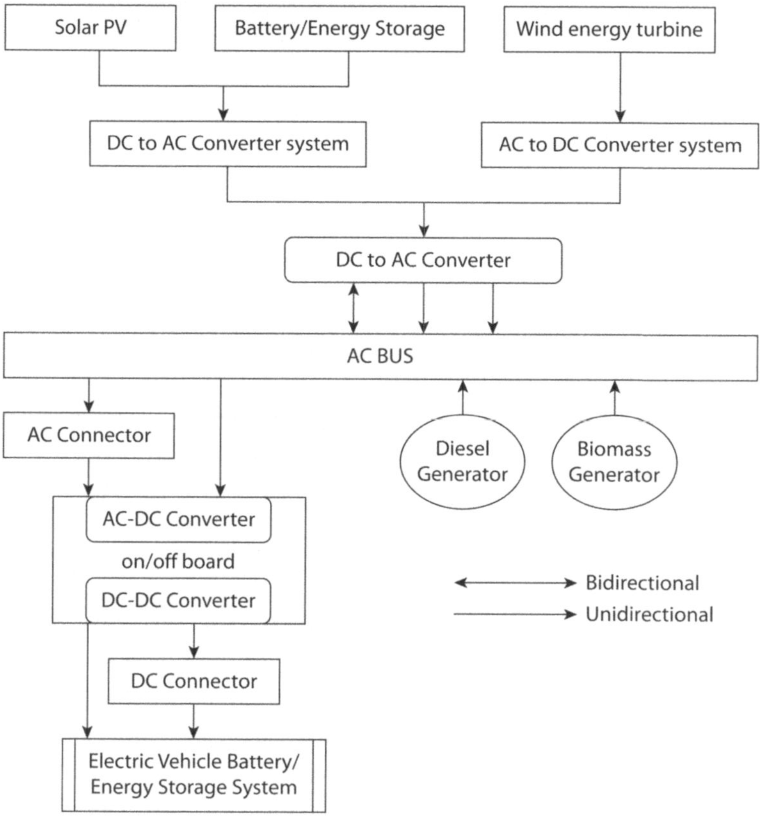

FIGURE 5.2 AC-based technique of off-grid charging system.

5.3.2.1 AC-Based Technique of Off-Grid Charging System

Figure 5.2 shows an AC-based technique for off-grid charging systems. AC power is used in this charging system, which is either generated by classical methods like diesel generators and biomass, or from new renewable sources like wind turbines. Sometimes, more than one method can also be used by the AC bus system to fulfill the required demands at charging stations [20].

Another renewable source of solar power generation is also used in such charging stations. Solar power generates DC power, which is first converted into AC using the converter, and then it is combined with the AC sources with the help of a transformer. This system requires a large area for the installation of the solar panel, so it is costly and requires a converter, but it is more compact compared to an AC generation system [21].

In the off-grid charging system, an energy storage system is used to provide an uninterrupted power supply when any outage arises in the main power supply system. We eliminated the DC-DC converter and just used a DC-AC converter in order to reduce the cost of the solar energy storage system. Control devices and local illumination were powered in tandem by storage batteries [21].

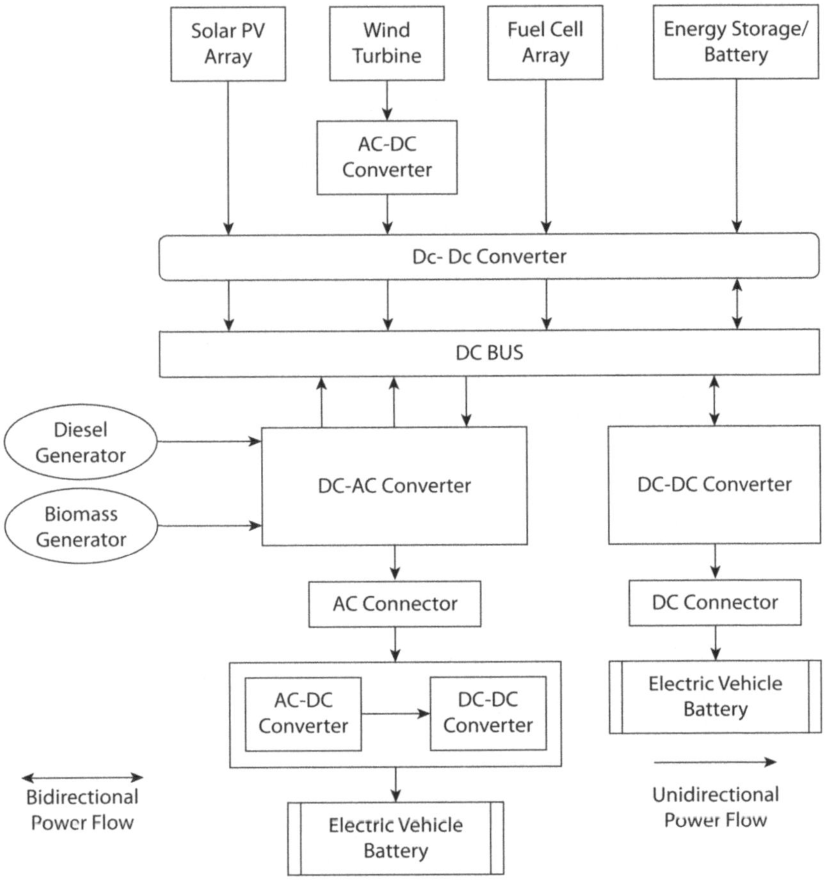

FIGURE 5.3 DC-based technique of off-grid charging system.

5.3.2.2 DC-Based Technique of Off-Grid Charging System

A DC-based method for off-grid charging systems is depicted in Figure 5.3. The DC bus facilitates the power exchange in this instance. Only a DC-DC converter is required to link PV, fuel cell systems, Energy storage systems (ESS), and EV batteries to a DC bus because they are all DC devices by nature. Stated otherwise, this architecture eliminates the requirement for a DC-AC converter, as opposed to those needed in the off-grid charging systems with an AC-based architecture [21].

Compared to an AC-based off-grid charging system, this is more effective and reliable because it takes less installation space and requires less uninterrupted power for the charging station. But if we can use diesel and biogas generators as power sources, then we require an AC-DC converter. In that case, a DC-based charging station requires more space and is more costly compared to off-grid AC EV charging stations. However, it is seen that if we can use diesel and biogas generators, it is more beneficial and more efficient than DC-based charging stations [22].

Similar to an AC-based off-grid charging system, a DC-based charging system also uses both AC and DC power supply systems. In an AC-based off-grid system, we required converters at the input stage from DC to AC and then output side AC-DC, whereas for DC-based charging stations, only a DC-DC converter is required [23].

5.3.2.3 AC-DC-Based Charging System

In this charging system, AC and DC powers are first generated by the different available energy sources, and then they are connected to AC and DC buses. The configuration diagram for an AC-DC-based charging station is shown in Figure 5.4. In AC-DC-based charging stations, loads are connected as per their requirements, meaning that if an AC load is available, it can connect to an AC bus, and if a DC load is available, it can connect to a DC bus. A bidirectional DC-AC converter is also used to transfer power between AC and DC, as seen in Figure 5.4 [24].

Such an AC-DC-based structure provides AC as well as DC charging by AC and DC buses simultaneously, so there is no need for extra power conversion devices. Due to such a good arrangement, this charging system is better than the off-grid AC-based charging system and the DC-based charging system. This charging system has a flexible charging setup, is reliable, and is more efficient compared to other charging systems.

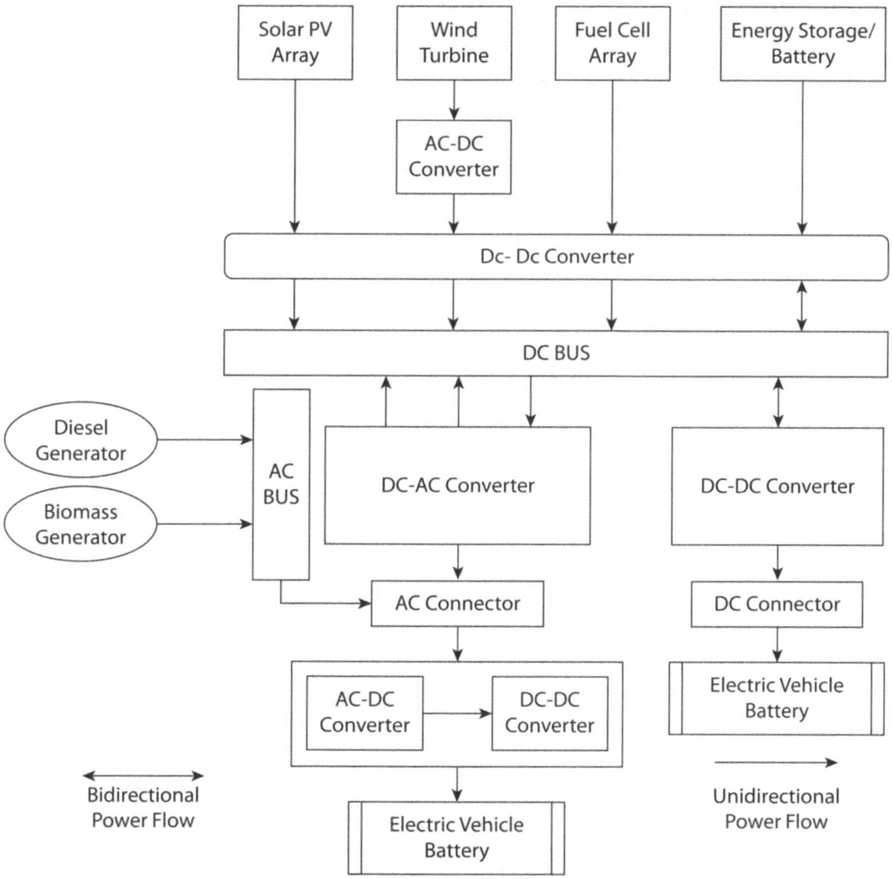

FIGURE 5.4 AC-DC-based charging system.

5.4 HYBRID POWER CHARGING TECHNIQUE

To charge electric cars, a hybrid power EV charging station integrates different energy sources as shown in Figure 5.5. To offer dependable and sustainable charging options, these stations usually combine grid electricity with RES, such as solar or wind turbines [25]. A hybrid power EV charging station might operate as follows:

Renewable Energy Generation: To produce power from renewable sources, the station uses wind turbines or solar panels. By utilizing solar or wind power to generate clean electricity, these energy sources lessen the need for fossil fuels and cut carbon emissions.

Energy Storage: Hybrid charging stations frequently incorporate energy storage devices like batteries or ultra-capacitors to guarantee constant power supply. When there is a spike in output or a dip in demand, these storage devices store extra energy produced by renewable sources and release it to augment charging [26].

Grid Connection: In situations where the output of renewable energy is insufficient to satisfy demand, hybrid charging stations stay linked to the electrical grid to offer a dependable power source. A grid connection guarantees continuous charging services, particularly in times of high demand or low supply of renewable energy.

Backup Power: To continue critical charging services in the event of a grid outage or emergency, hybrid charging stations may be equipped with diesel generators or fuel cells as backup power solutions. These backup systems offer further durability and dependability to guarantee continuous EV charging operations [27].

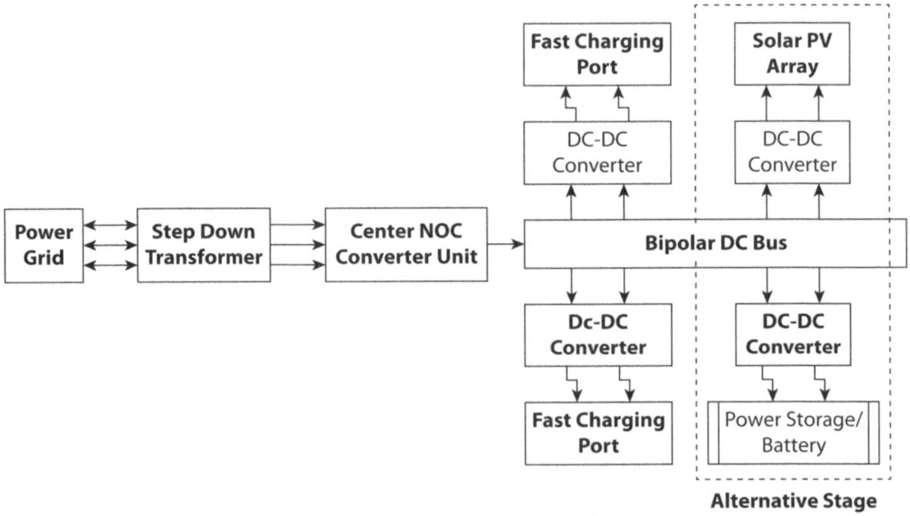

FIGURE 5.5 Architecture of hybrid charging station with balancing energy storage system.

5.5 ENERGY SOURCES

Based on system needs, Figure 5.6 shows how various energy sources—such as RES, fossil fuel-based energy sources, and the power grid—are employed in off-grid and HCSs. Unless RES are employed in the power grid, charging a high number of EVs through the grid will increase power demand and contribute to additional CO_2 emissions at the power plant for every kWh of electricity produced in the power system [28].

Therefore, in order to fulfill future charging demand while avoiding CO_2 emissions, it is advised that additional RES be incorporated into EV charging applications. Potential RES include wind, solar, and biomass energy.

Electricity is produced from wind energy using a wind turbine. Urban areas cannot provide the large amount of space needed for the installation of wind turbines, as they require a suitable site [29]. Rural locations can, nevertheless, provide these needs. However, compared to other RES, wind speed is a less desirable option due to its extreme fluctuations.

However, because biomass energy is easy to store and use when needed, it may provide a great substitute for wind energy. Wood, trash, and plants are the most often utilized biomass feedstock for the production of biomass energy [30]. However, the production of power from biomass feedstock releases greenhouse gas emissions into the atmosphere, severely polluting it. As such, it is unsuitable for locations with a high population density.

With solar PV arrays, solar energy is easily accessible in both rural and urban places. In addition to generating power, photovoltaic panels offer protection from sunlight and precipitation and don't require a dedicated installation area; they may be effortlessly installed on building roofs or in parking lots [18]. Additionally, selling extra power throughout the day and pulling electricity while the PV output is low are made possible by integrating with the UG [31].

5.6 ENERGY STORAGE SYSTEM

In order to run off-grid charging systems and continuously supply EV charging power, energy storage devices are essential. As seen in Figure 5.7 [22], it becomes crucial for supplying charging power in a hybrid charging station when it operates in standalone mode (from the grid-connected mode). Furthermore, to lessen the power fluctuation of RESs, an energy storage system in grid-connected mode is employed. A battery bank, hydrogen energy storage/fuel cell, ultra-capacitors, or a combination of these three can be installed in an energy storage system.

5.7 POWER ELECTRONICS CONVERTERS

For the conversion of electrical power, AC-DC, and DC-AC in any charging station, like off-grid or hybrid power charging stations, we required one of the main components called power electronics converters. As seen in Figure 5.8 [23], a variety of converter types, including DC-DC, DC-AC, and AC-DC, are frequently employed in these charging systems.

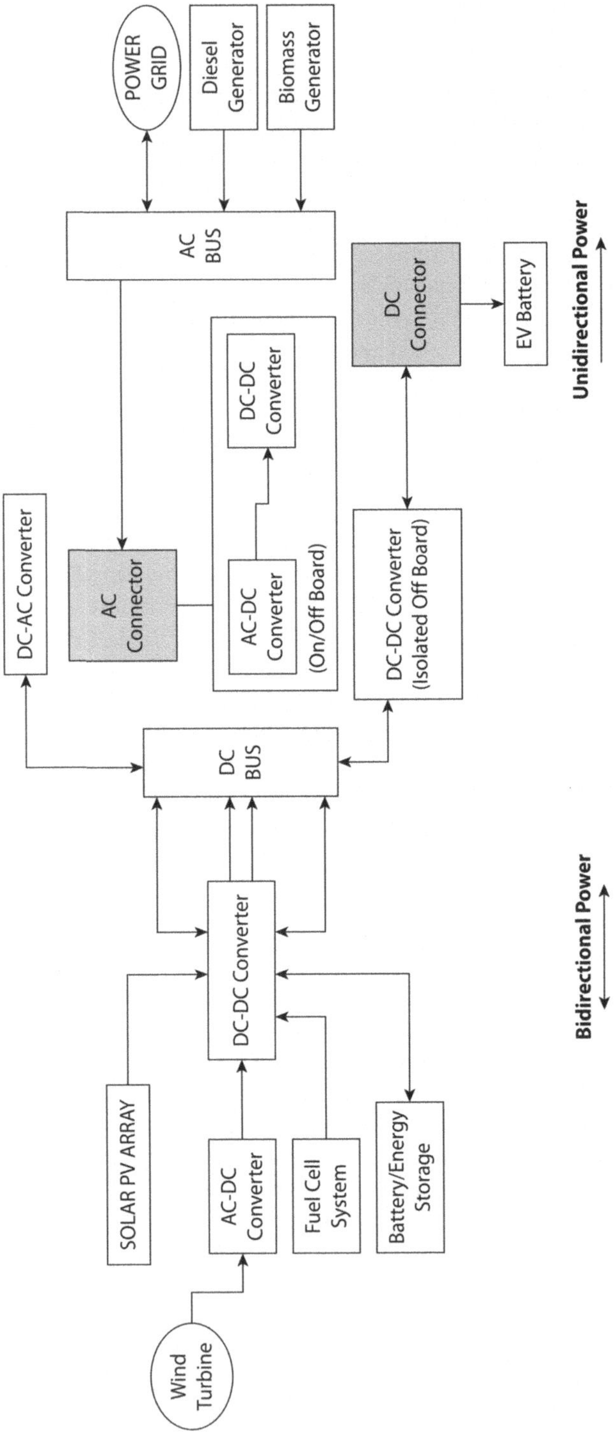

FIGURE 5.6 AC and DC-based architecture of hybrid charging station.

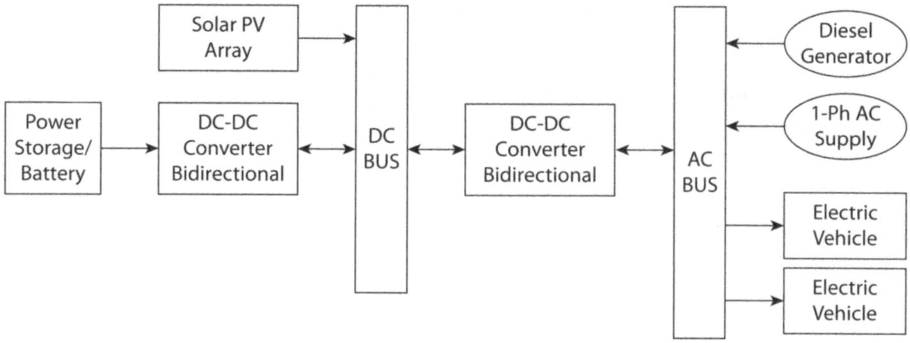

FIGURE 5.7 A HCS (AC-based) for EVs having PV array MPPT without a unidirectional DC-DC converter.

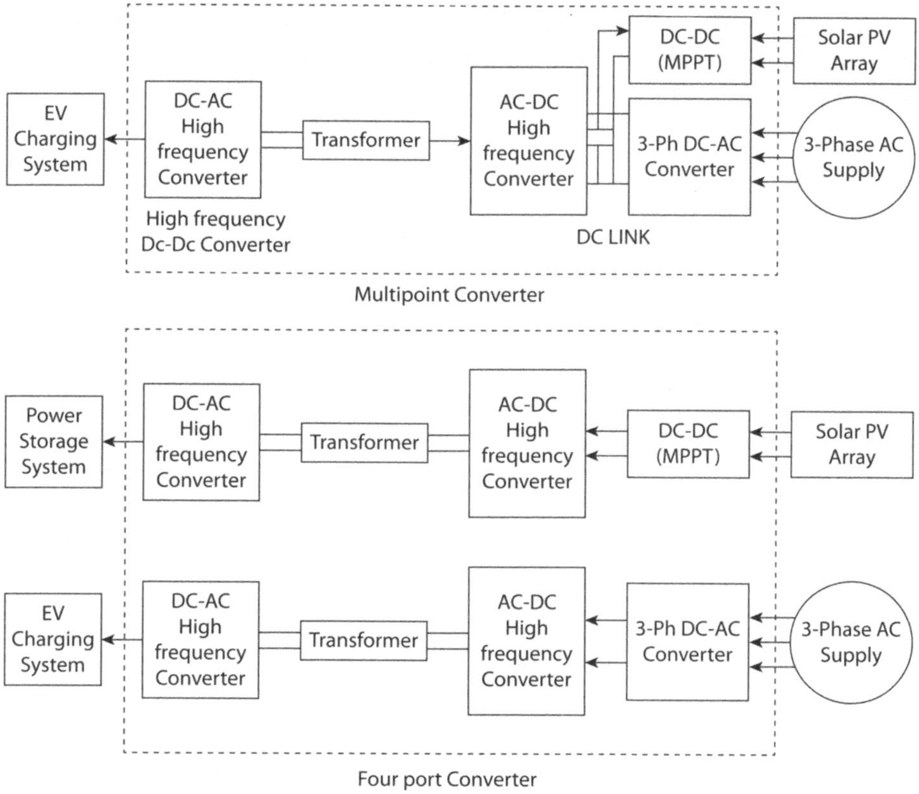

FIGURE 5.8 (a) A three-port converter using the standard DC connection voltage, (b) A four-port converter featuring power flows that are naturally decoupled.

5.8 ENERGY MANAGEMENT STRATEGY

Off-grid and HCSs need efficient functioning of several energy sources, energy storage systems, and loads in order to reduce charging costs, maximize system efficiency, and provide uninterruptible charging power [24]. The only objective of the AC and DC-based hybrid charging station architectures is to achieve uninterruptible charging at the lowest feasible cost while producing EMS, which is dependent on the energy balance of the DC bus [25].

5.9 EV CHARGING CONNECTORS, POWER LEVELS, AND STANDARDS

Electricity from AC or DC sources can now be used to charge EVs conductively. Both methods include converting AC from the UG to DC electricity in order to replenish the vehicle's battery [26, 27]. The EV battery in plug-in hybrid EVs is typically a few kWh, but the EV battery in plug-in EVs is in the tens of kWh range. The BMW i8 PHEV, which runs on a 7.1 kWh battery; the Brompton E-Bike M6L, which runs on a 300 Wh battery; the Tesla Model S vehicle PEV, which runs on an 85 kWh battery; the Hero Electric Photon HX scooter, which runs on a 72 V/26 Ah battery; the Ather 450X scooter, which runs on a 2.9 kWh battery; and so on are a few examples [28, 29].

5.10 CONCLUSIONS

The following are some possible findings that such a study might make:

- **Sustainability & Energy Independence:** There is a lot of room to reduce dependency on conventional fossil fuel-based energy sources by using off-grid and hybrid charging solutions. These systems support sustainability and energy independence by integrating renewable energy generation technologies like solar, wind, or hydropower. This aligns with international initiatives to mitigate climate change and reduce greenhouse gas emissions.
- **Flexibility and Adaptability:** By integrating several power generation techniques, such as RES and grid electricity, HCSs offer flexibility in energy sourcing. Because of its adaptability, EV charging infrastructure may be made more resilient and reliable by taking into account changing local resources, energy demand patterns, and environmental factors.
- **Cost-Effectiveness and Efficiency:** Research may reveal that hybrid and off-grid charging systems, especially in isolated or off-grid areas with expensive or limited grid connectivity, can provide economical EV charging options. Over the course of their lifetime, these systems can save operating costs and increase efficiency by utilizing RES and optimizing energy management techniques.
- **Possibilities for Innovation:** Studies may point to ways that off-grid and HCSs can innovate and advance technologically. Some examples include better energy storage technologies, enhanced grid integration capabilities, the creation of intelligent charging algorithms, and the optimization of

renewable energy harvesting methods. The performance, scalability, and affordability of off-grid and hybrid EV charging options should be significantly improved by these advancements.

REFERENCES

1. S. Habib, M. M. Khan, F. Abbas, L. Sang, M. U. Shahid, and H. Tang, "A comprehensive study of implemented international standards, technical challenges, impacts and prospects for electric vehicles," IEEE Access, vol. 6, pp. 13866–13890, 2018.
2. D. B. Richardson, "Electric vehicles and the electric grid: A review of modeling approaches, impacts, and renewable energy integration," Renew. Sustain. Energy Rev, vol. 19, pp. 247–254, 2013.
3. L. Liu, F. Kong, X. Liu, Y. Peng, and Q. Wang, "A review on electric vehicles interacting with renewable energy in smart grid," Renew. Sustain. Energy Rev., vol. 51, pp. 648–661, 2015.
4. N. Singh, "Optimization techniques used for design of economic electrical power distribution", Meta Heuristic Optimization Techniques Walter De Gruyter GmbH, Berlin, Germany, ebook, pp. 79–92, 2022.
5. A. R. Bhatti, Z. Salam, and R. H. Ashique, "Electric vehicle charging using photovoltaic based microgrid for remote islands," Energy Procedia, vol. 103, pp. 213–218, 2016.
6. A. Verma, and B. Singh, "A solar PV, BES, grid and DG set based hybrid charging station for uninterruptible charging at minimized charging cost," in: Proc. IEEE Ind. Appl. Soc. Annu. Meeting, pp. 1–8, 2018.
7. B. Singh, and A. Verma, "Control of solar PV and WEGS powered EV charging station," in: Proc. IEEE Transport. Electrific. Conf. Expo., pp. 609–614, 2020.
8. M. Nizam, and F. X. R. Wicaksono, "Design and optimization of solar, wind, and distributed energy resource (DER) hybrid power plant for electric vehicle (EV) charging station in rural area," in: Proc. 5th Int. Conf. Electric Veh. Tech., pp. 41–45, 2018.
9. S. Mishra, G. Dwivedi, S. Upadhyay, and A. Chauhan, "Modelling of standalone solar photovoltaic based electric bike charging," Mater. Today: Proc, vol. 49, pp. 473–480, 2021.
10. A. Verma, and B. Singh, "Energy management strategy of solar PV-battery and diesel generator based electric vehicle charging station," in: IEEE Energy Convers. Congr. Expo., pp. 1043–1050, 2018.
11. B. Sun, "A multi-objective optimization model for fast electric vehicle charging stations with wind, PV power and energy storage," J. Clean. Prod, vol. 288, pp. 1–16, 2021.
12. B. Singh, A. Verma, A. Chandra, and K. Al-Haddad, "Implementation of solar PV-battery and diesel generator based electric vehicle charging station," IEEE Trans. Ind. Appl, vol. 56, no. 4, pp. 4007–4016, 2020.
13. N. Singh, "Optimization of Hybrid Electric Power Generation System", Notion Press, vol.1, pp.1–158, 2022. https://notionpress.com/read/optimization-of-hybrid-electrical-power-generation-systems.
14. S. A. Singh, and S. S. Williamson, "Comprehensive review of PV/EV/grid integration power electronic converter topologies for DC charging applications," in: Proc. IEEE Transport. Electrific. Conf. Expo., pp. 1–5, 2014.
15. R. H. Ashique, Z. Salam, M. J. Bin Abdul Aziz, and A. R. Bhatti, "Integrated photovoltaic-grid DC fast charging system for electric vehicle: A review of the architecture and control," Renew. Sustain Energy Rev, vol. 69, pp. 1243–1257, 2017.
16. N. Singh, "Renewable power generation using asynchronous generator: A review," Springer Lect. Notes Electr. Eng., (Recent Advances in Power Systems), vol.688, pp. 205–215, 2020.

17. G. R. C. Mouli, P. Bauer, and M. Zeman, "Comparison of system architecture and converter topology for a solar powered electric vehicle charging station," in: Proc. 9th Int. Conf. Power Electron. ECCE Asia, pp. 1908–1915, 2015.
18. V. Sondhiya, K. Dwivedi, S. K. Almas, and N. Singh, Energy Management of Nanogrid through Flair of Deep Learning from IoT Environments, book Artificial Intelligence Techniques in Power Systems Operations and Analysis, Taylor& Francis, vol.1, pp 1–228, 2023.
19. A. R. Bhatti, Z. Salam, M. J. B. A. Aziz, and K. P. Yee, "A critical review of electric vehicle charging using solar photovoltaic," Int. J. Energy Res, vol. 40, no. 4, pp. 439–461, 2016.
20. M. K. Kumawar, and N. Singh, "Emerging Trends in Science & Technology", Bhumi Publisher, ISBN No. 978-81-951982-2-1, vol.1, pp.1–80, 2021. https://www.bhumi-publishing.com/wp-content/uploads/2021/06/Emerging-Trends-in-Science-and-Technology.pdf
21. C. Youssef, E. Fatima, E. S. Najia, and A. Chakib, "A technological review on electric vehicle DC charging stations using photovoltaic sources," IOP Conf. Ser.: Mater. Sci. Eng, vol. 353, pp. 1–12, 2018.
22. N. Singh, "Artificial Intelligence Techniques in Power System Operation and Analysis" CRC Press, vol.1, pp.1–175, 2023.
23. L. Olatomiwa, S. Mekhilef, M. Ismail, and M. Moghavvemi, "Energy management strategies in hybrid renewable energy systems: A review," Renew. Sust. Energ. Rev, vol. 62, pp. 821–835, 2016.
24. W. Xiao, M. S. El Moursi, O. Khan, and D. Infield, "Review of grid tied converter topologies used in photovoltaic systems," IET Renew. Power Gener, vol. 10, no. 10, pp. 1543–1551, 2016.
25. N. Singh, A. Mishra, M. Gupta, K. Dwivedi, and M. Swarnkar "Implementation of microgrid for rural electrification to cater on side load demand", in: 3rd (IEEE) International Conference on Energy, Power and Electrical Engineering (EPEE 2023), September 16, 2023.
26. N. Singh et al. "Analysis of optimum cost and size of the hybrid power generation system using optimization technique", in:12th IEEE International Conference CSNT-2023 held at Technocrats Institute of Technology (Excellence). Bhopal, Madhya Pradesh (India). 08 & 09 April, 2023. https://ieeexplore.ieee.org/document/10134720
27. L. Rubino, C. Capasso, and O. Veneri, "Review on plug-in electric vehicle charging architectures integrated with distributed energy sources for sustainable mobility," Appl. Energy, vol. 207, pp. 438–464, 2017.
28. L. S. A. Grande, I. Yahyaoui, and S. A. Gomez, "Energetic, economic and environmental viability of off-grid PV-bess for charging electric vehicles: Case study of Spain," Sustain. Cities Soc, vol. 37, pp. 519–529, 2018.
29. A. Al Wahedi, and Y. Bicer, "Assessment of a stand-alone hybrid solar and wind energy-based electric vehicle charging station with battery, hydrogen, and ammonia energy storages," Energy Storage, vol. 1, no. 5, pp. 23–29, 2019.
30. N. Singh, N. Priyadarshi, and N. Kumar. "Minimization of environmental emission and cost of generation using economic load dispatch with optimization techniques." AIP Conference Proceedings"2771, 020025, 2023.
31. H. T. Nguyen et al., "Review map of comparative designs for wireless high-power transfer systems in EV applications: Maximum efficiency, ZPA, and CC/CV modes at fixed resonance frequency independent from coupling coefficient," IEEE Trans. Power Electron., vol. 37, no. 4, pp. 4857–4876, 2022.

6 Mathematical Modelling and Analysis of DC-DC Converters for Voltage Regulation in DC Microgrids

Sureshkumar Alagarsamy, Pradeep Vishnuram, Mohit Bajaj, and Arvind R. Singh

6.1 INTRODUCTION

Renewable energy resources (RERs) such as wind turbines and photovoltaic (PV) solar panels have become much more popular in recent years. This is a big part of the plan to lessen our reliance on fossil fuels, which will help lower GHG emissions and stop global warming [1]. Direct current (DC) microgrids [2,3], alternating current (AC) microgrids [4,5], and distributed generation (DG) sources that are grid-connected are all examples of applications for RERs. A microgrid is a compact active distribution system. It comprises DGs (both renewable and non-renewable), batteries, and loads. It can work with the grid or on its own. The microgrid is more efficient, reliable, and has better power quality when it uses a variety of energy sources. It also loses less power. Also, the stand-alone microgrid is a good option for places too far away or isolated to get electricity through transmission lines.

Most electrical devices, like computers, cell phones, and media players, use a DC power source in a stand-alone microgrid that powers the buildings. In these microgrids, there are loads like lights and air conditioning that either DC or AC sources can power. The energy supplied and distributed in DC form is helpful because it reduces the number of switches needed, makes control easier than in an AC system where the frequency changes, and costs less. So, the stand-alone DC microgrid is a good choice for bringing electricity to remote places [6–8]. This made experts pay more attention to how voltage was controlled and how energy was used [9,10].

The stability of DC-bus voltages or energy-management-based control has been the subject of extensive study from various perspectives [6,7]. A study conducted by [11] provides a model predictive control approach for an AC-DC-AC converter with the aim of optimising the efficiency of a microgrid including a directly drive turbine and a system for storing energy. In addition, a proposed monitoring methodology has been put forth to efficiently synchronise the functioning of the turbine and the battery unit, notwithstanding the inherent unpredictable nature of converters and their pivotal contribution to voltage fluctuations. The objective of this technique is

DOI: 10.1201/9781003537823-8

to guarantee reliable operation across grid-connected and freestanding operational modes. In [12], a DC microgrid's DC bus voltage is stabilised using a distinct three-level control approach during various operational scenarios such as weight changes, production changes (such as islanding mode), grid reconnections (with each other and load shedding). Xu et al. [13] created a DC voltage regulator with an integrated energy storage system that functions well for the wind turbine system. All the works above employed theoretically perfect converters and ignored practical issues such converter movement control, switching activities, and duty cycle.

For the converters utilised in RERs or DC-voltage regulation-based management of energy, many control methods have been proposed, with an emphasis on the converters' switching activities and duty cycles [14,15]. In these studies, conversion models that were not perfect were used. Considering how these less-than-ideal processors change over time makes the study's results more realistic. The frequency-based approach outlined in reference [8] has been utilised to solve the power management in a DC microgrid. This microgrid consists of various components such as a wind turbine, solar panels, a diesel engine, and a storage device. A management technique based on polynomials has been devised for the purpose of efficiently managing the DC-bus voltage and current regulation in the context of using supercapacitors and battery as energy storage [16]. The simpler non-ideal conversion models were used in this work. In these studies, the supercapacitor quickly makes up for big changes in power. In a study conducted by [17], a flexible solar power system was developed, consisting of three DCs to DC converters for PV panels, two DC to AC converter devices, and a single DC-to-DC conversion for battery and local requirements. This strategy proposes a more decentralised control approach using DC-bus signalling to boost the system's dependability and efficiency.

In [18], researchers evaluated the effects of using "load-shedding" as a management strategy alongside a system of batteries and a turbine that generates electricity for a DC microgrid. Microgrid DC-to-DC converters, island modes, and DC-to-DC converters with batteries are all topics explored in the references [19–21]. Rashad et al. [22] have used mathematical modelling to evaluate the stability of a DC microgrid. They have also utilised a sliding feedback control in the course of this analysis. A new control technique utilising the principles of two rotating field theory has been presented for the management of a DC microgrid that includes single-phase static and dynamic alternating current loads [23]. Several researches have also examined several droop-based control methodologies for regulating the operation of the microgrid in question [24–27]. For coordinating voltage restoration and energy management in a DC microgrid, a droop control mechanism based on an imaginary voltage axis was developed in [27]. Reduced voltage fluctuations and enhanced power management are two benefits of using an imitation voltage line in the operating range of DC microgrids.

In [28], converter-based management of a DC microgrid using a battery-supercapacitor-based combined storage system for energy has been developed and analysed. In this research, the effect of the supercapacitor voltage changes and the best supercapacitor voltage on the safety of DC microgrids has been looked into. A study of stability and a method for stabilising a DC microgrid using several DC-to-DC converters linked in parallel is given in [29]. Capacity, inductance, peak power supply, and cable length constraints have been used to identify stable domains for DC microgrids in this method. To do so, we needed to identify the eigenvalues of the Jacobian matrix, and that we did by solving the corresponding matrix's system of linear equations.

In [30], nonlinear stability analysis was performed for a DC microgrid by lowering the number of differential equations to anticipate the behaviour of the microgrid. This was done to achieve the goal of accomplishing the prediction of the microgrid's behaviour. While acknowledging the significance of the previous study's findings, it is important to note that the research mentioned for the development of controllers has not specifically examined the independent analysis of the dynamic aspect of the converter's switching duties. When building controllers for switching functions, considering these dynamic aspects can lead to satisfactory outcomes in all settings, a suitable stable reaction in the steady state and a speedy transient-state response that can be attained despite variations in load and power. This motivates us to design a controller that separates the switch functionality from the dynamic effects of the converter's circuit's independent variables, thereby reducing current and voltage swings.

The design exhibits several noteworthy characteristics, including its emphasis on simplicity and ease of implementation, its ability to achieve rapid and efficient performance, its negligible impact on steady-state conditions, and its swift response during transient states. Additionally, the design incorporates a distributed management capacity that minimises the need for extensive communication between control systems. It demonstrates efficiency in both the multi-source and mono-source operational modes. Furthermore, the design features a flexible structure that enables microgrid integration without requiring significant controller reconfiguration.

The organisation of the remaining paper is as follows: Modelling of the multi-source DC microgrid system is explained in Section 6.2. Sections 6.3 and 6.4 explain the dynamic modelling of the individual system and overall system, respectively. The main conclusion of the paper is given in Section 6.5.

6.2 MODELLING OF THE MULTI-SOURCE DC MICROGRID SYSTEM

The multi-source DC microgrid configuration is shown in Figure 6.1. The system has two renewable sources and a lithium battery connected with a DC microgrid bus. The power generated from the solar and wind sources is given to the DC bus through a DC-DC converter. In this work, the DC bus voltage is assumed to be

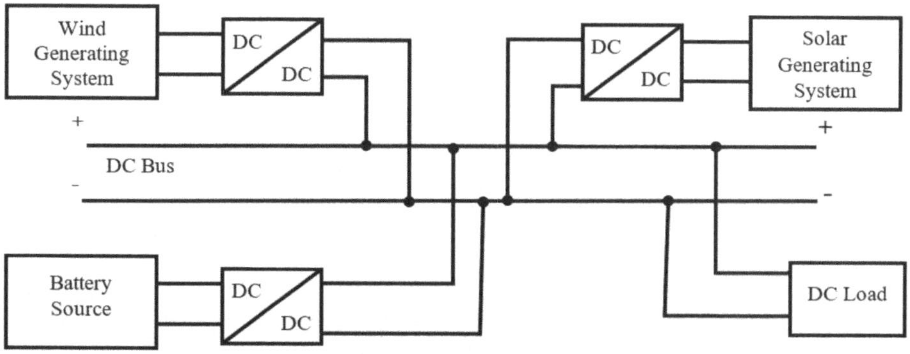

FIGURE 6.1 Multi-source DC microgrid configuration.

constant, and the steady-state mathematical modelling of the converter is analysed. Also, the stability of the converter topology is discussed by using various frequency domain approaches for the variation in the converter duty cycle. The detailed model is discussed in the subsection.

6.2.1 Modelling of DC-DC Converter

The sun's illumination is converted into DC voltage in the solar system. The DC voltage is stepped up using the DC-DC converter. Also, the wind energy is converted into electrical energy with the help of wind turbines and permanent magnet synchronous generators. The fluctuation in the wind output results in variable frequency AC waveforms; as a result, the rectifier with filter is converted into DC. Then, the DC-DC is varied with the help of a DC-DC converter. This work uses a buck converter to obtain the variable DC output per DC bus requirement.

The circuit diagram of the converter with renewable sources is illustrated in Figure 6.2. The mathematical modelling of the converter is obtained by analysing the modes of operation of the converter. When switch S_1 is turned on, the current flows through the inductor through the source to the DC bus. At this duration, the inductor and capacitor charges.

The mathematical equations governing the modes are given as follows:

$$\frac{dv_{in\,i}}{dt} = \frac{1}{C_{in\,i}}\left(i_{in\,i} - i_{L\,i}d_i\right) \tag{6.1}$$

$$\frac{di_{L\,i}}{dt} = \frac{1}{L_i}\left(v_{in\,i}d_i - V_{out\,i}\right) \tag{6.2}$$

$$\frac{dv_{out\,i}}{dt} = \frac{1}{C_{out\,i}}\left(i_{L\,i} - i_{out\,i}\right) \tag{6.3}$$

As two DC-DC converters are used for solar and wind renewable sources, the differentiation among them is done by considering the two different subscript values, i.e., $i = 1$ for solar and $i = 2$ for wind. Also, the input and output voltages are denoted as $v_{in\ i}$ and $v_{out\ I}$, respectively. $i_{in\ i}$ and $i_{L\ i}$ denote the input and inductor currents,

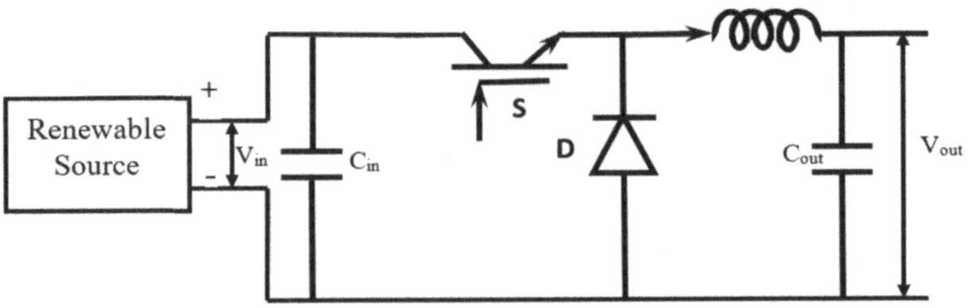

FIGURE 6.2 Buck converter with renewable sources.

respectively. $C_{in\,i}$ and $C_{out\,i}$ represent the input and output capacitors. The duty cycle is represented by d_i. For steady modelling, the state variables are: x_1 is considered as input voltage, x_2 is considered as inductor current, and x_3 is considered as output voltage. The state space matrix is given by:

$$\begin{bmatrix} \dot{x}_{1i} \\ \dot{x}_{2i} \\ \dot{x}_{3i} \end{bmatrix} = \begin{bmatrix} 0 & -\dfrac{d_i}{C_{in\,i}} & 0 \\ \dfrac{d_i}{L_i} & 0 & -\dfrac{1}{L_i} \\ 0 & \dfrac{1}{C_{out\,i}} & -\dfrac{1}{C_{out\,i} R_{grid}} \end{bmatrix} \begin{bmatrix} x_{1i} \\ x_{2i} \\ x_{3i} \end{bmatrix} + \begin{bmatrix} \dfrac{1}{C_{in\,i}} \\ 0 \\ 0 \end{bmatrix} \begin{bmatrix} i_{in} \end{bmatrix} \tag{6.4}$$

$$Y = \begin{bmatrix} 0 & 0 & 1 \end{bmatrix} \begin{bmatrix} x_{1i} \\ x_{2i} \\ x_{3i} \end{bmatrix} \tag{6.5}$$

The stability of the system is obtained by selecting the values for various parameters listed in Table 6.1. The transfer function of the system is given by:

$$G_i(s) = \frac{v_{out\,i}(s)}{v_{in\,i}(s)} = \frac{2.667 \times 10^{10}}{s^3 + 5000s^2 + 7.2 \times 10^6 s + 2.667 \times 10^9} \tag{6.6}$$

The frequency domain response of the system is obtained by using root locus and Nicholos plots as given in Figure 6.3. It is inferred from the root locus plot that all poles lie on the left hand side of the s-plane. It denotes that the system is stable. To confirm the closed-loop stability of the system, Nicholos plot is obtained by varying open loop gain and phase of the system. It noted that gain margin = 1.25 dB, phase margin = 7.1492°, gain crossover frequency = 2.6833 × 10^3 rad/sec, and phase crossover frequency = 2.4135 × 10^3 rad/sec. Also, for the various values of the gain, the obtained gain and phase margin is positive. It confirms the close loop stability of the system.

TABLE 6.1
System Parameters

Details	Parameters	Specifications
Bus voltage	Nominal value	960 V
Buck DC/DC converter	L	3 mH
	C_{out}	20 µF
	C_{in}	0.1 mF
Switching	Frequency	5 kHz
Buck – boost DC/DC converter	L	3 mH
	C_{out}	20 µF
DC grid power	Minimum	11.4 kW
	Maximum	52.25 kW

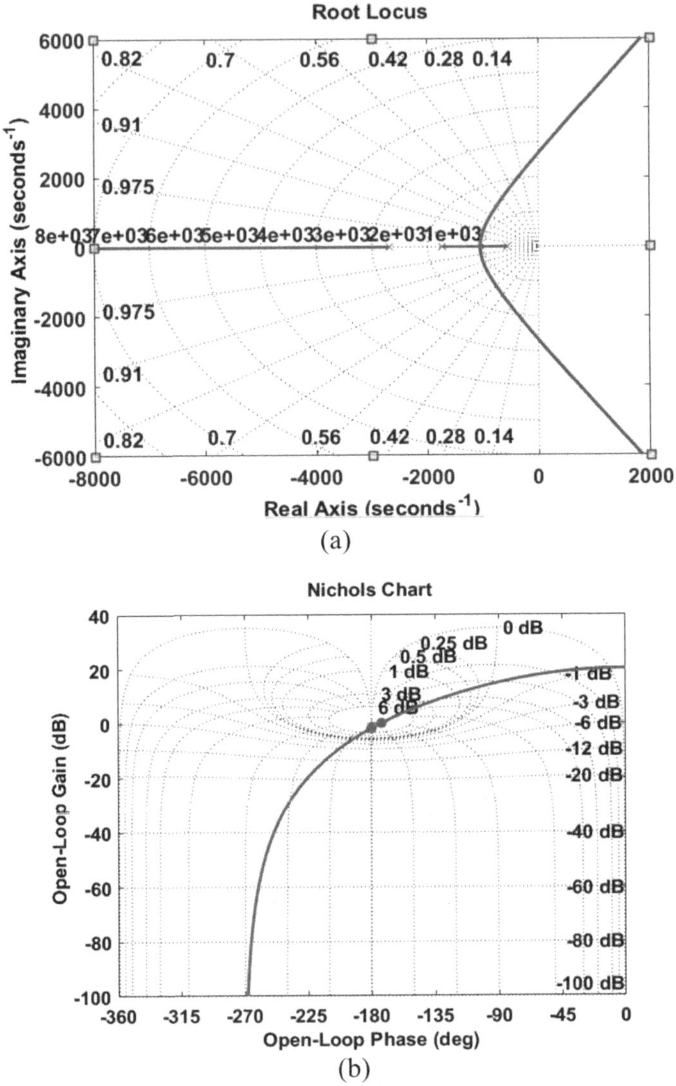

FIGURE 6.3 Stability analysis of renewable model system using (a) root locus and (b) Nicholos plot.

6.2.2 Modelling of Bidirectional Converter

The circuit diagram of the bidirectional converter is illustrated in Figure 6.4. In mode 1, the switch S_1 is on and S_2 is off. During this mode, the inductor gets charged by the supply voltage and capacitor discharges to the load. The mathematical equation governing the mode is given as follows:

$$\frac{di_{L3}}{dt}(on) = \frac{1}{L_3}\left(v_{in3} - V_{out3}\right) \tag{6.7}$$

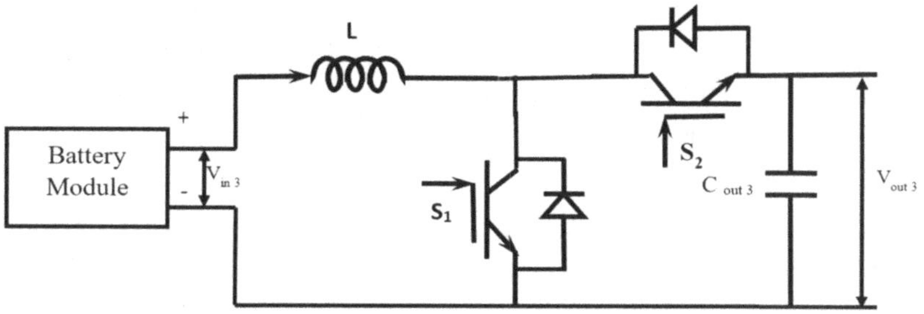

FIGURE 6.4 Bidirectional DC to DC converter.

$$\frac{dv_{out\,3}}{dt}(on) = \frac{1}{C_{out\,3}}\left(i_{L\,3} - i_{out\,3}\right) \tag{6.8}$$

In mode 2, the switch S_1 is off and S_2 is on. During this mode, the inductor gets discharged via S_2 and capacitor charges from the supply. The mathematical equation governing the mode is given as follows:

$$\frac{di_{L\,3}}{dt}(off) = \frac{v_{in\,3}}{L_3} \tag{6.9}$$

$$\frac{dv_{out\,3}}{dt}(off) = -\frac{i_{out\,3}}{C_{out\,3}} \tag{6.10}$$

The combined mathematical model of the bidirectional converter is as follows:

$$\frac{di_{L\,3}}{dt} = \frac{1}{L_3}\left(v_{in\,3} - d_3 V_{out\,3}\right) \tag{6.11}$$

$$\frac{dv_{out\,3}}{dt} = \frac{1}{C_{out\,3}}\left(d_3\, i_{L\,3} - i_{out\,3}\right) \tag{6.12}$$

For steady modelling, the state variables are: x_4 is considered as inductor current and x_5 is considered as output voltage. The state space matrix is given by:

$$\begin{bmatrix} \dot{x}_4 \\ \dot{x}_5 \end{bmatrix} = \begin{bmatrix} 0 & -\dfrac{d_3}{L_3} \\ \dfrac{d_3}{C_{out\,3}} & -\dfrac{1}{C_{out\,3} R_{grid}} \end{bmatrix} \begin{bmatrix} x_4 \\ x_5 \end{bmatrix} + \begin{bmatrix} \dfrac{1}{L_3} \\ 0 \end{bmatrix} \begin{bmatrix} v_{in\,3} \end{bmatrix} \tag{6.13}$$

$$Y = \begin{bmatrix} 0 & 1 \end{bmatrix} \begin{bmatrix} x_4 \\ x_5 \end{bmatrix} \tag{6.14}$$

The stability of the system is obtained by selecting the values for various parameters listed in Table 6.1. The transfer function of the bidirectional converter is given by:

$$G_3(s) = \frac{v_{out3}(s)}{v_{in3}(s)} = \frac{6.667 \times 10^6}{s^2 + 5000\,s + 2.667 \times 10^6} \tag{6.15}$$

The frequency domain response of the bidirectional converter is obtained by using root locus and Nicholos plots as given in Figure 6.5. The system is stable

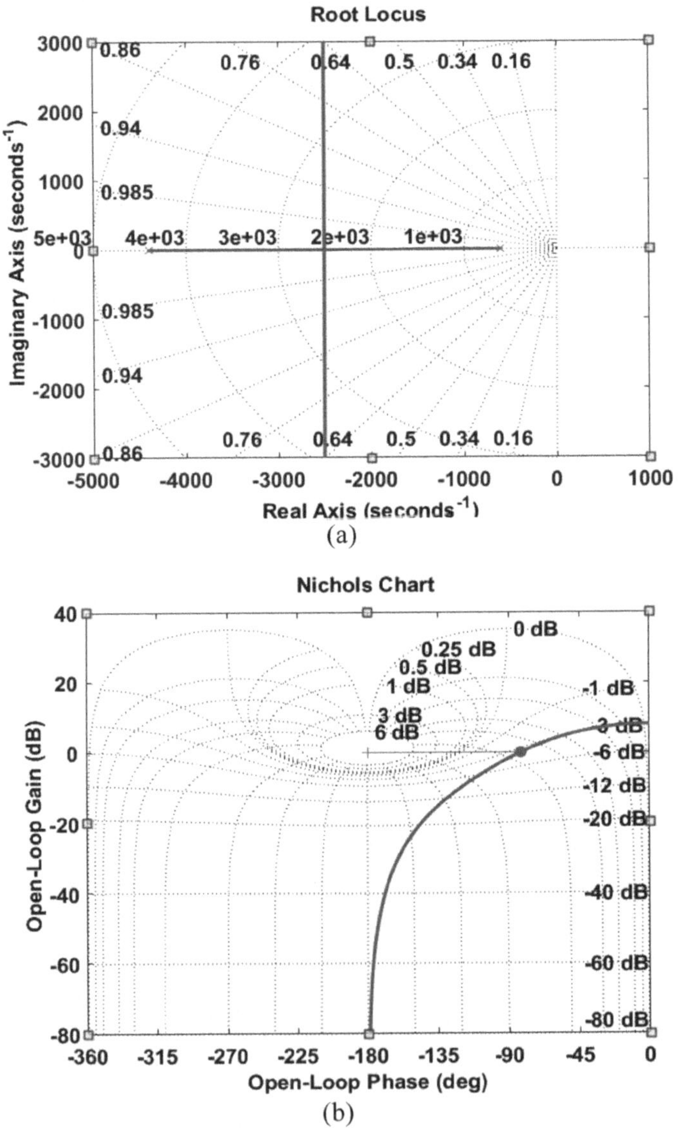

FIGURE 6.5 Stability analysis of bidirectional of DC-DC converter using (a) root locus and (b) Nicholos chart.

TABLE 6.2
Open Loop System's Time Domain Specification

Time Domain Specification	Wind/Solar System	Battery Module System
Rise time (ms)	4.3	3.7
Peak time (ms)	13.3	11.3
Settling time (ms)	8.0	6.7

as all poles lie on the left-hand side of the s-plane. From the Nicholos plot, gain margin and phase margin values are infinity and 97.9603°, respectively. It confirms the close loop stability of the system as the values are positive. Open loop system's time domain specification is listed in Table 6.2. Step response of open loop system is illustrated in Figure 6.6. It is understood from the figure that the system results in better time domain response.

6.3 DYNAMIC MODELLING OF THE SYSTEM

Dynamic modelling plays a pivotal role in the design and study of control systems. The process entails the development of mathematical models that depict the temporal behaviour of a system, incorporating the system's dynamic reaction to various inputs and disturbances. The consideration of stability is an essential component in the design of control systems. The utilisation of dynamic modelling facilitates the assessment of the stability of a closed-loop control system, hence guaranteeing the absence of undesired oscillations or instability. This section dealt the stability of the perturbed system.

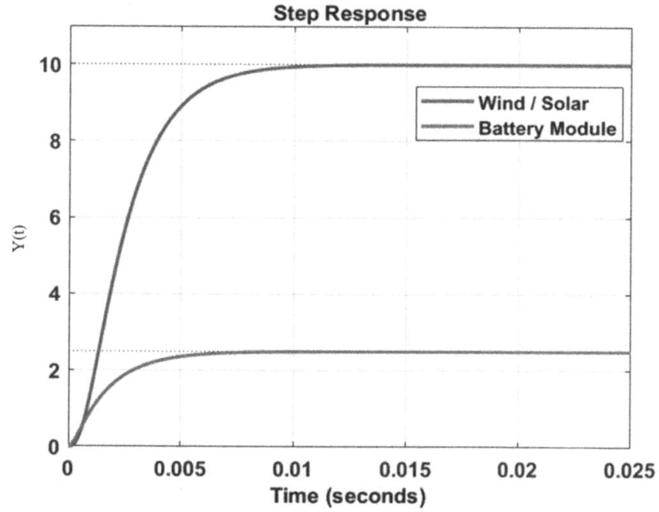

FIGURE 6.6 Step response of open loop system.

6.3.1 Perturbed Model of DC-DC Converter

For dynamic modelling of the system, the perturbed state variables are given as: $\hat{x}_1$ is considered as change in input voltage, $\hat{x}_2$ is considered as inductor current, $\hat{x}_3$ is considered as output voltage, $\hat{d}$ denoted the change in duty cycle, and $\hat{I}_{in}$ is the change in input current. The perturbed state space matrix of the DC-DC converter is given by:

$$\begin{bmatrix} \hat{\dot{x}}_{1i} \\ \hat{\dot{x}}_{2i} \\ \hat{\dot{x}}_{3i} \end{bmatrix} = \begin{bmatrix} 0 & -\frac{d_i}{C_{in\ i}} & 0 \\ \frac{d_i}{L_i} & 0 & \frac{1-2d_i}{L_i} \\ 0 & \frac{d}{C_{out\ i}} & \frac{-1+2d_i}{C_{out\ i}\ R_{grid}} \end{bmatrix} \begin{bmatrix} \hat{x}_{1i} \\ \hat{x}_{2i} \\ \hat{x}_{3i} \end{bmatrix} + \begin{bmatrix} \frac{V_{out\ i}}{C_{in\ i}\ R_{grid}} \\ \frac{V_{out\ i}}{L} \\ \frac{I_{L\ i}}{C_{out}} - \frac{2\ V_{out\ i}}{C_{out\ i}\ R_{grid}} \end{bmatrix} \begin{bmatrix} \hat{d} \end{bmatrix} + \begin{bmatrix} \frac{d}{C_{in\ i}} \\ 0 \\ 0 \end{bmatrix} \begin{bmatrix} \hat{I}_{in} \end{bmatrix} \tag{6.16}$$

$$Y = \begin{bmatrix} 0 & 0 & 1 \end{bmatrix} \begin{bmatrix} x_1 \\ x_2 \\ x_3 \end{bmatrix} \tag{6.17}$$

The perturbed transfer function of the DC-DC converter is:

$$\hat{G}_i(s) = \frac{\hat{v}_{out\ i}(s)}{\hat{d}(s)} = \frac{3.167\times10^5 s^2 + 4.222\times10^9 s + 5.067\times10^{11}}{s^3 + 133.3s^2 - 3.556\times10^5 + 7.111\times10^7} \tag{6.18}$$

The perturbed frequency domain response of the system is obtained by using root locus and Nicholos plots as given in Figure 6.7. As all the poles lies on the left half of the s-plane, the system is stable. The perturbed closed-loop response is obtained using Nicholos chart by varying open loop gain and phase of the system. It noted that gain margin = 3.1958×10^{-4} dB, phase margin = 87.6152°, gain crossover frequency = 996.8896 rad/sec, and phase crossover frequency = 3.1695×10^5 rad/sec. Also, for the various values of the gain, the obtained gain and phase margin is positive. It confirms the close loop stability of the system.

6.3.2 Perturbed Bidirectional Converter

For dynamic modelling of the bidirectional converter, the perturbed state variables are given as: $\hat{x}_4$ is considered as the change in inductor current and $\hat{x}_5$ is considered as change in output voltage. The state space matrix is given by:

$$\begin{bmatrix} \hat{\dot{x}}_4 \\ \hat{\dot{x}}_5 \end{bmatrix} = \begin{bmatrix} 0 & -\frac{d_3}{L_3} \\ \frac{d_3}{C_{out\,3}} & \frac{1-d}{C_{out\,3}\ R_{grid}} \end{bmatrix} \begin{bmatrix} \hat{x}_4 \\ \hat{x}_5 \end{bmatrix} + \begin{bmatrix} \frac{2V_{in\,3} - V_C}{L} \\ -\frac{V_{out\,3}}{C_{out\ i}\ R_{grid}} \end{bmatrix} \begin{bmatrix} \hat{d} \end{bmatrix} + \begin{bmatrix} \frac{1}{L} \\ 0 \end{bmatrix} \begin{bmatrix} \hat{V}_{in} \end{bmatrix} \tag{6.19}$$

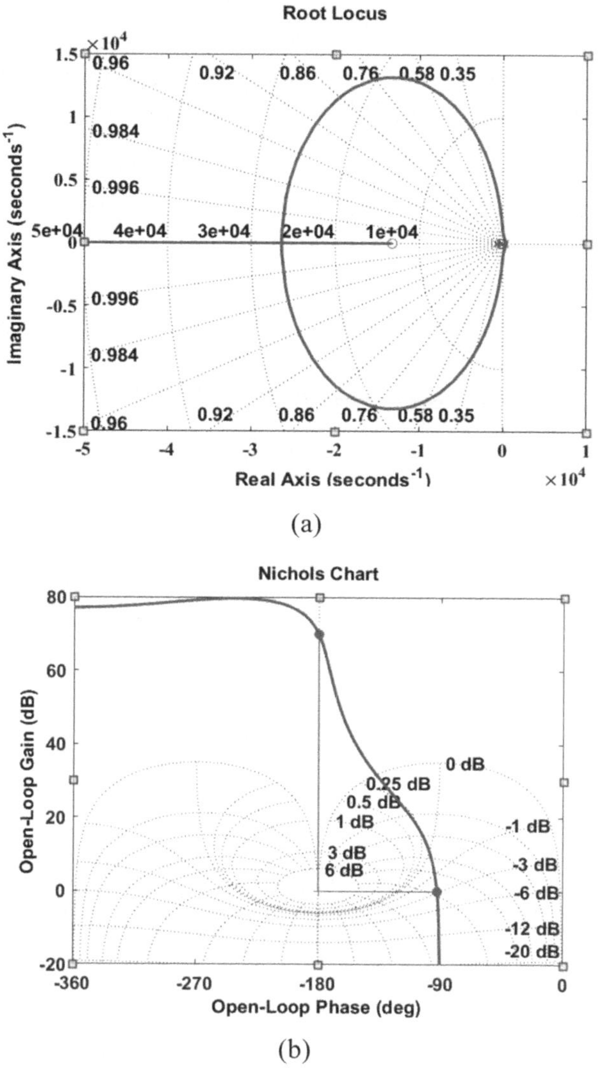

FIGURE 6.7 Perturbed stability analysis of the DC-DC converter using (a) root locus and (b) Nicholos chart.

$$Y = \begin{bmatrix} 0 & 1 \end{bmatrix} \begin{bmatrix} x_4 \\ x_5 \end{bmatrix} \tag{6.20}$$

The perturbed transfer function of the DC-DC converter is given by:

$$\hat{G}_3(s) = \frac{\hat{v}_{out\,3}(s)}{\hat{d}(s)} = \frac{-2.639\times10^6\, s + 2.097\times10^{10}}{s^2 + 2.639 s + 7.707\times10^6} \tag{6.21}$$

The perturbed frequency domain response of the bidirectional converter is obtained by using root locus and Nichols plots as given in Figure 6.8. The system is stable as all poles lie on the left-hand side of the s-plane. From the Nicholos plot, gain margin = 1 dB, phase margin = $1.5717 \times 10^{-6°}$, gain crossover frequency = 1.4483×10^5 rad/sec and phase crossover frequency = 1.4483×10^5 rad/sec. It confirms the closed-loop stability of the system as the values are positive. Closed-loop system time domain specification is tabulated in Table 6.3 and perturbed step response is shown in Figure 6.9. The closed-loop system results in better time domain response.

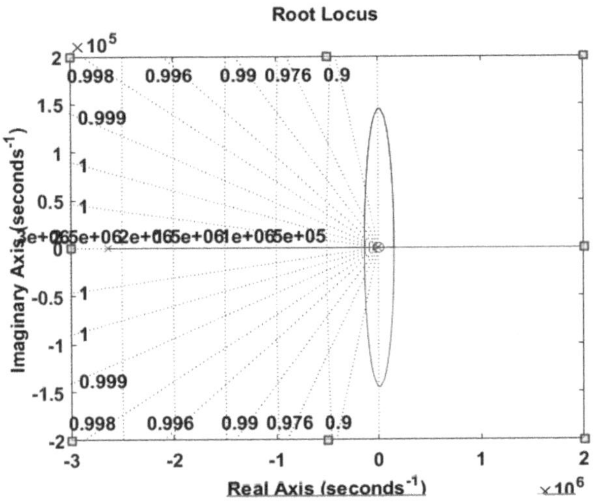

(a)

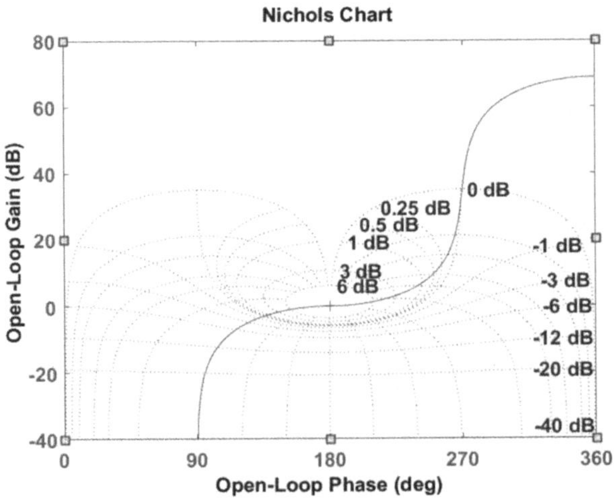

(b)

FIGURE 6.8 Perturbed stability analysis of bidirectional DC to DC converter using (a) root locus and (b) Nicholas chart.

TABLE 6.3
Closed-Loop System Time Domain Specification

Time Domain Specification	Wind/Solar System
Rise time (μs)	6.22
Peak time (μs)	21.4
Peak overshoot (%)	3.4
Settling time (μs)	63.6

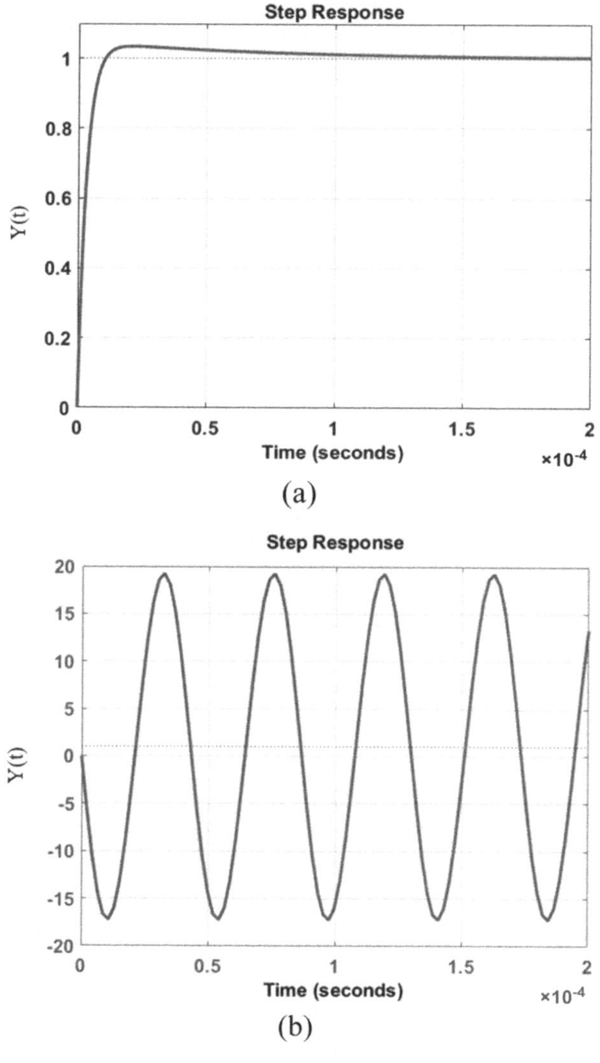

FIGURE 6.9 Perturbed step response of (a) wind/solar module and (b) battery module.

The perturbed step response of the wind/solar module and battery module is given in Figure 6.9. From the response, it is inferred that the system response of the wind/solar module is stable as illustrated in Figure 6.9(a), whereas the step response becomes unstable for battery-operated module as shown in Figure 6.9(b).

6.4 DYNAMIC MODELLING OF THE OVERALL SYSTEM

Figure 6.10 enumerates the schematic of a centralised n-dimensional multivariable closed-loop control system with negative feedback. The system input is represented as $a_1, a_2, a_3, \ldots a_n$, $b_1, b_2, b_3, \ldots b_n$ represents the controller output. The process output of the multivariable system is represented as $c_1, c_2, c_3, \ldots c_n$.

The transfer function of the plant $G(s)$ is given by:

$$G(s)=\begin{bmatrix} r_{11}(s) & r_{12}(s)\cdots & r_{1n}(s) \\ r_{21}(s) & r_{22}(s)\cdots & r_{2n}(s) \\ \vdots & \vdots & \vdots \\ r_{n1}(s) & r_{n2}(s) & r_{nn}(s) \end{bmatrix} \tag{6.22}$$

The controller $C(s)$ is given by:

$$C(s)=\begin{bmatrix} p_{11}(s) & p_{12}(s)\cdots & p_{1n}(s) \\ p_{21}(s) & p_{22}(s)\cdots & p_{2n}(s) \\ \vdots & \vdots & \vdots \\ p_{n1}(s) & p_{n2}(s) & p_{nn}(s) \end{bmatrix} \tag{6.23}$$

The closed transfer function of the system is given by:

$$M_i(s)=\frac{r_{i1}c_{1i}+r_{i2}c_{2i}+\cdots+r_{in}c_{ni}}{1+r_{i1}c_{1i}+r_{i2}c_{2i}+\cdots+r_{in}c_{ni}} \tag{6.24}$$

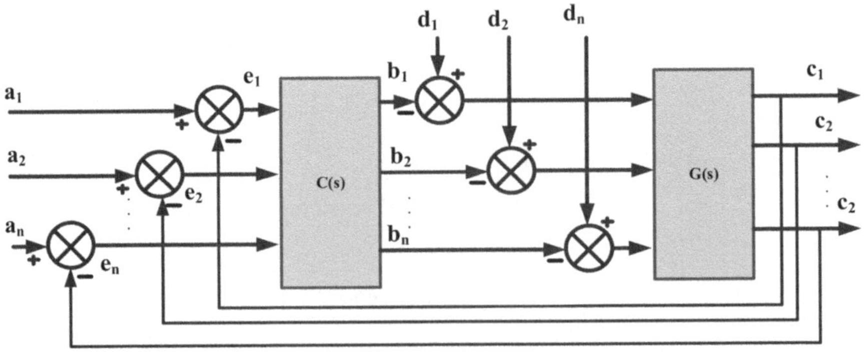

FIGURE 6.10 Block diagram of the multiple-input multiple-output (MIMO) system.

$$M_j(s) = \frac{r_{j1}c_{1j} + r_{j2}c_{2j} + \cdots + r_{jn}c_{nj}}{1 + r_{j1}c_{1j} + r_{j2}c_{2j} + \cdots + r_{jn}c_{nj}} \tag{6.25}$$

Where the system is defined for *i*th input and *k*th output.

A key requirement of a system using multiple-input multiple-output (MIMO) is to get the required output for each input while minimising interaction with the other inputs, preferably to zero. The ideal closed-loop matrix for a MIMO system should have off-diagonal entries set to 0 and diagonal elements representing the required input-to-output transfer functions.

The required transfer function from the *i*th input to the *i*th output, without any interaction among the loops, is determined by the forward route transfer function from the *i*th input to the *i*th output, denoted as $g(s)$.

Considering all the points indicated, the necessary closed-loop transfer function matrix can be constructed as a diagonal matrix $M(s)$:

$$M(s) = \begin{bmatrix} m_{11}(s) & 0\cdots & 0 \\ 0 & m_{22}(s)\cdots & 0 \\ \vdots & \vdots & \vdots \\ 0 & 0 & m_{nn}(s) \end{bmatrix} \tag{6.26}$$

The transfer function of the MIMO system is given by:

$$M(s) = \frac{G(s)C(s)}{1 + \left(G(s)C(s)\right)^{-1}} \tag{6.27}$$

$$M(s)\left(1 + \left(G(s)C(s)\right)^{-1}\right) = G(s)C(s) \tag{6.28}$$

$$M(s)\left(1 + \left(G(s)C(s)\right)^{-1}\right)G(s)^{-1} = C(s) \tag{6.29}$$

In the given controller equation, $M(s)$ is a diagonal matrix, allowing for the convenient evaluation of the term $(1 - M(s))^{-1}$. Yet, due to the off-diagonal parts of $G(s)$ being transfer functions, the inversion of $G(s)$ is challenging, especially for systems with non-minimum phase zeros and numerous time delays.

The following solution is suggested to prevent such difficulty. All transfer function matrices are converted into the frequency domain as specified:

$$C(j\omega) = G^{-1}(j\omega)M(j\omega)\left(1 + M(j\omega)\right)^{-1} \tag{6.30}$$

Where $C(j\omega)$ can be obtained for various values of ω.

The expression for the Proportional-Integral (PI) controller as the function of the controller gain is given by:

$$C_{PI}(s) = K_P + \frac{K_I}{s} \tag{6.31}$$

$$C_{PI}(j\omega) = K_P + \frac{K_I}{j\omega} \tag{6.32}$$

The PI controller parameters are determined using straightforward matrix calculations.

$$K_P + j\frac{K_I}{\omega_0} = G^{-1}(j\omega_0)M^{-1}(j\omega_0)\left(1 - M(j\omega_0)\right)^{-1} \tag{6.33}$$

The DC microgrid comprises a solar PV unit, a wind-turbine unit, and an energy storage system. The energy storage technology utilised in the DC microgrid for maintaining a constant DC-bus voltage during steady-state operation or in response to load changes is referred to as a lithium-battery unit. The power-generating systems ought to account for the fluctuations in the DC bus voltage that arise from variations in load inside the DC microgrid. The overall perturbed state space matrix is given by:

$$\hat{x} = \begin{bmatrix} A_1 & 0 & 0 \\ 0 & A_2 & 0 \\ 0 & 0 & A_3 \end{bmatrix} \hat{x} + \begin{bmatrix} B_1 & 0 & 0 \\ 0 & B_2 & 0 \\ 0 & 0 & B_3 \end{bmatrix} \hat{d} \tag{6.34}$$

$$Y = \begin{bmatrix} C_1 & 0 & 0 \\ 0 & C_2 & 0 \\ 0 & 0 & C_3 \end{bmatrix} x \tag{6.35}$$

Where, A_1, A_2, A_3 and B_1, B_2, B_3 are state matrix of PVU, WTU, and bidirectional converter, respectively. C_1, C_2, C_3 are input matrix of PVU, WTU, and bidirectional converter. The perturbed MIMO system transfer functions are shown in Table 6.4

TABLE 6.4
MIMO Transfer Function

		Input $\left(\hat{v}_{outi}(s)\right)$	
Output $\left(\hat{d}_i(s)\right)$	U_1	U_2	U_3
Y_1	$\frac{3.16\times10^5 s^2 + 4.21\times10^9 s + 5.05\times10^{11}}{s^3 + 133s^2 - 3.53\times10^5 s + 7.09\times10^7}$	0	0
Y_2	0	$\frac{3.16\times10^5 s^2 + 4.21\times10^9 s + 5.05\times10^{11}}{s^3 + 133s^2 - 3.53\times10^5 s + 7.09\times10^7}$	0
Y_3	0	0	$\frac{-2.6\times10^5 s + 2.09\times10^{10}}{s^3 + 2.63\times10^6 3 s^2 s + 7.70\times10^6}$

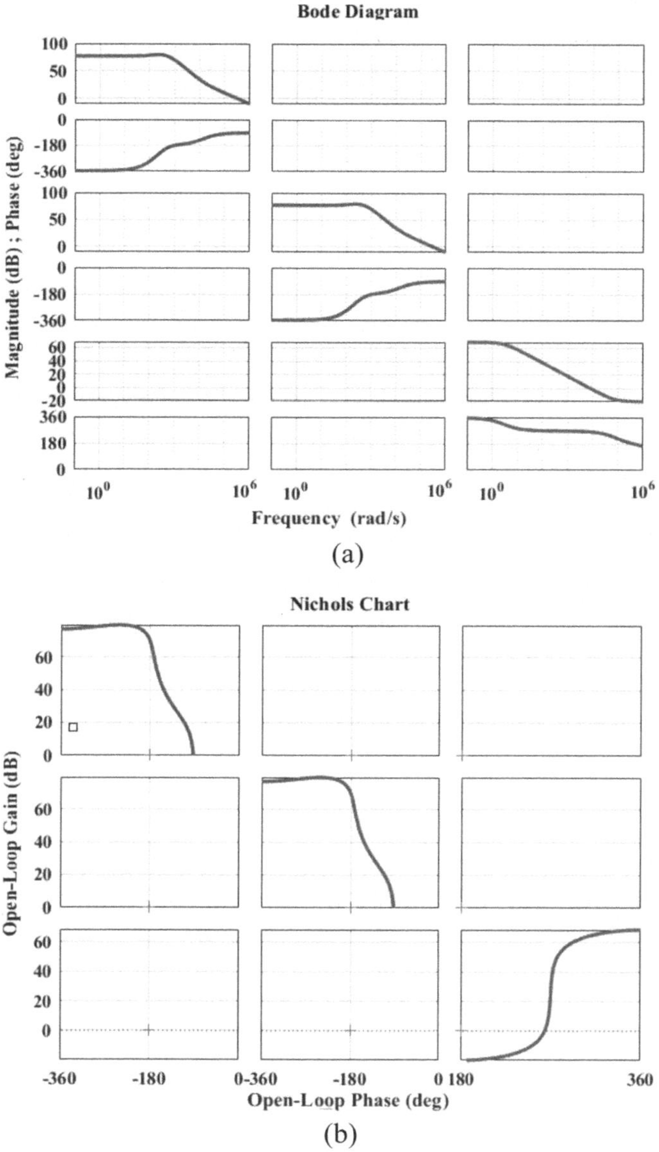

FIGURE 6.11 MIMO system frequency response plot (a) root locus and (b) Nicholos chart.

The frequency domain response of the overall system is given in Figure 6.11. The figure represents the response plot of the different input with respect to different output. The system operated based on the different individual renewable energy system connected to the microgrid. The overall frequency response plot is equal to the individual system response plot. Use the MATLAB commend (isstable(sys)) to check the dynamic system model is stable or unstable. In this system, the logical zero value shown is unstable.

6.5 ROBUST PI CONTROLLER DESIGN

Figure 6.12 displays the block diagram of a system with multiplicative input uncertainty. In the figure, $M(s)$ is the nominal plant, M_C(s), the PI controller; M_Δ, the perturbed plant; and M_T is the multiplicative weight.

The plant transfer function can be expressed as:

$$M(j\omega) = M_R(\omega) + jM_I(\omega) \tag{6.36}$$

By expressing the transfer function of the proportional-integral-derivative (PID) controller in the frequency domain, one can get the following equation:

$$M_C(j\omega) = K_P + \frac{K_I}{j\omega} + K_D(j\omega) \tag{6.37}$$

In light of multiplicative input, the expression for uncertainty weight (MT) is given by dividing the real part (RT) by the imaginary part (IT).

$$M_T(j\omega) = R_T(\omega) + jI_T(\omega) \tag{6.38}$$

The goal is to determine the PID controller parameters to ensure that the system depicted in Figure 6.12 remains stable under the condition that $|\Delta I(j\omega)| \leq 1$. The system is considered robustly stable if the nominal system is stable and satisfies the requirements.

$$\left\| M_T(j\omega) T(j\omega) \right\|_\infty = \leq \gamma \tag{6.39}$$

$T(j\omega)$ represents the complementary sensitivity function and γ equals 1. Complementary sensitivity define $T(j\omega)$ as:

$$T(j\omega) = \frac{M(j\omega) M_C(j\omega)}{1 + M(j\omega) M_C(j\omega)} \tag{6.40}$$

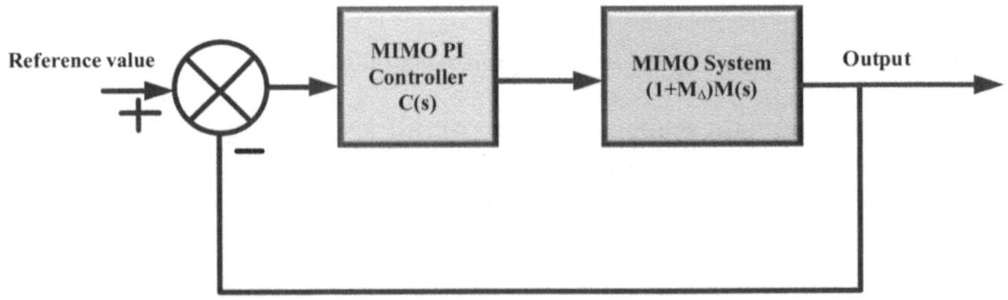

FIGURE 6.12 Block diagram of the closed-loop MIMO system.

The restrictions, namely equation (6.39), can be formulated for a single-input single-output decoupled system as follows:

$$\left\| M_T(j\omega)T(j\omega) \right\| \le \gamma \forall \omega \tag{6.41}$$

Additionally, equation (6.41) can be expressed in terms of magnitude and phase angle as follows:

$$\left\| \left| M_T(j\omega)T(j\omega) \right| e^{j\angle M_T(j\omega)T(j\omega)} \right\| \le \gamma \forall \omega \tag{6.42}$$

By dividing the real and imaginary components:

$$1 + \left(K_P + \frac{K_I}{j\omega} + j\omega K_D \right)\left(M_R(\omega) + jM_I(\omega) \right)$$
$$-\frac{1}{\gamma}\left(R_T(\omega) + jI_M(\omega) \right)\left(K_P + \frac{K_I}{j\omega} + j\omega K_D \right)\left(R_e(\omega) + jI_M(\omega) \right)\left(\cos\theta_T + jsin\theta_T \right) = 0$$

(6.43)

The above equation is simplified as:

$$X_{RP}K_P + X_{RI}K_I + X_{RD}K_D = -\omega \tag{6.44}$$

$$X_{IP}K_P + X_{II}K_I + X_{ID}K_D = 0 \tag{6.45}$$

The unknown parameters X_{RP}, X_{IP}, X_{RI}, X_{II}, X_{RD}, and X_{ID} are given by:

$$X_{RP} = \omega\left(M_R(\omega)\left(1 - \alpha R_T(\omega) + \beta_T(\omega) \right) + M_I(\omega)\left(\beta A_T(\omega) + \alpha\beta_T(\omega) \right) \right) \tag{6.46}$$

$$X_{RI} = M_R(\omega)\left(-\beta R_T(\omega) + \alpha\beta_T(\omega) \right) + I_M(\omega)\left(1 - \alpha A_T(\omega) + \beta\beta_T(\omega) \right) \tag{6.47}$$

$$X_{RP} = \omega^2\left(M_R(\omega)\left(\beta R_T(\omega) - \propto \beta_T(\omega) \right) + I_M(\omega)\left(\alpha A_T(\omega) - \beta\beta_T(\omega) - 1 \right) \right) \tag{6.48}$$

$$X_{IP} = \omega M_R\left((\omega)\left(-\beta A_T(\omega) \right) - \alpha\beta_T(\omega) \right) + I_M(\omega)\left(1 - \alpha A_T(\omega) + \beta I_T(\omega) \right) \tag{6.49}$$

$$X_{LI} = M_R(\omega)\left(\alpha A_T(\omega) - \beta I_T(\omega) - 1 \right) + R_M(\omega)\left(-\beta A_T(\omega) - \alpha I_T(\omega) \right) \tag{6.50}$$

$$X_{ID} = \omega^2\left(M_R(\omega)\left(1 - \alpha A_T(\omega) - \beta I_T(\omega) \right) + M_I(\omega) - \alpha I_T(\omega) \right) \tag{6.51}$$

Upon solving equations (6.45) and (6.46) (6.51) with $K_D = \widehat{K_D}$, the parameters of the PID controller are determined as follows:

$$\begin{bmatrix} X_{RP} & X_{RI} \\ X_{LP} & X_{LI} \end{bmatrix} \begin{bmatrix} K_P \\ K_I \end{bmatrix} = \begin{bmatrix} -\omega - XR_D\hat{K}_D \\ -XI_D\hat{K}_D \end{bmatrix} \tag{6.52}$$

The values of the controller gain can be obtained by equation (6.52).

6.6 CONTROLLER DESIGN OF OVERALL SYSTEM

The grid should maintain constant output voltage irrespective to the variation in the input as it fed from renewable energy source. The closed-loop response of the system is illustrated in Figure 6.13. It is noted from the response that the solar/wind system possesses the better stability, whereas the output converter has an oscillating response. Hence, a closed-loop controller is required to maintain the system stability.

The block diagram of the closed-loop system is given in Figure 6.12. The output of the system is compared with the required reference and the error is given to the controller. The PI controller is frequently utilised to improve stability and minimise steady-state error in control systems. The proportional term (K_p) responds to the current mistake, while the integral term (K_i) works to eliminate any steady-state

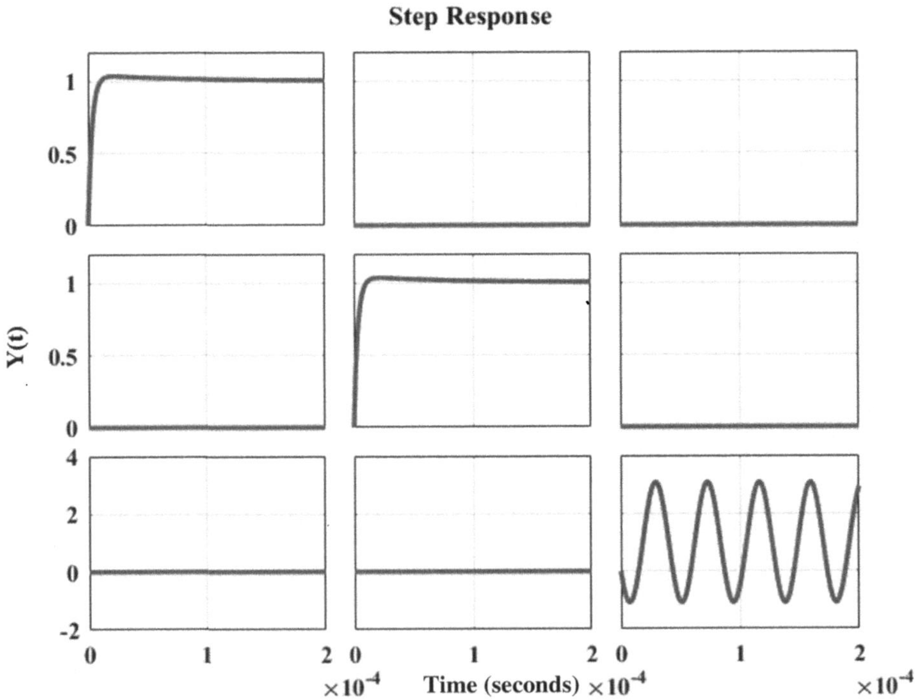

FIGURE 6.13 Closed-loop step response of MIMO system.

error gradually. Robust MIMO PI controller produces the control signal given to the MIMO system.

The robust controller gain produces the multiple controllers gain as given in the equation (6.53). This is to note that the system will be stable for all the values of the selected controller gains. By confirming the stability of the system, the response of the system is obtained.

$$\begin{bmatrix} K_1 \\ K_2 \\ K_3 \end{bmatrix} = \begin{bmatrix} 0.692 & 4.32\times10^4 \\ 0.692 & 4.32\times10^4 \\ 0.006 & 0.118 \end{bmatrix} \begin{bmatrix} 1 \\ 1/s \end{bmatrix} \tag{6.53}$$

Figure 6.14 shows the step response of the PI controller MIMO system. The stability of the system for the chosen values of K_p and K_i is advantageous, given the critical nature of stability in the design of control systems. Stability is a critical factor in preventing the system response from escalating uncontrollably and culminating in undesirable behaviour.

In short, the closed-loop control system incorporating the PI controller demonstrates commendable stability, promptness, and efficacy in mitigating errors. This indicates that the control system for the specified application is precise and

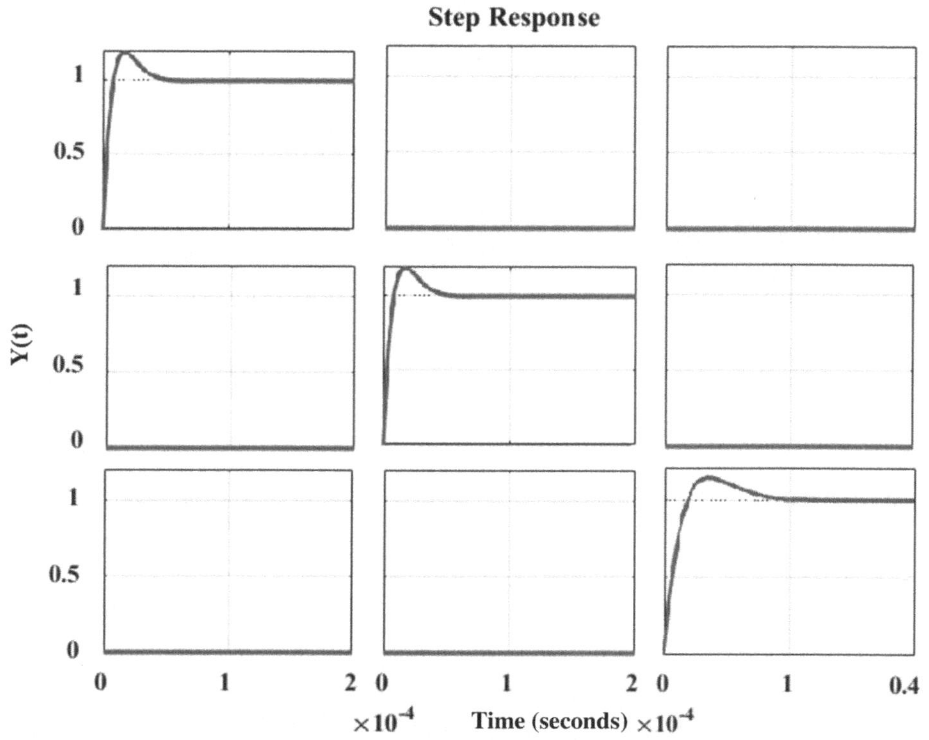

FIGURE 6.14 Step response of the MIMO system with PI controller.

dependable. Additional examination and experimentation might be required in order to optimise the controller for various operational circumstances or to fulfil particular performance criteria.

6.7 CONCLUSION

The present work proposes a recommended system for the regulation of the DC-bus voltage in an autonomous DC microgrid. The microgrid is composed of a solar unit, a wind producing unit, a power storage unit, and a load regulating unit. The proposed strategy utilises a study of system based on distributed load management to regulate the DC-bus voltage under varying load and power generation conditions. This study employs an extensive dynamic model to assess the electrical power conversion of a system for storing energy and its producing units. The proposed methodology investigates both steady-state and dynamic scenarios inside the dynamic model in order to address fluctuations caused by variations in load and the operation of generation units. The study's main advantages encompassed the capacity to produce separate elements for the dynamic characteristics of converter switching processes and the incorporation of all dynamic zeros identified in converter-based equations. Moreover, this study exhibits notable attributes such as simplicity, straightforward implementation, and limited necessity for inter-controller communication. It is anticipated that by the rate of convergence and DC-DC converters' transients response, state parameters will be significantly enhanced by implementing an additional closed-loop-based control system that achieves the suggested control technique's dynamic factors while also considering the dynamic errors of the mechanically driven variables.

REFERENCES

1. W. Alharbi, K. Raahemifar (2015) Probabilistic coordination of microgrid energy resources operation considering uncertainties. Electric Power Systems Research 128: 1–10. https://doi.org/10.1016/j.epsr.2015.06.010
2. A.T. Elsayed, A.A. Mohamed, O.A. Mohammed (2015) DC microgrids nd distribution systems: An overview. Electric Power Systems Research 119: 407–417. https://doi.org/10.1016/j.epsr.2014.10.017
3. Z. Wang, B. Chen, J. Wang, (2015) Decentralized energy management system for networked microgrids in grid-connected and islanded modes, IEEE Transactions on Smart Grid 7(2): 1097–1105. https://doi.org/10.1109/TSG.2015.2427371
4. M. Mehrasa, et al., (2015) A control plan for the stable operation of microgrids during gridconnected and islanded modes. Electric Power Systems Research 129: 10–22. https://doi.org/10.1016/j.epsr.2015.07.004
5. E. Pouresmaeil, M. Mehrasa, J.P. Catalão (2015) A multifunction control strategy for the stable operation of DG units in smart grids. IEEE Transactions on Smart Grid 6(2): 598–607. https://doi.org/10.1109/TSG.2014.2371991
6. V. Chapparya, A. Dey, S.P. Singh (2023) A novel non-isolated boost-zeta interleaved DC-DC converter for low voltage bipolar DC micro-grid application. IEEE Transactions on Industry Applications 59(5): 6182–6192. https://doi.org/10.1109/TIA.2023.3284810.
7. A. Rajamallaiah, S.P.K. Karri, Y.R. Shankar (2024) Deep reinforcement learning based control strategy for voltage regulation of DC-DC buck converter feeding CPLs in DC microgrid. IEEE Access 12: 17419–17430. https://doi.org/10.1109/ACCESS.2024.3358412

8. P. Shafiee, Y. Khayat, Y. Batmani, Q. Shafiee, J.M. Guerrero (2022) On the design of event-triggered consensus-based secondary control of DC microgrids. IEEE Transactions on Power Systems 37(5): 3834–3846. https://doi.org/10.1109/TPWRS.2021.3136679
9. Y. Kuang, et al. (2016) A review of renewable energy utilization in islands. Renewable Sustainable Energy Reviews 59: 504–513. https://doi.org/10.1016/j.rser.2016.01.014
10. Y.-S. Kim, E.-S. Kim, S.-I. Moon (2016) Frequency and voltage control strategy of standalone microgrids with high penetration of intermittent renewable generation systems. IEEE Transactions on Power Systems 31(1) 718–728. https://doi.org/10.1109/TPWRS.2015.2407392.
11. J. Han, S.K. Solanki, J. Solanki (2013) Coordinated predictive control of a wind/battery microgrid system. IEEE Journal of Emerging and Selected Topics in Power Electronics 1(4): 296–305. https://doi.org/10.1109/JESTPE.2013.2282601.
12. D. Chen, L. Xu, L. Yao (2013) DC voltage variation based autonomous control of DC microgrids, IEEE Transactions on Power Delivery 28(2): 637–648. https://doi.org/10.1109/TPWRD.2013.2241083.
13. G. Xu, et al., (2012) Coordinated DC voltage control of wind turbine with embedded energy storage system. IEEE Transactions on Energy Conversion 27(4): 1036–1045. https://doi.org/10.1109/TEC.2012.2220361.
14. M. Mehrasa, et al., (2014) Passivity-based control technique for integration of DG resources into the power grid. International Journal of Electrical Power & Energy Systems 58 281–290. https://doi.org/10.1016/j.ijepes.2014.01.034
15. N. Eghtedarpour, E. Farjah (2014) Distributed charge/discharge control of energy storages in a renewable-energy-based DC micro-grid, IET Renewable Power Generation 8(1): 45–57. https://doi.org/10.1049/iet-rpg.2012.0112
16. M.B. Camara, B. Dakyo, H. Gualous (2012) Polynomial control method of DC/DC converters for DC-bus voltage and currents management—Battery and supercapacitors. IEEE Transactions on Power Electronics 27(3): 1455–1467. http://doi.org/10.1109/TPEL.2011.2164581
17. K. Sun, et al., (2011) A distributed control strategy based on DC bus signaling for modular photovoltaic generation systems with battery energy storage. IEEE Transactions on Power Electronics 26(10): 3032–3045. http://doi.org/10.1109/TPEL.2011.2127488
18. L. Xu, D. Chen (2011) Control and operation of a DC microgrid with variable generation and energy storage. IEEE Transactions on Power Delivery 26(4): 2513–2522. http://doi.org/10.1109/TPWRD.2011.2158456
19. M. Kumar, S. Srivastava, S. Singh (2015) Control strategies of a DC microgrid for grid connected and islanded operations. IEEE Transactions Smart Grid 6(4): 1588–1601. http://doi.org/10.1109/TSG.2015.2394490
20. A. Deihimi, M.E.S. Mahmoodieh (2017) Analysis and control of battery-integrated dc/dc converters for renewable energy applications. IET Power Electron 10(14): 1819–1831. https://doi.org/10.1049/iet-pel.2016.0832
21. B. Banerjee, W.W. Weaver (2014) Generalized geometric control manifolds of power converters in a DC microgrid. IEEE Transactions on Energy Conversion 29(4): 904–912. http://doi.org/10.1109/TEC.2014.2351621
22. M. Rashad, et al., (2018) Mathematical modeling and stability analysis of DC microgrid using SM hysteresis controller. International Journal of Electrical Power & Energy Systems 95: 507–522. https://doi.org/10.1016/j.ijepes.2017.09.001
23. M. Kumar, et al., (2018) Development of a new control strategy based on two revolving field theory for single-phase VCVSI integrated to DC microgrid. International Journal of Electrical Power & Energy Systems 98 290–306. https://doi.org/10.1016/j.ijepes.2017.11.042

24. M. Baharizadeh, H.R. Karshenas, J.M. Guerrero (2018) An improved power control strategy for hybrid AC-DC microgrids. International Journal of Electrical Power & Energy Systems 95:364–373. https://doi.org/10.1016/j.ijepes.2017.08.036
25. N. Yang, et al., (2015) Compensation of droop control using common load condition in DC microgrids to improve voltage regulation and load sharing. International Journal of Electrical Power & Energy Systems 64: 752–760. https://doi.org/10.1016/j.ijepes.2014.07.079
26. V. Mortezapour, H. Lesani (2017) Hybrid AC/DC microgrids: A generalized approach for autonomous droop-based primary control in islanded operations. International Journal of Electrical Power & Energy Systems 93: 109–118. https://doi.org/10.1016/j.ijepes.2017.05.022
27. B.S. Ko, G.-Y. Lee, K.-Y. Choi, R.-Y. Kim, (2019) A coordinated droop control method using a virtual voltage axis for power management and voltage restoration of DC microgrids. IEEE Transactions on Industrial Electronics, 66(11): 9076–9085. https://doi.org/10.1109/TIE.2018.2877135.
28. S. Kotra, K. Mishra Mahesh (2019) Design and stability analysis of DC microgrid with hybrid energy storage system. IEEE Transactions on Sustainable Energy 10(3): 1603–1612. https://doi.org/10.1109/TSTE.2019.2891255.
29. M. Su, Z. Liu, Y. Sun, H. Han, X. Hou (2018) Stability analysis and stabilization methods of DC microgrid with multiple parallel-connected DC–DC converters loaded by CPLs. IEEE Transactions Smart Grid 9(1): 132–142. http://doi.org/10.1109/TSG.2016.2546551
30. A.P.N. Tahim, D.J. Pagano, E. Lenz, V. Stramosk (2015) Modeling and stability analysis of island-ed DC microgrids under droop control. IEEE Transactions on Power Electronics 30(8): 4597–4607. http://doi.org/10.1109/TPEL.2014.2360171

Section III

Intelligent Technologies in Energy Systems

7 The Impact of IoT on Performance of Electric Vehicle Charging Station Network

Sneha Nagar, Ashutosh Kashiv, and Nagendra Singh

7.1 INTRODUCTION

A new era of environmentally friendly transportation has been initiated by the emergence of electric vehicles (EVs) that offer lower emissions, less dependency on petroleum and other petroleum products, and increased energy efficiency. With the worldwide community turning progressively towards electronic transportation, there is an urgent need for reliable and easily accessible infrastructure for charging. The charging infrastructure is the main obstacle to the adoption of EVs. With the integration of the Internet of Things (IoT), this transformation is possible [1].

The IoT is a network of sensors that can collect, transmit, and analyse real-time data. IoT provides an acceptable solution to a number of major problems with respect to EV charging. IoT totally changes the way that we charge and communicate with EVs by integrating sensors into charging infrastructure [2].

This study explores the potential and many consequences of IoT-enhanced EV charging infrastructure (EVCI). It is important to understand how IoT technology will impact transportation in the future when it is adopted into the e-transportation industry. This study aims to figure out the effects of IoT on EVCI in various aspects, including reliability, availability, energy efficiency, environmental friendliness, and customer satisfaction [3].

7.1.1 Background and Significance

7.1.1.1 Electric Vehicles and Eco-Friendly Transportation

The adoption of EVs is now a popular way to address the urgent problems of energy sustainability and pollution. EVs, with their no emissions from the tailpipe and renewable energy source possibilities, offer a promising way to reduce the negative effects of traditional fossil fuel powered mobility [4].

DOI: 10.1201/9781003537823-10

7.1.1.2 Infrastructure Challenges for EV Charging

The move to electronic mobility isn't without challenges, though. One of the greatest challenges is creating an efficient system for charging EVs in order to fulfil the requirements of the growing demand for such automobiles. For those considering purchasing an EV, concerns about range, excessive charging times, and the accessibility of charging stations have been major issues [5].

7.1.1.3 IoT as a Game-Changer

The development of IoT technology offers the potential to provide a comprehensive solution to these issues. IoT makes it possible to build intelligent, integrated charging infrastructures that can predict maintenance requirements, enhance charging procedures, and improve the overall user experience. IoT promises to improve the accessibility, reliability, and efficiency of EVs by using real-time information and connectivity among vehicles, charging stations, and electrical grids [6].

7.1.1.4 Research Needs and Objectives

Although the infrastructure for charging EVs is increasingly utilizing the IoT, further study is necessary to figure out the exact impact of IoT on various elements of that infrastructure. By analysing the effects of IoT-enhanced EVCI on the rate of adoption, station availability, dependability, customer satisfaction, and the larger environmental and economic landscape orientation, this study aims to close this knowledge gap [7].

7.1.2 IoT in EV Charging

IoT comprises three major elements: charging equipment, a mobile app, and a charging management platform, as shown in Table 7.1. Figure 7.1 shows the electrical vehicle charging station setup with the IoT system.

TABLE 7.1
Major Elements of IoT Enabled EVCI

Charging Equipment Hardware Unit	Mobile App	Management Station
It contains sensors, charging protocols, and several connectivity options.	An app or web-based application that connects electric vehicle users to charging infrastructure.	A platform that provides analysis of data collected from sensors and devices through the cloud.
It provides a physical connection between vehicles and the power grid and conducts important charging functions such as fault detection, energy metering, port engagement and disengagement, and charging start and stop status.	It gives users authentication and management for charging functions, like managing charging, starting and stopping, charging the on boarding process, exploring a charging network, booking a slot, payment, and billing.	A platform that operates in the cloud collects information from sensors and devices and processes it. In the role of monitoring regular tasks like load balancing, dashboards, firmware updates, reminders, notifications, and remote monitoring.

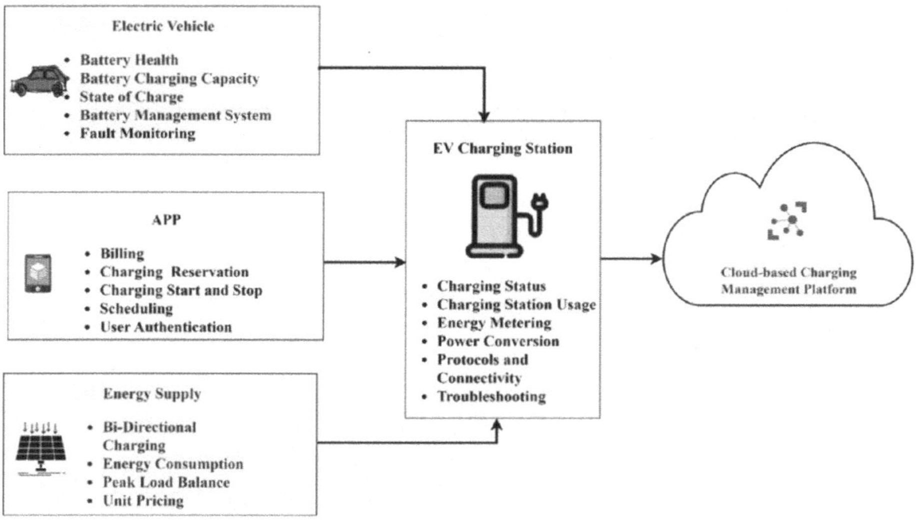

FIGURE 7.1 IoT based EV charging system.

7.1.3 Schematic of IoT Enabled Electric Vehicles Charging Infrastructures (EVCI)

This section mentions the schematic of IoT-enabled EVCI. Figure 7.2 shows the Schematic of IoT-enabled EVCI.

IoT Devices: It includes sensors, actuators, and controllers. These devices monitor and control various characteristics of the charging procedure.

EV Charging Stations: This represents the physical charging infrastructure where EVs are charged. IoT devices can monitor charging status, power levels, and availability within these stations.

Cloud-Based Platform: It offers a central point for data analytics and remote control. It collects data from the IoT devices, then processes the data and stores it securely.

Data Analytics: Data analytics helps in optimising charging operations, predicting maintenance needs, and improving energy efficiency. It represents the analytics tools and algorithms used to process the data collected by IoT devices.

User Interfaces: It include two types of user interfaces:

- The user-facing interface on the EV (like a mobile app).
- A web-based interface.

These interfaces permit users to check charging status, locate charging stations, and manage their charging sessions remotely.

Grid Integration: It ensures that the EVCI is well-integrated with the electrical grid, allowing for demand management, load balancing, and grid stability.

Payment and Billing: It handles payment processing and billing for users, whether through subscription models, pay-as-you-go, or other payment methods.

Security: Security is topmost in IoT-enabled systems, mainly in critical infrastructure like EV charging. It includes encryption, verification, and access control.

Regulatory Compliance: It shows the need for observing legal and regulatory standards in the operation of EVCI.

Impact Analysis: It represents the evaluation of the impact of IoT-enhanced EVCI on energy consumption, emissions reduction, cost savings, and user experience.

This study includes an in-depth overview of the development of IoT-enhanced EVCI using a combination of case studies and real-world challenges. There are several insights provided regarding the potential and challenges. Proposals are made for innovative solutions and a framework that increases the effectiveness, dependability, and sustainability of EVCI. The objective of this work is to encourage the global shift towards less polluted, environmentally friendly transportation and to enhance electric mobility.

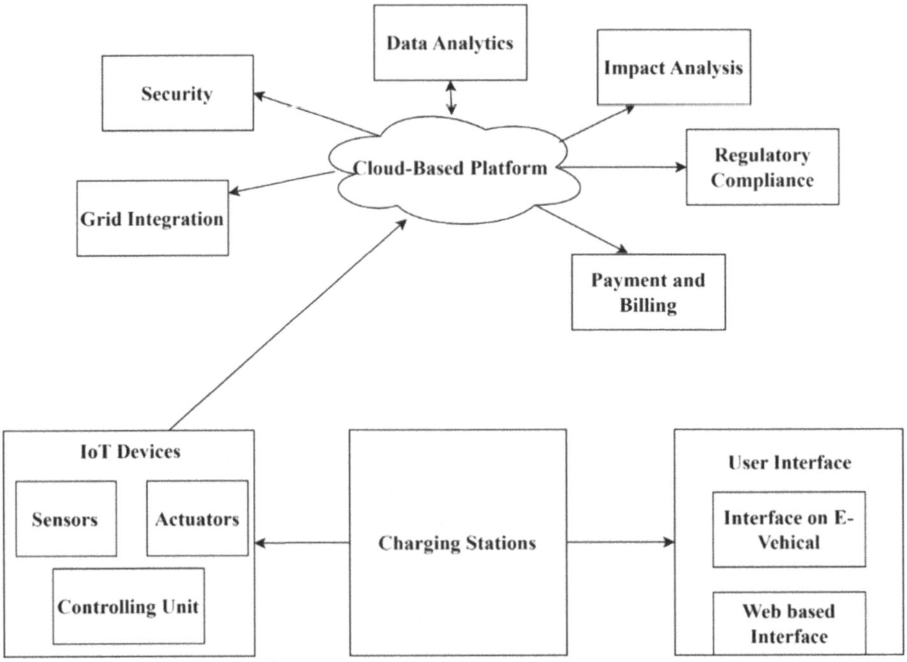

FIGURE 7.2 Schematic of IoT-enabled electric vehicles charging infrastructures (EVCI).

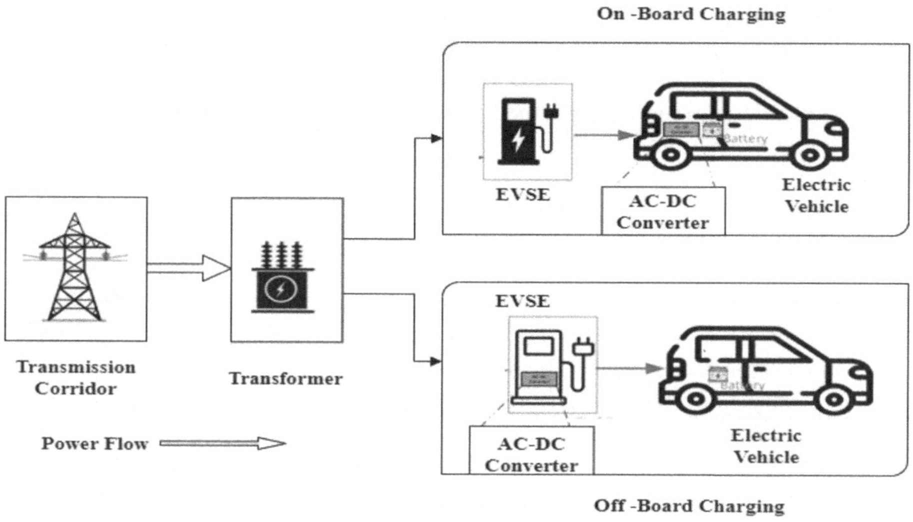

FIGURE 7.3 Schematic layout of integration of IoT-Enhanced EV charging infrastructure with electric power system.

7.1.4 Integration of IoT-Enabled EVCI with Existing Electric Power System

The installation of infrastructure for EV charging is essential to speeding up the transition to EVs. Researchers and industry partners have adopted the IoT as an innovative technology to fulfil the changing requirements of electric vehicle improving the sustainability, reliability, and efficiency of EVCI [8]. Figure 7.3 shows the IoT-based EVCI.

7.2 LITERATUREREVIEW

1. **IoT-Enabled Data Analytics:** Real-time data analytics is one of the main goals of IoT-enhanced EVCI. A study by [5] and [8] highlighted the significance of IoT sensors in collecting information on user behaviour, energy consumption, and the availability of charging infrastructure. Predictive maintenance, demand prediction, and intelligent energy distribution are all made possible by these data-driven insights, optimising the charging procedure for grid operators and consumers together.
2. **Grid Integration and Energy Management:** Effective grid integration is a key element of the EVCI. Artilce [6] point out that intelligent grid integration and demand response systems are two ways that IoT may help with grid synchronization. As a result of this, charging infrastructure will be able to function in collaboration with the electricity system, reducing stress during high demand and enhancing grid stability.

3. **Convenience and User Experience:** The widespread acceptance of EVs depends mainly on customer satisfaction. Many studies, including the work of [6] and [7], show how IoT can enhance the consumer's experience. Mobile applications, integrated payment solutions, and reservation platforms are examples of IoT-driven features that have been suggested to optimise and improve the charging process.
4. **Environmental Sustainability:** Environmental sustainability is also addressed with IoT-enhanced EVCI. Authors of [9] explain how IoT helps with load balancing, reducing carbon emissions, and utilising renewable energy sources. IoT enhances the sustainability and environmental friendliness of the transportation ecosystem through energy optimisation and the adoption of renewable energy sources.
5. **Security and Privacy:** IoT devices are an important component of the charging infrastructure. Addressing cybersecurity and protecting the privacy of users. A study done by [10] and [11] explores the significance of strong privacy protections and cybersecurity measures in reducing possible risks and weaknesses related to IoT-enabled systems.

In summary, research on IoT-enhanced EVCI shows that it has the potential to completely transform electric transportation. Scholars from several fields have acknowledged the important role of IoT in solving issues with grid integration, data analytics, security, user experience, and the sustainability of the environment. Our study paper, which works to present additional information, propose innovative approaches, and lay out a structure for maximizing the efficiency as well as the effectiveness of EVCIs in the IoT era, depends upon this comprehensive literature review.

7.3 SENSORS, PROTOCOLS, AND SECURITY USED IN IOT-ENABLED EVCI

This section elaborates on different sensors, devices, protocols, and analytics tools that can be used in IoT-enabled EVCI [12]. All these sensor protocols can be classified into two types– independent variables and dependent variables.

7.3.1 Independent Variables

Independent variables consist of sensors, communication protocols, and data analytics tools.

7.3.1.1 Types of IoT Sensors and Devices

Different types of sensors and devices that are most commonly used and can be useful in designing IoT-enabled EVCI are listed below:

Current Sensors: Current sensors monitor the electric current flowing through the charging cable, ensuring safety and preventing overloads.

Voltage Sensors: Voltage sensors measure the voltage provided to the charging station, helping to maintain the right charging voltage levels.

Power-Metres: These sensors measure the electrical power consumption of the EV charging process, helping to track energy usage accurately.

Temperature Sensors: Temperature sensors monitor the temperature of charging connectors and cables, ensuring safe and efficient charging.

Occupancy Sensors: These sensors detect whether a vehicle is present at the charging station, helping manage station availability.

RFID or NFC Readers: These devices are used for user authentication, access control, and billing.

Load Balancing Devices: IoT-enabled load balancers distribute power efficiently among multiple charging stations, optimising energy use.

Connection Status Monitors: These devices track the status of charging connections to ensure proper engagement and minimise user inconvenience.

Environmental Sensors: These sensors monitor environmental conditions, such as air quality and temperature, ensuring the safety and comfort of users.

7.3.1.2 Communication Protocols

Various types of available communication protocols that are used and can be useful to design IoT-enabled EVCI are enlisted below.

OCPP: OCPP (Open Charge Point Protocol) is a widely used protocol for communication between charging stations and central management systems. It allows for control, monitoring, and data exchange between EVCI components [13].

MQTT: Message Queuing Telemetry Transport is a lightweight publish-subscribe protocol that is used for efficient real-time data transfer between sensors, charging stations, and management systems.

HTTP/HTTPS: Standard web protocols are often used for transmitting data between the charging station and cloud-based analytics platforms.

Modbus: Modbus is a common protocol for reading data from power sensors and other industrial equipment used in charging stations.

Zigbee and Z-Wave: These wireless communication protocols are used for connecting various IoT devices within the charging station, such as temperature sensors and access control systems.

7.3.1.3 Data Analytics Tools

Different types of data analytics tools that are available and can be useful to design IoT-enabled EVCI are listed below:

Cloud Platforms: Cloud-based platforms like AWS, Azure, or Google Cloud are used for storing and analysing data from charging stations. These platforms offer scalability and data processing capabilities.

Data Analytics Software: Tools like Tableau, Power BI, or custom analytics software can be used to analyse charging data, monitor station performance, and create reports.

Machine Learning and AI: Advanced analytics techniques, including machine learning and artificial intelligence, can be applied to predict maintenance needs, optimise charging schedules, and improve energy efficiency.

Energy Management Systems: These systems are used to manage power distribution and load balancing among multiple charging stations.

User Interface Software: Charging station operators often use custom software to provide user interfaces, allowing EV owners to find and access charging stations and monitor their charging sessions [14].

The specific sensors, devices, protocols, and analytics tools used can vary depending on the manufacturer, location, and requirements of the EV charging station. These technologies collectively enable efficient and reliable EV charging while providing data for on-going management and optimization.

7.3.2 Dependent Variables

EV adoption rates, reliability, user satisfaction, energy efficiency, charging station availability, and environmental impact are the dependent variables. EV adoption has been steadily increasing in recent years, driven by advancements in technology, environmental concerns, and government incentives [15]. This surge in adoption has led to changes and improvements in various aspects of the EV ecosystem.

EV Adoption Rates: EV adoption rates have been on the rise globally, with more consumers opting for electric cars due to increased awareness of environmental issues, improved range and performance, and government incentives. However, the adoption rate may vary by region and depend on factors like infrastructure development and economic considerations.

Charging Station Availability: The availability of charging stations has improved significantly to support the growing number of EVs. Many governments and private companies have invested in expanding charging infrastructure, making it more accessible for EV users. This expansion includes public charging stations, home charging options, and workplace charging.

Charging Station Reliability: Charging station reliability has also improved with advancements in IoT technology. Real-time monitoring, predictive maintenance, and remote management have helped reduce downtime and ensure that charging stations are operational, contributing to a more reliable charging experience [16].

User Satisfaction: User satisfaction with EVs and charging infrastructure has generally increased as technology has evolved. Real-time availability information, ease of use, and improved user interfaces have contributed to higher satisfaction levels among EV owners.

Energy Efficiency: EVs are inherently more energy-efficient than traditional internal combustion engine vehicles. However, improvements in battery technology and charging infrastructure have further enhanced the energy efficiency of EVs. Smart charging, load balancing, and integration with renewable energy sources have reduced energy waste in the charging process.

Environmental Impact: EVs have a lower environmental impact compared to gasoline or diesel vehicles. They produce zero tailpipe emissions and reduce greenhouse gas emissions when charged with clean energy sources. As the electricity grid becomes greener and more renewable energy sources are integrated, the environmental benefits of EVs continue to grow [17].

7.3.3 Impact of IoT-enhanced Charging Infrastructure on the Adoption and Practical Use of Electric Vehicles

The impact of IoT-enhanced EVCI is discussed in Table 7.2.

7.3.4 Evaluation of the Effects of IoT on Energy Efficiency, Availability, and Dependability of Charging Stations

In the context of EVs, the IoT has a significant influence on the availability, dependability, and energy efficiency of charging infrastructure. To improve these factors and enhance the practicality and effectiveness of electric transportation, IoT technology is essential [18]. This is an evaluation of its effect on each.

TABLE 7.2
Different Impact of IoT-Enabled EVCI

S. No	Aspects	Summary
1	Monitoring and management in real-time	Permits charging station monitoring in real-time, improving fast problem identification and reducing maintenance.
2	Data-oriented	Collects and evaluates data to optimize the location and power of charging infrastructure
3	Enhancing user experience	Offers current information on charging expenses, available stations, and expected charging times.
4	Remote monitoring	Reduces the need for in-person visits by enabling remote control of stations for maintenance and software upgrades.
5	Predictive maintenance	Improves the overall dependability of the infrastructure for charging by using sensors to calculate future maintenance requirements.
6	Load balancing	Lowers energy costs and increases efficiency by balancing grid load through the optimization of charging plans.
7	Protection and verification	Provides safe access and authentication techniques to prevent unauthorized usage and ensure accurate billing.
8	Connectivity to smart grids	Effortless communication with smart grids, which is providing stability for the grid bidirectional power flow.
9	The management of fleets	Provides information for effective fleet management in both public and commercial transportation.

7.3.4.1 Availability of Charging Stations

- **Positive Impact:** The IoT makes it possible to monitor and control charging stations in real time. Operators can identify problems like failures, breaks, or limited access quickly and then take immediate action to resolve them. Through this, charging station availability and working are ensured, which reduces maintenance and raises satisfaction among consumers [19].
- **Enhanced User Experience:** IoT provides real-time data on station availability, which is accessible through mobile applications or in-car displays. Users can easily find available charging stations.
- **Data-Driven Optimization:** By carefully placing charging infrastructure to match user demand and improve overall availability, IoT data may be utilized to enhance their location.

7.3.4.2 Reliability of Charging Stations

- **Predictive Maintenance:** IoT sensors can predict maintenance needs by continuously monitoring the condition of equipment and systems. This enables proactive maintenance and minimizes unexpected breakdowns, ultimately improving the reliability of charging stations [12].
- **Remote Management:** IoT allows operators to remotely manage and control charging stations, reducing the need for physical visits. This not only saves time and resources but also ensures that stations are well-maintained and reliable.
- **Enhanced Security:** IoT enhances security through secure access and authentication methods, preventing unauthorized use and ensuring the reliability of the charging infrastructure.

7.3.4.3 Energy Efficiency

- **Load Balancing:** IoT technology optimizes charging schedules based on the availability of renewable energy sources and off-peak periods. This reduces energy costs and promotes energy efficiency by aligning charging with the grid's capacity.
- **Integration with Smart Grids:** IoT-enabled charging infrastructure can be integrated with smart grids, allowing bidirectional power flow. This means EVs can feed excess energy back into the grid when needed, contributing to grid stability and improving overall energy efficiency.
- **Data-Driven Insights:** IoT data analytics provide insights into energy consumption patterns, enabling adjustments to improve energy efficiency and reduce overall electricity consumption.

In summary, IoT has a highly positive impact on charging station availability, reliability, and energy efficiency for EVs. It allows for real-time monitoring, predictive maintenance, load balancing, and integration with smart grids, all of which collectively enhance the effectiveness and sustainability of electric mobility. As the IoT technology continues to advance, its impact in these areas is likely to become even more pronounced, further supporting the growth of EV adoption.

7.4 IoT-BASED EV CHARGING STATION

EV charging stations require cutting-edge technologies that enhance efficiency and automatic monitoring. IoT-based EV charging stations provide remote monitoring, smart scheduling, and real-time updates on charging status. They also provide information about the stations' availability and energy consumption. IoT systems also support secure payments and integrated operation with renewable energy sources.

7.4.1 Potential and Benefits of IoT-Enhanced E-Vehicle Charging

Figure 7.4 shows the Potential and benefits of IoT-Enhanced EV Charging system. The key findings of a research study on IoT-Enhanced EVCI are as follows:

Improved Charging Availability: IoT-enhanced charging infrastructure significantly improves the availability of charging stations. Real-time monitoring and data analytics enable better station management, reducing instances of station unavailability due to maintenance or congestion [19].

Enhanced Reliability: The integration of IoT sensors and predictive maintenance algorithms has a positive impact on charging station reliability. It reduces unplanned downtime, ensuring that EV owners can rely on the infrastructure for their charging needs.

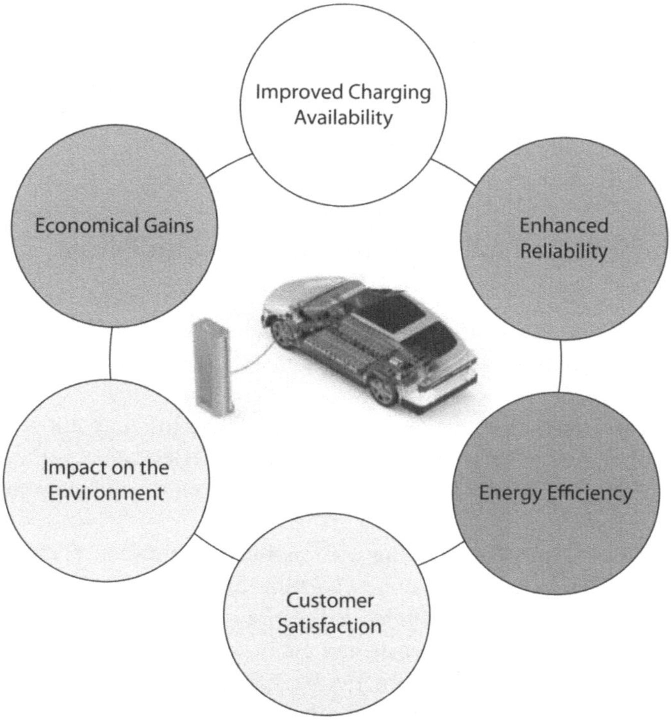

FIGURE 7.4 Potential and benefits of IoT-enhanced EV charging.

Energy Efficiency: By varying charging costs grid situations, the availability of renewable energy sources, and the needs of particular vehicles, IoT-enabled infrastructure minimizes the use of energy. As a result, the grid experiences less strain during peak hours and gains in energy efficiency [20].

Customer Satisfaction: Users of EVs show a higher degree of satisfaction with the IoT-enhanced charging infrastructure. A more pleasant charging experience is made possible by features like user-friendly interfaces, mobile applications for remote tracking and payments, and real-time station availability updates [21].

Impact on the environment: IoT-based charging systems have the potential to reduce emissions of greenhouse gases. Such systems support the use of renewable energy sources and optimize consumption of energy, which helps in their connection with environmental sustainability [22].

Economical Gains: According to the study, IoT-enhanced charging infrastructure could improve economic growth by attracting additional owners of EVs, increasing investment in the charging sector, and generating opportunities for employment in the support and maintenance sectors [23].

7.4.2 Challenges of IoT in Electric Vehicles

Privacy and security: Introducing IoT in EVs enhances the possibility of cyberattacks, compromising the security and safety of the vehicle and those inside it.

Communication: The lack of communication between different IoT platforms and devices can create integration challenges, making it difficult to create an effortless and connected user experience.

Cost: Implementing the IoT can be costly, particularly for lower-cost models of EVs. This may result in a more limited acceptance of the technology.

7.4.3 Possibilities for IoT in Electric Vehicles in the Future

Implementation of IoT technology in EVs is presently in its early stages, and there is lots of scope for innovation and development shown in Figure 7.5. These are a few possible applications of the IoT in EVs:

Intelligent driver assistance systems (IDAS): The IoT may allow the application of IDAS, which can increase safety and effectiveness. Examples of these technologies include self-driving automobiles and communication between vehicles.

Intelligent charging: By enabling communication between EVs and charging infrastructure, enabling smart charging, and reducing charging periods.

Integration of EVs with the grid: IoT can assist with vehicle-to-grid integration as it lowers energy costs and enhances the reliability of energy by enabling EVs to deliver more energy back into the system.

Preventative management: With the help of sensors, we have to collect various information, identify problems before they arrive, and optimise maintenance schemes. The IoT helps with preventative management.

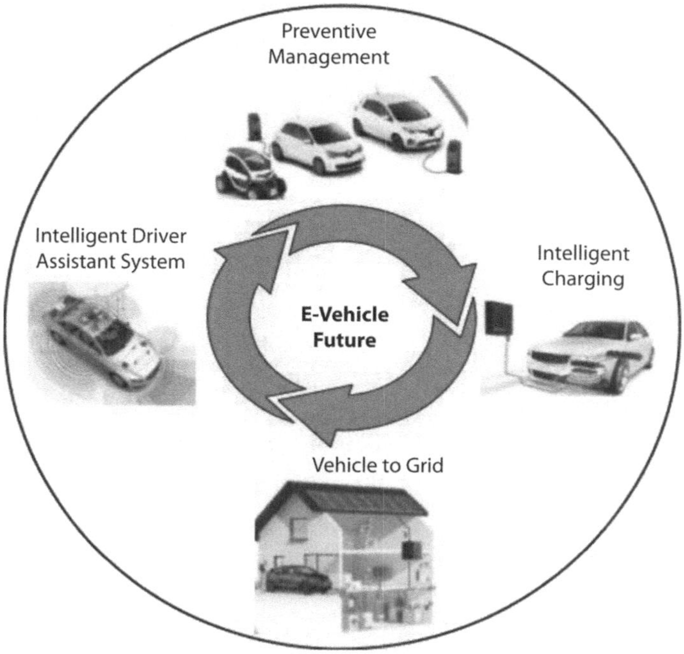

FIGURE 7.5 Future of IoT-enabled electric vehicles.

7.5 A CASE STUDY OF USER SATISFACTION AND PREFERENCES

7.5.1 User Satisfaction

A survey was conducted to assess user satisfaction with IoT-enabled charging solutions for EVs shown in Figure 7.6. [3] The following results were obtained.

The majority of users expressed satisfaction with IoT-enabled charging solutions, with 70% reporting being either "Satisfied" or "Very Satisfied."

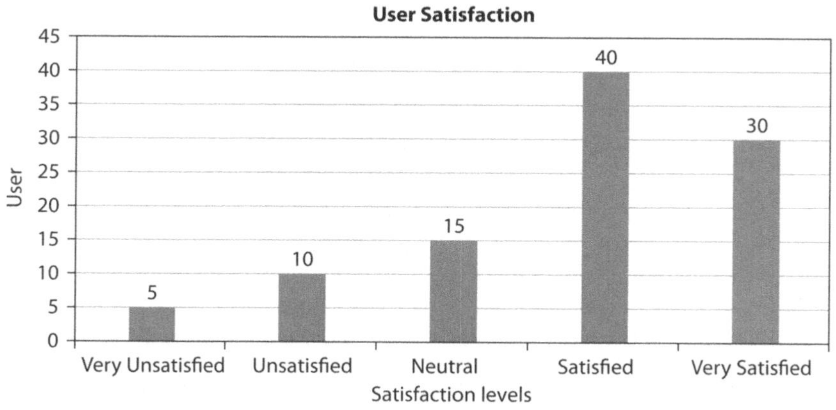

FIGURE 7.6 User satisfaction.

7.5.2 Preferred Features

Users were asked to rank their preferred features in IoT-enabled charging solutions shown in Figure 7.7. The results are as follows:

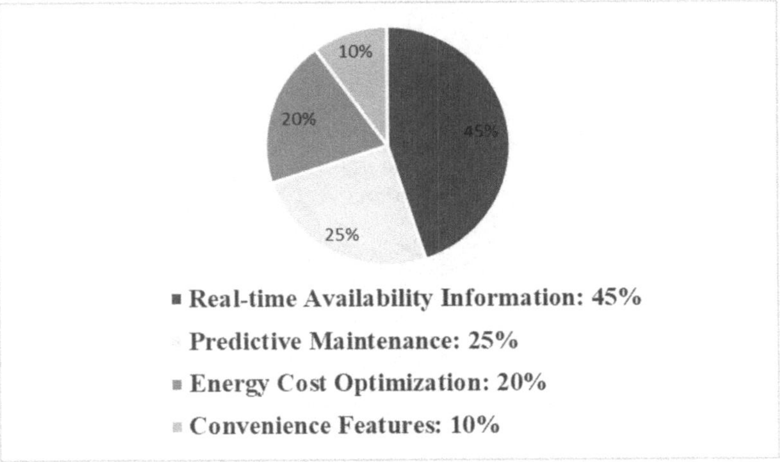

FIGURE 7.7 Preferred features.

Users chose "Real-time Availability Information" as their favourite feature, with 45% of them choosing it as their first option.

7.5.3 Reasons for Satisfaction

Users who reported satisfaction with IoT-enabled charging solutions were asked to specify the reasons for their satisfaction shown in Figure 7.8. The following results were obtained:

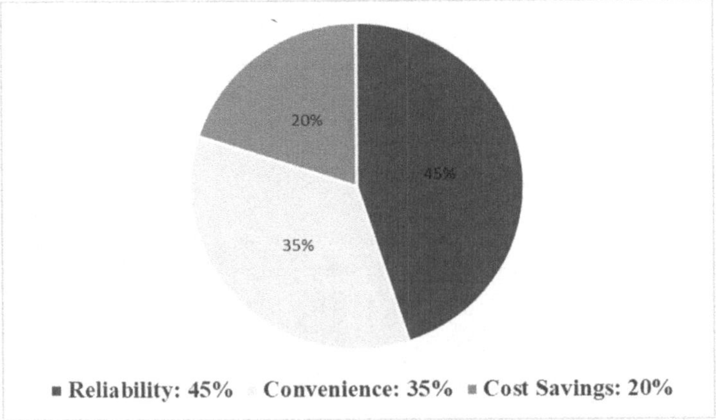

FIGURE 7.8 Reasons for satisfaction.

Overall, "reliability" and "convenience" are the most frequently identified reasons for satisfied consumers.

The two main factors contributing to user satisfaction are "convenience" and "reliability." These components stand out as the core elements of satisfying user experiences, highlighting the importance of a strong and stable charging infrastructure. The large number of users who cite dependability as a primary reason for their pleasure is indicative of the value that users place on IoT-enabled products. Furthermore, the focus on convenience highlights the need for intuitive user interfaces, smooth experiences, and effective charging procedures.

The overall conclusions obtained from the case study of the survey are given below:

1. The majority of users expressed satisfaction with IoT-enabled charging solutions, with 70% reporting being either "satisfied" or "very satisfied."
2. "Real-time Availability Information" is the favourite feature of 45% of the respondents.
3. "Reliability" and "convenience" are the most frequently identified reasons by 45% of satisfied consumers.

By keeping in mind the above-mentioned conclusion of the survey, the design engineer of the IoT-enhanced EVCI should plan the design of the system for optimum utilization of the facilities.

7.6 CONCLUSIONS

In conclusion, the impact of IoT-enhanced EVCI is a transformative force in the world of electric mobility. IoT technology has redefined the way EVs are charged, offering a multitude of benefits that encompass convenience, reliability, energy efficiency, and sustainability. Through real-time monitoring and management, IoT ensures that charging stations remain operational and available, reducing downtime and addressing user concerns about range anxiety. Data-driven optimization empowers decision-makers to strategically position charging stations and understand user behaviours, further improving accessibility. Predictive maintenance, remote management, and enhanced security provide robust reliability, allowing users to trust the charging infrastructure. IoT's load balancing and integration with smart grids optimize energy use, contributing to energy efficiency and grid stability. IoT enhances the entire charging experience as well as offers immediate updates on station availability, which greatly improves satisfaction among users. The user-driven requests for IoT-enabled systems are highlighted by preferences for features like cost reduction, maintenance planning, and real-time availability. An even more reliable, effective, and environmentally friendly future for electric transportation will be assured by the IoT technology's combination with EVCI as it gradually improves. The IoTs' positive impacts on the infrastructure required for charging EVs highlight how important it is to increase the acceptance and efficiency of EVs, which in turn promotes more environmentally friendly transportation.

REFERENCES

1. R. Manivannan, "Research on IoT-Based Hybrid Electrical Vehicles Energy Management Systems Using Machine Learning-Based Algorithms", Elsevier Sustainable Computing: Informatics and Systems, 41, 10094, 2024.
2. J. Debadarshini, R.M. Vardhan, S. Saha and C.N. Bhende, "iCoord: Efficient Charging of the Electric-Vehicles Through IoT-Assisted Coordination With the Charging-Infrastructure", Ad Hoc Networks, 153, 103345, 2024.
3. N.V. Emodi, U.B. Akuru, M.O. Dioha, P. Adoba, R.J. Kuhudzai and O. Bamisile, "The Role of Internet of Things on Electric Vehicle Charging Infrastructure and Consumer Experience", MDPI Journal Energies, 16(10), 4248, 2023.
4. Y.T. Chai, H.S. Che, C. Tan, W.-N. Tan, S.-C. Yip and M.-T. Gan, "A Two-Stage Optimization Method for Vehicle to Grid Coordination Considering Building and Electric Vehicle User Expectations", International Journal of Electrical Power & Energy Systems, 148, 108984, 2023.
5. N. Singh, A. Sharma, S. Prabhudand and K. Chadaga, "Article: IoT-Based Greenhouse Technologies for Enhanced Crop Production: A Comprehensive Study of Monitoring, Control, and Communication Techniques", Systems Science & Control Engineering, Vol. 12, Issue 1, 30 Jan 2024.
6. M.S. Mastoi, S. Zhuang, H.M. Munir, M. Haris, M. Hassan, M. Usman, S.S.H. Bukhari and J.-S. Ro, "An in-Depth Analysis of Electric Vehicle Charging Station Infrastructure, Policy Implications, and Future Trends", Elsevier Energy Reports, 8, 11504–11529, 2022.
7. Y. Wang, Y. Chi, J.H. Xu and Y. Yuan, "Consumers' Attitudes and Their Effects on Electric Vehicle Sales and Charging Infrastructure Construction: An Empirical Study in China", Energy Policy, 165, 112983, 2022.
8. N. Singh, "Meta Heuristic Techniques Used To Improve The Performance Of Static Converters" International Conference on Innovative Development in Virtual and Rapid Prototyping (IDVRP-2022). AIP Conference Proceedings 2771, 010001 2023. https://doi.org/10.1063/5.0152913.
9. N. Singh et al. "IOT Enabled Hybrid Model With Learning Ability for E-Health Care Systems", Measurement: Sensors (Science Direct), vol. 24, pp.1–14, Dec. 2022.
10. Z. Teimoori and A. Yassine, "A Review on Intelligent Charging Management System for Future Electric Vehicle Transportation", Sustainability, 14, 14100, 2022.
11. M.S. Mastoi, S. Zhuang, H.M. Munir, M. Haris, M. Hassan, M. Usman, S.S.H. Bukhari and J.-S. Ro, "An in-Depth Analysis of Electric Vehicle Charging Station Infrastructure, Policy Implications, and Future Trends", Energy Reports, 8, 11504–11529, 2022.
12. M. Nour, S. Sengupta and C. Mahapatra, "Internet of Things (IoT) for Electric Vehicle Charging Infrastructure: A Comprehensive Survey", Computers & Electrical Engineering, 89, 106923, 2021.
13. N. Singh et al. "Analysis of Optimum Cost and Size of the Hybrid Power Generation System Using Optimization Technique", 12th IEEE International Conference CSNT-2023 held at Technocrats Institute of Technology (Excellence), Bhopal, Madhya Pradesh (India). 08 & 09 April, 2023. https://ieeexplore.ieee.org/document/10134720
14. N. Singh, "Artificial Intelligence Techniques In Power System Operation And Analysis", Book CRC Press, vol.1, pp.1–185, 2023.
15. M.A. Khan and R. Hasan, "Internet of Things (IoT)-Based Smart Electric Vehicle Charging Infrastructure: A Comprehensive Review", IEEE Access, 8, 175103–175122, 2020.
16. S. Zheng, J. Hu, C. Zhang and Y. Xia, "A Smart and Scalable Charging Infrastructure for Electric Vehicles in IoT Environments", IEEE Internet of Things Journal, 7(9), 8234–8244, 2020.

17. Z. Wang and V.K. Sood, "Internet of Things (IoT)-Enabled Intelligent Electric Vehicle Charging Infrastructure: Challenges and Opportunities", Energies, 13(22), 6024, 2020.
18. Y. Chen, K. Zhang and Z. Zhang, "Intelligent Electric Vehicle Charging Infrastructure: A Comprehensive Survey", IEEE Transactions on Industrial Informatics, 16(6), 4325–4337, 2019.
19. V. Sondhiya, K. Dwivedi, S.K. Almas and N. Singh, Energy Management of Nanogrid through Flair of Deep Learning from IoT Environments, Artificial Intelligence Techniques in Power Systems Operations and Analysis, Taylor & Francis, Vol.1, pp. 1–20, July 2023.
20. J. Zhang, W. Gao and L. Tian, "IoT-Based Intelligent Electric Vehicle Charging System: A Comprehensive Survey", IEEE Internet of Things Journal, 4(5), 1449–1466, 2017.
21. U. Samarakoon, A.Q. Huang and K. Zhang, "Smart Charging for Electric Vehicles: A Review on the Vehicle to Grid Integration and the Applications in the Internet of Energy", Applied Energy, 225, 592–605, 2018.
22. N. Singh, "Renewable Energy For Sustainable Rural Development And Challenges," IJRAR, 6(1), 185–187, February 2019.
23. M. Kumawat and N. Singh, "Emerging Trends In Science & Technology", Bhumi, ISBN No. 978-81-951982-2-1, vol.1, pp. 1–175, June 2021.

8 Prediction Forecasting and Optimization of Energy Consumption

An AI-ML-based Approach

Preeti Khanwalkar and Swetha Ravikanti

8.1 INTRODUCTION

The increasing economic growth, industrial demands, smart city/building initiatives, automated electronic appliances, increasing indoor activities, air conditioners, air Fresheners, etc., have increased the demand of energy consumption. Energy consumption is increasing day by day for both residential and commercial buildings and thus requires handling energy resources. Fulfilling supply and demands of energy consumption is essential for economic growth and sustainable development [1, 2]. To optimize energy consumption analysis, prediction and forecasting of energy consumption is required, which enables to take suitable action and to reduce unnecessary energy consumption. With the advent in technology, smart sensors, Internet of Things (IoT) devices, energy meters, etc., are continuously capturing real-time data, which will be useful for energy suppliers, consumers, and management systems to analyze and recognize energy consumption demands [3]. Moreover, artificial intelligence (AI) and machine learning (ML) being a data analysis based tool provide a promising way of optimal decision-making, analyzing, and reducing energy consumption and hence enhance the efficient energy utilization and long-term reliability of the system [4–6].

For effective energy utilization, the most important factor to consider is to alleviate energy waste by identifying anomalies during energy consumption [7, 8]. Anomaly detection in energy consumption also includes faults, failures, and other irregular conditions, where the data samples differ from the usual samples collected on a daily, weekly, monthly, or yearly basis. Timely anomaly identification not only supports effective energy utilization but also recognizes any faulty equipment, and thus eliminates the wastage of resources. The energy consumption in residential and commercial buildings is affected by many aspects like weather conditions, such as cloudy weather, summer, festival seasons, etc. Even in a few cases these aspects may not directly affect the energy consumption but may be highly correlated to energy consumption. So, all these aspects need to be taken into consideration for anomaly detection [9].

In this work, AI-ML-based algorithms are used to predict and forecast any anomaly situation in energy consumption, which maintains reliable and efficient energy utilization among energy suppliers and consumers and reduces energy wastage. The contribution of our work is as follows: (i) real-time large-scale energy anomaly

 DOI: 10.1201/9781003537823-11

detection (LEAD) dataset available from Kaggle [10] is considered. (ii) Dataset cleaning-preprocessing and analysis is conducted with ML algorithms (iii) Anomalous situation that detects faults, failures, and other irregular conditions are predicted. (iv) Next, different ML models are evaluated and results are compared.

The rest of the chapter is organized as follows. The related work is discussed in Section 8.2, Section 8.3 introduces the proposed methodology, Section 8.4 describes the simulation and results, and Section 8.5 presents the conclusion and future works.

8.2 RELATED WORKS

The recent proliferation of AI and ML tools has shown a great impact on the prediction, forecasting, and optimization of energy consumption in both residential as well as industrial sectors. A literature survey on AI-based methods for load forecasting, identification of anomaly, and demand response are discussed in [11] with methods focused on the demand-side applications of the future energy systems. Authors in [12] have used a support vector machine (SVM), artificial neural network (ANN), and K-nearest neighbors (KNNs) model to identify the prediction of data. In another work, authors in [13] have discussed the ML-based method for the identification of the consumption of electricity. To forecast the power usage linear regression, KNNs, XGBOOST, Random Forest (RF), and ANN models are used.

In [14], the authors have discussed the RF and decision tree (DT) based model to predict the power anomaly consumption. The authors have used the Irish Social Science Data; and the Isolation Forest algorithm is used to label the readings of the smart meter electricity consumption as normal or abnormal. In [15] Amir Mosavi and Abdullah Bahmani have discussed the literature survey of ML approaches for energy consumption prediction. For identifying the consumption of energy, authors in [16, 17] have used regression models and performed analysis using support vector regression (SVR), RF, DT, and KNN.

Himeur et al. [18, 20] have presented a rule-based model using a deep neural network for classification and detection of anomalies. Sujay Kumar et al. in [19] under the PM-Gati-Shakti Initiative have discussed the forecasting methods to identify the seasonal patterns from the day-to-day information of maximum demand for the Union Territory of Delhi. The authors have discussed the auto-regressive integrated moving average (ARIMA) model for time series prediction to identify the patterns.

8.3 PROPOSED METHODOLOGY

In this work, the ML algorithms are used for the prediction and forecasting of anomalies of a LEAD dataset available from Kaggle [10]. The data is analyzed to detect faults, failures, and other irregular conditions. ML-based predictive analysis enables to analyze the potential problems beforehand, helps to minimize the failure of critical resources and abrupt shutdowns, and reduces the operative costs of energy consumption.

The proposed methodology is shown in Figure 8.1. Since real-time data consists of missing values, duplicate values, and different types of data. Thus, the first data is cleaned and preprocessed with missing data imputed with the mean value. Further feature importance is considered for different values. With data ranges in different scales, the feature scaling is performed through a standard scaler, and to handle

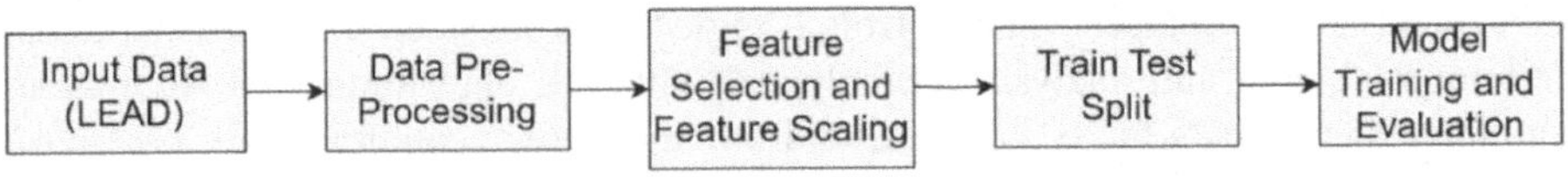

FIGURE 8.1 Proposed methodology.

imbalance data handling, synthetic minority oversampling technique (SMOTE) analysis is performed. After preprocessing and feature scaling the data, the data is fed to the learning algorithm. Different features are decided to be used in the prediction, and based on predictions, data is divided into features and target variables. Further, data is partitioned into training, testing, and validation groups.

In our work, four ML models are used, which include RF, SVMs, AdaBoost, and ANN. Further, these models are evaluated based on the R-square results.

8.4 SIMULATION AND RESULTS

The real-time data for the study is considered as LEAD dataset available from Kaggle [10], which consists of data samples of 200 buildings. The training dataset consists of 1,749,494 rows and 57 columns. The dataset consists of attributes such as "building id", "meter reading", "anomaly", "primary use", "square feet", "year built", "floor count", "air temperature", "cloud coverage", "dew temperature", "weekday", "month", "year", "is holiday", etc. Each data sample is annotated with anomalous (1) or normal (0) usage. With experiments, the accurate anomaly detection model is designed on the training dataset and then predicts and forecasts the meter reading in the test dataset from the other 206 buildings that it is anomalous '1' or normal '0'. The experimental results of the data preprocessing and exploratory data analysis are as shown below. Figure 8.2 indicates the anomaly pattern detected in different types of buildings. The education building with higher energy consumption shows a higher anomaly as compared to other buildings like religious worship.

Figure 8.3 indicates the heatmap that is used to determine the correlation between different variables. Anomaly patterns are highly correlated with building type,

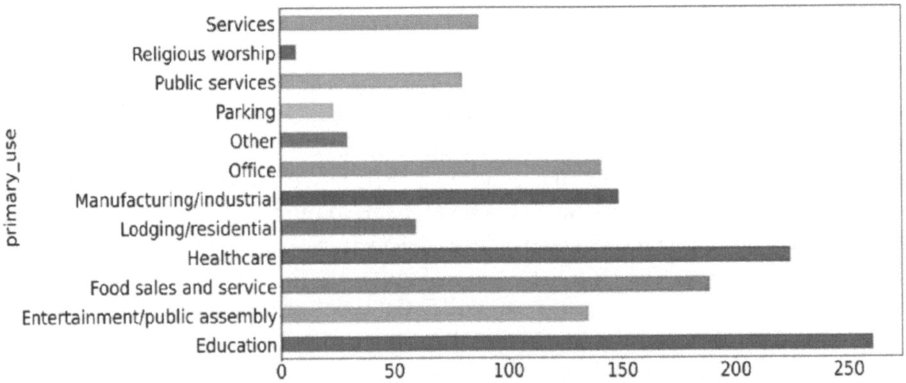

FIGURE 8.2 Energy consumption anomaly concerning different types of buildings.

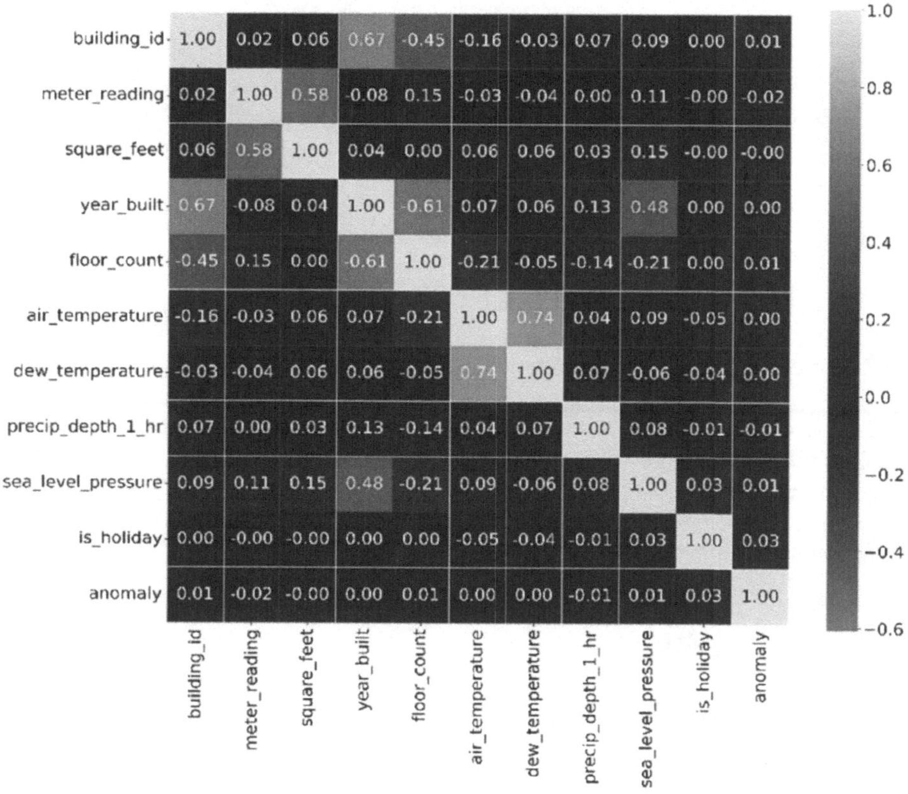

FIGURE 8.3 Correlation among different attributes of energy consumption.

holiday season, and weather conditions such as sea pressure; such highly co-related variables are removed to build an appropriate anomaly detection model.

Further, based on the data points anomalies are also predicated in different types of buildings as shown in Figure 8.4. Anomaly in a random sample of buildings is depicted in Figure 8.5.

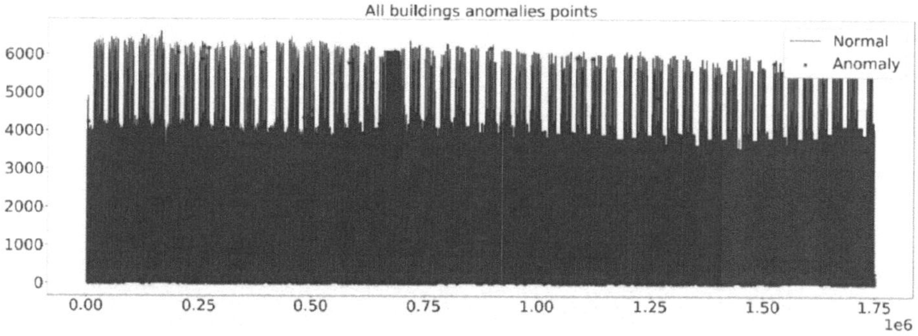

FIGURE 8.4 Predicted anomaly in different types of buildings.

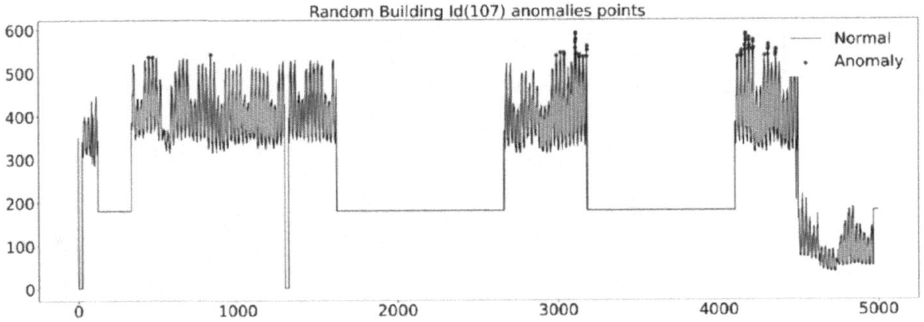

FIGURE 8.5 Predicted anomaly detected in one of the buildings.

The several ML models are evaluated based on their *R*-square results. Table 8.1 displays the comparative *R*-square scores of different ML models.

As shown in Figure 8.6, AdaBoost model outperforms the other SVM, ANN, and RF with 95.5% accuracy, and it indicates that for the given dataset AdaBoost model predicts the anomaly and optimizes efficient energy utilization.

TABLE 8.1
Comparative Results of Train and Validation R^2 Scores of Different Models

Models	Train R^2 Score	Validation R^2 Score
SVM	0.228	0.115
ANN	0.626	0.364
Random Forest	0.867	0.613
AdaBoost	0.955	0.649

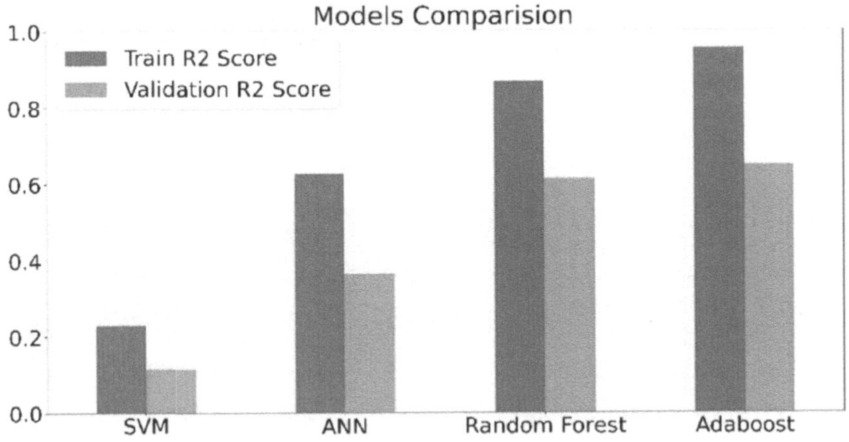

FIGURE 8.6 Comparison of model predictions.

8.5 CONCLUSION AND FUTURE WORKS

This chapter discussed the real-time data analysis for prediction, forecasting anomaly detection, and thus optimized energy consumption. The LEAD data is collected from the Kaggle data sources. Further, the data is preprocessed and analyzed using ML algorithms to detect anomalies and to provide reliable and optimized energy consumption. For the predictive analysis of anomaly and energy consumption, different ML models are used, including RF, SVMs, Adaboost, and ANN, and then compared based on *R*-square results. The simulation results show that the Adaboost model outperforms other models in anomaly detection of the considered data. In the future, a multiclass classification model for different categories of energy consumption with short- and long-term prediction and forecasting can be considered.

REFERENCES

1. X. Wang, and S.-H. Ahn, Real-time prediction, and anomaly detection of electrical load in a residential community, Applied Energy, Volume 259, 2020, https://doi.org/10.1016/j.apenergy.2019.114145
2. L. Sharma, The rise of internet of things and smart cities, in: Towards Smart World, Chapman and Hall/CRC, 2020, pp. 3–14. https://doi.org/10.1201/9781003056751-1.
3. N. Mazlan, N.A. Ramli, L. Awalin, M. Ismail, A. Kassim, and A. Menon A smart building energy management using internet of things (IoT) and machine learning, TEST Engineering & Management, Volume 83, 8083–8090, 2020.
4. X. Wang, Z. Yao, and M. Papaefthymiou A real-time electrical load forecasting and unsupervised anomaly detection framework, Applied Energy, Volume 330, 120279, 2023.
5. S.S. Ali, and B.J. Choi State-of-the-art artificial intelligence techniques for distributed smart grids: A review, Electronics, Volume 9, 1030, 2020. https://doi.org/10.3390/electronics9061030
6. D. Karthika, and K. Karthikeyan. "Forecasting electricity consumption in India by using univariate time-series analysis." Advances and Applications in Mathematical Sciences, Mili Publications, Volume 22, pages 71–78 (2022).
7. L. Lei, B. Wu, X. Fang, L. Chen, H. Wu, and W. Liu (2023). A dynamic anomaly detection method of building energy consumption based on data mining technology, Energy, 263, 125575. https://doi.org/10.1016/j.energy.2022.125575
8. M. Shapi, N.A. Ramli, and L. Awalin Energy consumption prediction by using machine learning for smart building: Case study in Malaysia, Developments in the Built Environment, Volume 5, 100037, 2020.
9. A. Malki, E.-S. Atlam, and I. Gad. (2022). Machine learning approach of detecting anomalies and forecasting time-series of IoT devices, Alexandria Engineering Journal, Volume 61. 10.1016/j.aej.2022.02.038.
10. https://www.kaggle.com/code/abdallahmohamedamin/large-scale-energy-anomaly-detection-lead
11. G. Reddy, L. Aitha, C. Poojitha, A. Shreya, D. Reddy, and G. Meghana, Electricity Consumption Prediction Using Machine Learning, E3S Web of Conferences, Volume 391, 2023, 10.1051/e3sconf/202339101048
12. A. Zainab, S.S. Refaat, and O. Bouhali, Ensemble-based spam detection in smart home IoT devices time series data using machine learning techniques, Information, Volume 11, 2020 https://doi.org/10.3390/info11070344.

13. C. Xu, and H. Chen. (2020). A hybrid data mining approach for anomaly detection and evaluation in residential buildings energy data, Energy and Buildings, Volume 215, 109864. 10.1016/j.enbuild.2020.109864.
14. H. Rahman, I. Selvarasan, and J. Begum A Short-term forecasting of total energy consumption for India-a black box based approach, Energies, Volume 11, 3442, 2018. https://doi.org/10.3390/en11123442
15. R. El-Hadad, Y.-F. Tan, and W.-N. Tan Anomaly prediction in electricity consumption using a combination of machine learning techniques, International Journal of Technology, Volume 13(6), 1317, 2022.
16. A. Mosavi, and A. Bahmani Energy consumption prediction using machine 3 learning; A review, MDPI Journal of Energies, Volume 11, 2018.
17. A. González-Briones, G. Hernández, J.M. Corchado, S. Omatu, and M.S. Mohamad, "Machine learning models for electricity consumption forecasting: A review," 2019 2nd International Conference on Computer Applications & Information Security (ICCAIS), Riyadh, Saudi Arabia, 2019, pp. 1–6, doi: 10.1109/CAIS.2019.8769508.
18. Y. Himeur, A. Alsalemi, F. Bensaali, and A. Amira Smart power consumption abnormality detection in buildings using micro-moments and improved k-nearest neighbors, International Journal of Intelligent Systems, Volume 36(6), 2865–2894, 2021.
19. S. K. Reddy M, and G. G, PM-Gati Shakti: Advancing India's energy future through demand forecasting – A case study, arXiv, 2308.07320, Economics, 2023
20. Y. Himeur, A. Alsalemi, F. Bensaali, and A. Amira A novel approach for detecting anomalous energy consumption based on micro-moments and deep neural networks, Cognitive Computation, Volume 12(6), 1381–1401, 2020.

9 Cyber Physical Systems Related Issues and Digital Twins for Power and Energy Systems

Shekinah P and S R Mani Sekhar

9.1 INTRODUCTION

Cyber-physical systems (CPS) integrate and compute physical operations and networking. These are systems where physical processes interact with near, real-time computing [1, 2]. In the context of power and energy systems, CPS is the fusion of digital technology with electricity infrastructure to enhance efficiency, dependability, and safety [3]. Digital twins are virtual representations of actual assets, processes, or configurations. They mimic the movements of their real-world counterparts by continually collecting data from sensors, analyzing it using algorithms, and providing insights into performance and behavior [4, 5]. In power and energy systems, digital twins are useful instruments for monitoring, regulating, optimizing, and carrying out predictive maintenance [6]. Numerous significant elements are involved in the concerns of digital twins and CPS in power and energy systems. This means ensuring seamless communication across the many CPS framework systems and parts, safeguarding infrastructure against cyberattacks to prevent disruptions, and ensuring dependability and resilience against failures and disasters [7, 8]. Sustaining strict guidelines for data privacy and quality is essential for obtaining accurate insights from sensors and other sources that are utilized in digital twins and CPS applications [9]. Scalability and flexibility are also critical for CPS architectures and digital twin models to accommodate expanding infrastructure and adapt to shifting operating requirements [10]. Meeting regulatory compliance requirements for cybersecurity, data privacy, and interoperability is necessary to ensure legal and safe deployment. In order to guarantee the resilience, safety, and security of vital infrastructure, it is crucial to solve connected challenges. Notwithstanding these challenges, the application of CPS and digital twins has a lot of potential to enhance the sustainability, dependability, and efficiency of power and energy systems. We examine the state-of-the-art research on CPS in power and energy systems in this chapter, outlining important ideas and implementing management strategies. We also examine how new technologies, such as digital twins, might solve problems with security and scalability and improve system performance.

DOI: 10.1201/9781003537823-12

9.2 CURRENT KNOWLEDGE AND RESEARCH ON CPS IN THE CONTEXT OF POWER AND ENERGY SYSTEMS

9.2.1 Introduction to CPS in Power and Energy Systems

9.2.1.1 Cyber-Physical Systems (CPS) and Their Significance in Power and Energy Systems

Linking the digital and physical worlds, CPS provide increased monitoring, control, optimization, and automation capabilities, driving a revolution in power and energy systems. Sensing, computing, communication, and control technologies are integrated by CPS in the context of power and energy systems to improve the sustainability, dependability, and efficiency of energy generation, transmission, distribution, and consumption processes. Being able to monitor and control crucial infrastructure components in real-time is one of the main benefits of CPS in power and energy systems [11, 12]. CPS can gather a large amount of data about ambient elements, equipment health, energy usage, and grid conditions by distributing sensors and actuators across the power system. By providing the basis for sophisticated analytics and algorithms for making decisions, this data enables operators to identify irregularities, anticipate malfunctions, and enhance system efficiency almost instantly. CPS can also provide increased flexibility and resilience in the grid by facilitating the integration of demand-side resources, energy storage devices, and renewable energy sources [13–17]. By utilizing sophisticated coordination and control algorithms, CPS may dynamically modify patterns of generation, storage, and consumption to align with variations in supply and demand [18]. This improves grid stability and lessens dependency on fossil fuels. Through the use of smart grid technology and demand response programs, CPS can make it possible for customers to take an active role in energy management and conservation initiatives [8]. CPS can lower peak demand, promote more energy-efficient consumption habits, and lessen grid stress during times of heavy load by giving real-time feedback and incentives [19]. CPS are essential for improving power and energy systems' resilience and cybersecurity. Cyber risks and assaults are becoming a major worry as essential infrastructure becomes more connected and digitized [8, 18]. Strong security features like intrusion detection, authentication, and encryption are incorporated into CPS systems to protect against hostile activity and guarantee the availability and integrity of the energy infrastructure. Additionally, CPS make it possible to quickly identify and address cybersecurity problems, reducing the likelihood of interruptions and lessening the effect on the distribution and supply of electricity.

9.2.1.2 Integration of Computing, Communication, and Control Technologies in Physical Systems

CPS are centered on the integration of computer, communication, and control technologies with physical systems, which propels innovation and change in a number of disciplines. The smooth communication between digital and physical processes is made possible by this integration, which improves the capabilities of automation, control, analysis, and monitoring [20, 21]. CPS also use communication technologies to enable information interchange, computer power to analyze data, and control systems to actuate changes in the physical environment based on computational insights. Many application sectors, such as manufacturing, transportation, healthcare, energy, and infrastructure, benefit greatly from this integration [22–24]. In manufacturing, for example, CPS enable the deployment of smart factories where interconnected sensors, actuators, and production systems communicate and collaborate in real-time to optimize production processes, improve quality, and reduce downtime. CPS make it possible to establish "smart factories" in the manufacturing sector, where networked sensors, actuators, and production systems work together in real-time to enhance output, lower downtime, and enhance quality [25–27]. In transportation, CPS facilitate the development of autonomous vehicles and intelligent transportation systems, where onboard computers, communication networks, and control algorithms work together to enhance safety, efficiency, and sustainability. CPS aids in the creation of intelligent transportation systems and driverless cars, combining communication networks, control algorithms, and onboard computers to improve sustainability, efficiency, and safety [28, 29]. In healthcare, CPS enable remote patient monitoring, personalized treatment delivery, and predictive analytics, leveraging interconnected medical devices, wearable sensors, and data analytics platforms to improve patient outcomes and healthcare delivery [30–32]. Moreover, in energy and infrastructure systems, CPS empower the creation of smart grids, smart buildings, and smart cities, where interconnected sensors, meters, and control systems enable real-time monitoring, optimization, and management of energy resources and infrastructure assets. CPS enabled development of smart grids, smart buildings, and smart cities in the domain of energy and infrastructure systems. These systems include of networked sensors, meters, and control systems that facilitate real-time monitoring, optimization, and administration of energy resources and infrastructure assets [33–35].

9.2.1.3 Benefits of Cyber-Physical Systems

CPS have many potential applications, but they are especially useful for improving power grid resilience, efficiency, and dependability [36]. CPS provide a wide variety of enhanced monitoring, control, and optimization capabilities that have the potential to completely transform the way power systems function by fusing digital intelligence with physical infrastructure. Increased efficiency is one of the main advantages of CPS in power networks. By continuously monitoring grid conditions in real time, CPS is able to spot inefficiencies like voltage swings or line losses and dynamically modify operating settings to maximize energy flow and reduce waste. Furthermore, by continually assessing the condition of grid assets and anticipating any breakdowns before they arise, CPS enable predictive maintenance plans, which lower downtime and boost overall system efficiency. By offering improved situational awareness and control capabilities, CPS also help to strengthen the stability of electricity systems [37]. Operators may minimize the impact on service dependability and lower the possibility of cascade failures by using CPS to identify and respond in real-time to grid disruptions or equipment breakdowns [38]. Additionally, by diversifying the energy supply and lowering reliance on centralized generation facilities, CPS improve system resilience by facilitating the inclusion of distributed energy resources and renewable energy sources into the grid. CPS can minimize the consequences of variations in renewable energy generation, optimize grid functioning, and dynamically balance supply and demand using sophisticated coordination and control algorithms. Increased resilience is a key advantage of CPS in power networks. CPS enables quick reaction to disturbances, including natural disasters or cyberattacks, and helps restore service in impacted regions by utilizing enhanced monitoring and control capabilities. In order to maintain service continuity and reduce downtime, CPS also supports self-healing techniques by autonomously rearranging grid topology and rerouting power flows to avoid damaged or crowded locations. CPS further strengthen cybersecurity by putting strong authentication, encryption, and intrusion detection systems (IDS) in place, protecting vital infrastructure from cyberattacks and guaranteeing its availability and integrity [37, 39].

9.2.2 Key Challenges and Research Areas

9.2.2.1 Challenges Associated with CPS Deployment in Power and Energy Systems

The deployment of CPS in power and energy systems brings about significant opportunities for innovation and improvement, there are also a number of issues that must be resolved. These difficulties include interoperability problems, cybersecurity risks, and system complexity, each of which presents a different barrier to the effective application of CPS technology in the energy industry [40]. Cybersecurity concerns are a major obstacle to the introduction of CPS in power and energy systems. These systems are more susceptible to cyberattacks as a result of their increased connectivity and digitization, which can seriously compromise data integrity, safety, and grid dependability. Vulnerabilities in CPS components might be used by malicious actors to obtain unauthorized access, interfere with operations, or steal confidential data. Furthermore, because CPS is linked, there are additional attack avenues available, as compromising one system might have repercussions for the whole energy grid. Strong security features, including as encryption, authentication, access control, and intrusion detection, must be incorporated at every stage of the CPS architecture in order to reduce cybersecurity risks [41–44]. Continuous monitoring, exchange of threat data, and incident response skills are necessary for quickly identifying and reducing cyberthreats. Another major obstacle to the introduction of CPS in power and energy systems is interoperability. The energy infrastructure comprises a wide range of devices, protocols, and standards, making it difficult to integrate and communicate among the many components of the CPS with ease. The incompatibility of new CPS technologies with older systems can limit system interoperability, obstruct data transmission, and make it more difficult to realize benefits like improved control, monitoring, and optimization. Industry stakeholders need to work together to create and implement standardized data formats, interfaces, and communication protocols that allow CPS components to plug and play together in order to overcome interoperability difficulties. The creation of frameworks for interoperability testing and certification schemes can aid in guaranteeing adherence to established norms and streamline the incorporation of CPS solutions into the current energy infrastructure. Another barrier to the introduction of CPS in power and energy systems is system complexity. System design, implementation, and maintenance may become more difficult as a result of the integration of CPS technologies, which adds new layers of complexity involving heterogeneous devices, dispersed computing systems, and linked networks. Complex CPS designs require specialized knowledge in a number of areas, including data analytics, cybersecurity, networking, and control theory [45–48]. CPS implementation and operation are further complicated by the dynamic nature of energy systems, which is typified by varying demand, integration of renewable energy, and instability in the grid. Industry stakeholders need to make investments in workforce development programs, research, and development projects aimed at streamlining CPS designs, boosting automation capacities, and strengthening system resilience in order to handle system complexity. The deployment of CPS technologies in power and energy systems may be facilitated and complexity reduced by using modular, scalable CPS solutions and model-based design methodologies.

9.2.2.2 Research Areas Addressing Challenges in CPS Deployment

Research efforts on CPS aim to address power and energy system challenges by concentrating on different aspects of system improvement in many key sectors. These include real-time monitoring and control, energy management optimization techniques, and flaw identification and diagnostics [49, 21].

Within CPS for power and energy systems, real-time control and monitoring are important study areas. Researchers in this field concentrate on creating sophisticated sensor technologies, data collection techniques, and analytical algorithms to allow accurate and ongoing monitoring of a variety of energy infrastructure factors. CPS can provide proactive decision-making and quick responses to changing grid circumstances by utilizing real-time data streams. In order to gather important data points throughout the grid, research in this field integrates cutting-edge sensing technologies, including distributed sensor networks, smart meters, and synchrophasors. Additionally, in order to improve grid operations in real-time and guarantee stability, dependability, and efficiency, researchers investigate cutting-edge control techniques, such as distributed control algorithms and model predictive control.

Another crucial area of research for power and energy systems within CPS is fault diagnosis and detection [50–52]. This area of study focuses on creating methods and algorithms for quickly identifying, locating, and diagnosing errors or anomalies in the energy infrastructure. Researchers want to find early warning signs of possible breakdowns or deterioration in grid components, including transformers, transmission lines, and substations by utilizing data analytics, machine learning (ML) approaches, and pattern recognition algorithms. Moreover, innovative monitoring systems and sensor networks are being used as part of fault detection and diagnostic research to provide continuous surveillance of vital infrastructure, supporting early intervention, and preventative maintenance operations. An important field of study that aims to maximize the sustainability, dependability, and efficiency of power and energy systems is optimization strategies for energy management. To optimize the processes involved in energy generation, transmission, distribution, and consumption, researchers in this field create optimization models, algorithms, and decision support systems. These methods develop optimal solutions that balance supply and demand, minimize costs, and lessen environmental effect by taking into account a number of variables, including energy demand, generating costs, grid limits, and environmental objectives. In order to improve grid flexibility and resilience, research in energy management optimization also incorporates demand response programs, energy storage devices, and renewable energy sources.

9.2.2.3 Importance of Interdisciplinary Approaches Addressing CPS Challenges

To address the issues posed by CPS, an interdisciplinary strategy involving cybersecurity, computer science, and electrical engineering is required [53, 54]. To effectively handle complexities, these systems that fuse the digital and physical domains need a multimodal approach that draws on expertise from several disciplines. The many domains provide unique perspectives, methods, and assets to this collaborative structure, enabling the development of comprehensive solutions that enhance the efficiency, stability, and safety of CPS in various applications, such as energy and power systems. Fundamental to CPS is the integration of digital intelligence with physical processes, necessitating a deep understanding of both domains. Electrical engineering provides the understanding of power systems, control theory, and hardware design principles that are required to comprehend the dynamical and physical components of CPS settings. Electrical engineers are adept in designing, optimizing, and maintaining the infrastructure that underpins energy and power systems, including transformers, distribution networks, and generators. This understanding also includes the principles of feedback control, which enable the dynamic changing of system parameters in response to sensor input and computer analysis, and thus facilitate CPS operation.

Computer science offers essential insights into data analytics, algorithms, and software development, which enhance CPS and are necessary to effectively exploit this expertise. In the discipline of CPS, computer scientists develop complicated algorithms and software for the real-time monitoring, control, and optimization of complex systems. Computer scientists employ techniques such as distributed computing, optimization algorithms, and ML to empower CPS to analyze massive amounts of data, extract meaningful insights, and reach logical conclusions on its own. Moreover, software engineering expertise ensures the interoperability, scalability, and robustness of CPS applications, facilitating a seamless integration into various settings and existing infrastructure.

However, the convergence of physical and digital systems presents new challenges, particularly in the area of cybersecurity. As CPS becoming increasingly digital and networked, they become more vulnerable to cyberthreats such as malware, denial-of-service attacks, and hacking. Cybersecurity experts are crucial to defending CPS against these threats because they leverage their knowledge of encryption, network security, and risk management to build robust defenses. Cybersecurity professionals use IDS, access control, authentication, and encryption to fortify CPS against unwanted access, data breaches, and cyber-physical attacks. By doing this, the availability, confidentiality, and integrity of crucial infrastructure are guaranteed.

Interdisciplinary cooperation fosters innovation and adaptation in the development of CPS, enabling the seamless integration of these diverse perspectives and methodologies. Academics and practitioners may leverage the unique capabilities of each area to address CPS issues by breaking down barriers and encouraging collaboration and communication across disciplines. Electrical engineers and computer scientists may work closely together to build smart sensors and actuators with integrated intelligence and communication capabilities, enabling real-time data collection and control in CPS situations. Cybersecurity experts collaborate with both disciplines to develop secure communication protocols, encryption strategies, and IDS that are especially tailored for CPS applications in order to guarantee the confidentiality and integrity of data sent inside the system.

In addition to technological domains, transdisciplinary research in CPS also addresses broader ethical and socioeconomic challenges. As CPS permeates essential infrastructure, transportation, healthcare, and other industries, academics must address privacy, autonomy, and equity problems. By consulting with experts in fields such as ethics, law, and social sciences, interdisciplinary teams can anticipate and mitigate risks and unintended consequences associated with the use of crisis psychology services. This comprehensive approach ensures the technological stability of CPS while also promoting social acceptance of these technologies and responsible innovation.

9.2.3 Case Studies and Applications

9.2.3.1 Case Studies and Applications of CPS in Power and Energy Systems

CPS are a new wave of disruptive technologies in power and energy systems that provide creative answers to the problems facing today's energy infrastructure. Power production, transmission, distribution, and consumption are all being revolutionized by CPS's improved monitoring, control, optimization, and automation capabilities, which are made possible by the seamless integration of digital intelligence with physical components. In this section, we examine case studies and CPS applications in power and energy systems, with an emphasis on demand response management, smart grids, and the integration of renewable energy sources. One of the most well-known uses of CPS in the energy industry is smart grids, which provide a comprehensive strategy for updating and streamlining the electrical system.

These intelligent networks facilitate the integration of demand-side resources, energy storage devices, and renewable energy sources while enhancing grid resilience, efficiency, and reliability through the use of CPS technologies. The Pacific Northwest Smart Grid Demonstration Project is a noteworthy case study that exemplifies the application of smart grid technology in the Pacific Northwest. It is a large-scale program supported by the U.S. Department of Energy [55]. The project showcased the advantages of real-time monitoring, control, and optimization in improving grid stability, decreasing energy losses, and allowing dynamic pricing schemes through the use of smart meters, sophisticated sensors, and grid automation systems. A more sustainable and resilient energy future may be achieved by utilities and grid operators by utilizing CPS to improve grid operations, balance supply and demand, and lessen the impact of intermittent renewable energy sources. Another important area of CPS use in power and energy systems is renewable energy integration, which addresses issues related to the erratic and intermittent nature of renewable energy sources like solar and wind [56]. Real-time monitoring, forecasting, and management of renewable energy output is made possible by CPS technologies, which also preserve system stability and dependability and enable a smooth integration of renewable energy generation into the grid. Deploying CPS-enabled microgrids in off-grid or distant locations where renewable energy sources are the main source of electricity is one example of a case study that demonstrates the concept.

In the lack of conventional grid infrastructure, these microgrids use energy storage technologies, energy management systems, and smart inverters to optimize energy consumption, balance supply and demand, and maintain grid stability. Remote villages and island nations may minimize their environmental effect while lowering energy prices, increasing energy availability, and reducing their dependency on fossil fuels by utilizing CPS.

A third use case for CPS in power and energy systems is demand response management, which lets utilities and grid managers dynamically modify electricity use in response to supply-demand imbalances or grid limitations. Real-time communication, monitoring, and control of energy usage across several sectors—residential, commercial, and industrial—are made possible by CPS technology. The implementation of CPS-enabled demand response programs by utilities and energy aggregators to encourage customers to adjust their energy use during peak hours or times of high demand is a noteworthy case study [57, 58]. Smart appliances, automated control systems, and thermostats allow users to take part in demand response events, which lowers their energy costs and contributes to grid stabilization. Demand response systems help utilities and customers by increasing grid flexibility, reducing the need for expensive infrastructure changes, and encouraging energy conservation and efficiency through the use of CPS. The case studies and CPS applications in power and energy systems show how these technologies may revolutionize the way energy infrastructure is optimized and modernized. Demand response management, smart grid integration, and renewable energy integration are just a few of the ways that CPS help utilities, grid operators, and customers increase the sustainability, efficiency, and dependability of the production, transmission, distribution, and use of electricity. CPS will become more and more important in determining the future of energy systems as they develop and mature, propelled by advancements in sensing, communication, control, and analytics technologies. They will foster innovation, resilience, and sustainability in the shift toward a cleaner and more decentralized energy landscape.

9.2.3.2 Implementation of CPS Technologies in Real-World Scenarios

The practical use of CPS technology has yielded noteworthy achievements, imparted invaluable knowledge, and indicated auspicious prospects for the future. Applications for CPS have revolutionized operations, increased productivity, and boosted safety in a number of industries, including manufacturing, energy, transportation, and healthcare. Here, we address future prospects for CPS implementation, reflect on lessons learned, and highlight a few noteworthy cases.

Intelligent transportation systems, linked infrastructure, and driverless cars have all been made possible by CPS technologies, which have completely transformed the automobile sector. Autonomous driving systems with CPS capability have been implemented by firms such as Tesla, Waymo, and Uber. These systems use sophisticated algorithms, cameras, and sensors to safely and effectively traverse roadways. Road safety, traffic congestion, and the mobility of those with impairments or restricted mobility have all improved as a result of these deployments. But as we've seen from events like autonomous car accidents, regulatory monitoring, stringent testing, and validation are crucial to guaranteeing the dependability and safety of CPS technology in the transportation sector [59–62]. CPS applications have improved patient outcomes and healthcare delivery by enabling individualized therapy delivery, remote patient monitoring, and predictive analytics. For instance, wearable technology with wireless connectivity and biosensors permits real-time monitoring of health metrics and vital signs, enabling medical professionals to act quickly in the event of an emergency or a decline in patient condition. Furthermore, by enabling remote consultations, diagnosis, and treatment, CPS-enabled telemedicine platforms reduce healthcare inequities and increase access to healthcare for underprivileged areas. To safeguard patient information and guarantee adherence to healthcare laws, strong cybersecurity measures and regulatory frameworks are necessary, as evidenced by issues with data privacy, security, and regulatory compliance.

By providing real-time monitoring, control, and optimization of production processes, CPS technologies in manufacturing have converted conventional factories into intelligent, networked manufacturing systems. In order to improve production, quality, and efficiency, businesses like Siemens, Bosch, and General Electric have implemented CPS-enabled industrial automation systems that include sensors, actuators, and ML algorithms. Significant cost savings, less downtime, and more flexibility in adapting to shifting market needs have resulted from these deployments. Lessons from cyberattacks on industrial control systems, however, highlight how crucial cybersecurity is to protecting vital infrastructure and guaranteeing the availability and integrity of production processes.

In the energy sector, CPS technologies have been essential in modernizing and streamlining the processes involved in electricity generation, transmission, distribution, and consumption. CPS-facilitated smart grid installations have improved the resilience, efficiency, and dependability of the grid while facilitating the integration of demand-side resources and renewable energy sources. Demand response initiatives, improved metering infrastructure, and grid automation technologies have all been successfully implemented to enhance grid performance and customer satisfaction in initiatives like the U.S. Smart Grid Investment Grant Program.

To overcome these obstacles and realize the full potential of CPS in the energy sector, stakeholders must work together to address interoperability, cybersecurity, and regulatory framework difficulties. In order to further improve the capabilities and robustness of CPS applications, future directions for CPS implementation include developments in sensor technologies, communication protocols, control algorithms, and data analytics approaches. Moreover, tackling issues with cybersecurity, privacy, ethics, and regulatory compliance will need multidisciplinary cooperation and collaborations between the public and private sectors. CPS technologies can continue to spur innovation, resilience, and sustainability across a range of disciplines by utilizing the combined knowledge and resources of multiple stakeholders, eventually helping society as a whole.

9.2.3.3 Role of Data Analytics, Machine Learning, and Artificial Intelligence in Optimizing CPS-enabled Power Systems

Artificial intelligence (AI), ML, and data analytics play a major role in optimizing CPS-enabled power systems by providing cutting-edge capabilities to increase sustainability, dependability, and efficiency. These technologies use the vast amount of data produced by CPS components to forecast, gain insights, and improve system performance in real time. Here, we examine the particular roles that AI, ML, and data analytics play in improving CPS-enabled power systems [51, 63]. Through the processing, analysis, and interpretation of the massive volumes of data produced by sensors, meters, and other monitoring equipment, data analytics plays a critical role in CPS-enabled power systems. Through methods including data mining, statistical analysis, and visualization, data analytics gives stakeholders important insights into equipment health, energy consumption trends, and grid operations. For instance, utilities and grid operators may enhance the generation, transmission, and distribution processes by using data analytics to find patterns, anomalies, and correlations in energy data [64]. Data analytics minimizes maintenance costs and downtime by recognizing early signs of equipment breakdowns or deterioration, hence enabling predictive maintenance plans. Based on past data, ML algorithms may learn patterns, anticipate future events, and adjust to changing conditions, providing tremendous capabilities for improving CPS-enabled power systems. ML algorithms may be used for a variety of tasks in the context of power systems, such as energy management, defect detection, and load forecasting. For instance, historical load data may be used to train supervised learning algorithms like support vector machines (SVMs) and neural networks, which can then be used to reliably predict future power demand [65–67]. Analogously, unsupervised learning techniques like clustering and anomaly detection can spot odd trends or occurrences in energy data, indicating possible grid malfunctions or disruptions. Energy management solutions may be made more efficient and reliable by constantly modifying control settings through the use of reinforcement learning techniques. AI is a more general term for a collection of methods and strategies that allow computers to carry out activities that normally require human intellect, such as ML, expert systems, and natural language processing. AI approaches may be used for a variety of purposes in CPS-enabled power systems, including as control, optimization, and decision-making. Expert systems have the capability to encode rules and domain knowledge to facilitate decision-making in grid operations. This aids operators in fault diagnosis, maintenance planning, and emergency response. Furthermore, textual data like incident reports, maintenance logs, and customer comments may be analyzed using natural language processing techniques to glean insightful information and guide decision-making.

9.3 KEY CONCEPTS, THEORIES, AND METHODOLOGIES RELEVANT TO CPS IMPLEMENTATION AND MANAGEMENT IN THE POWER AND ENERGY SECTOR

9.3.1 Concepts and Frameworks for CPS Implementation

9.3.1.1 System Integration, Interoperability, and Cyber-Physical Interactions in Power and Energy Systems

The core ideas of contemporary energy infrastructure are cyber-physical interactions, interoperability, and system integration. System integration is the process of integrating various parts, subsystems, and technologies together in a smooth manner to create an integrated energy ecosystem that can produce, transmit, distribute, and use power in an efficient manner [68, 69]. It integrates hardware, software, and data at several levels to make it easier for different parts of the energy grid to work together and coordinate. In order to create a reliable and linked grid network, system integration at the hardware level entails physically connecting electricity production facilities, substations, transformers, and distribution networks. These physical elements, which constitute the foundation of the energy infrastructure, cooperate to guarantee the efficient and dependable flow of power from its source to end consumers. The software component of system integration includes advanced control systems, energy management programs, and data analytics tools that coordinate grid activities in real time. Grid operators may make educated judgments and improve system performance by using data from sensors, meters, and other monitoring devices to monitor, optimize, and regulate energy flows with the help of these software solutions.

System integration is the process of incorporating new technologies into the grid to increase its resilience, flexibility, and sustainability. Examples of these technologies include energy storage systems, demand-side management techniques, and renewable energy sources. Another crucial idea, interoperability, emphasizes how crucial it is for various systems and components in the energy ecosystem to interact and communicate with one another and with one another in a smooth manner. Interoperability across different devices, networks, and applications is made possible by ensuring that disparate technologies, protocols, and standards can coexist and share information in an efficient manner. Greater flexibility, efficiency, and resilience in the grid are made possible by the integration of distributed energy resources, smart grid technology, and renewable energy sources, all of which depend on interoperability.

Additionally, interoperability promotes innovation, competitiveness, and teamwork in the energy industry by facilitating the creation of open and standardized communication protocols and interfaces. A key component of contemporary energy systems is cyber-physical interactions, which illustrate the complex interplay between digital and physical components in the grid. These exchanges take place at the nexus of physical systems, which include assets for power generation, transmission, distribution, and consumption, and cybernetic systems, which include computer, communication, and control technologies. By utilizing this synergy, CPS may provide sophisticated monitoring, control, automation, and optimization capabilities, turning energy infrastructure into networks that are intelligent, flexible, and responsive. Grid operators may anticipate and minimize possible breakdowns, manage energy flows, and react quickly to shifting supply and demand situations by using CPS to monitor grid conditions in real-time. However, these cyber-physical connections also bring with them new concerns about privacy, cybersecurity, and dependability. To protect against cyberthreats and guarantee the availability and integrity of vital infrastructure, strong security measures and resilience methods are also required. Modern power and energy systems are based on core ideas like system integration, interoperability, and cyber-physical interactions, which also influence how energy infrastructure is developed, used, and evolved in the digital age.

9.3.1.2 Theoretical Frameworks for Understanding CPS Deployment

The investigation of theoretical frameworks for comprehending the deployment of CPS spans several disciplines, each providing distinct perspectives on the dynamics, control, and interconnections of these intricate systems. System dynamics, network theory, and control theory are three important theoretical frameworks that are utilized in the analysis and optimization of CPS implementation. The foundation for comprehending how to control and modify the behavior of dynamic systems, such as CPS, is provided by control theory. The fundamental goal of control theory is to create controllers that can adjust system inputs to produce desired results while taking uncertainties and disruptions into consideration. Control theory makes it possible to create control algorithms and methods to govern the behavior of interconnected physical and cyber components in the context of CPS deployment.

Control theory, for instance, is used to create algorithms in smart grid applications that manage power flow, voltage, and frequency inside the system to maintain stability and dependability even in the face of fluctuating renewable energy sources and dynamic demand patterns. Furthermore, control theory makes it easier to create feedback control mechanisms that let CPS adjust in real time to shifting operational constraints, system dynamics, and environmental changes.

System dynamics provides an alternative viewpoint by emphasizing the analysis of feedback loops, causal connections, and dynamic interactions in order to comprehend the behavior of complex systems across time. System behavior may be analyzed under a variety of settings and scenarios, thanks to system dynamics models, which represent the interdependencies between the many parts and subsystems of CPS. System dynamics models facilitate decision-making and optimization efforts by offering insights into system performance, robustness, and stability by modeling the dynamic behavior of CPS.

System dynamics models, for example, can simulate how market dynamics, policy changes, and technological advancements affect energy production, consumption, and distribution in energy systems. This allows stakeholders and policymakers to assess various approaches and interventions for reaching resilience and sustainability objectives.

Another perspective through which to view CPS deployment is network theory, which focuses on the composition, interconnectivity, and dynamics of linked networks. Network theory in CPS offers a framework for dissecting the information flow, communication protocols, and topology of CPS. Network theory allows the investigation of network features like resilience, efficiency, and connectedness by representing CPS as networks of interconnected nodes and edges. This modeling sheds light on the propagation of information and how it affects system behavior.

Network theory, for instance, is used in smart grid applications to assess communication networks' resilience, spot security flaws, and enhance network architecture to guarantee dependable and safe data transfer. Additionally, network theory makes it easier to investigate complex behaviors and emergent phenomena that result from interactions between the physical and cyber components of CPS, offering insights into vulnerabilities and system-wide dynamics.

9.3.2 Cybersecurity and Resilience in CPS

9.3.2.1 Theories and Methodologies for Cybersecurity in CPS-Enabled Energy Systems

Protecting critical infrastructure from cyberattacks and vulnerabilities requires theories and methods for maintaining cybersecurity in CPS-enabled energy systems. These techniques cover a variety of methods for identifying, stopping, and lessening cyberattacks, such as anomaly and intrusion detection as well as the creation of secure communication protocols.

IDS are essential cybersecurity tools for CPS-enabled energy systems. They are made to keep an eye on system activity and network traffic in order to identify and stop malicious activity or unauthorized access. Conventional IDS use signature-based detection methods to find known attack patterns, such particular malware or network signatures. But as cyberattacks have grown more sophisticated, anomaly-based IDS have proliferated. These systems use statistical analysis and ML to identify departures from typical system behavior. IDS are essential for spotting suspicious activity in CPS-enabled energy systems, such illegal access attempts or odd data transmission patterns, so that possible risks may be quickly addressed and mitigated [70–72]. By spotting odd or unexpected behavior in CPS-enabled energy systems, which might point to a security breech or malicious activity, anomaly detection techniques support intrusion detection. These methods find anomalies in typical patterns of activity by examining sensor data, system logs, and other information sources. Since ML techniques can learn from past data and adjust to changing system circumstances, they are frequently employed for anomaly detection. Examples of these algorithms include neural networks, SVMs, and clustering algorithms. Anomaly detection, which enables proactive reaction and mitigation, aids in the discovery of new risks in CPS-enabled energy systems, such as insider threats or zero-day assaults that may elude conventional security mechanisms.

Ensuring the secrecy, integrity, and validity of data transmitted within energy systems allowed by CPS requires the use of secure communication protocols. These protocols authenticate users and devices, encrypt data being communicated between devices, and guard against spoofing, tampering, and eavesdropping threats. Transport Layer Security (TLS), Secure Shell (SSH), and Virtual Private Networks (VPNs), which offer encryption and authentication procedures to protect data transfer over networks, are examples of frequently used secure communication protocols. Secure communication protocols safeguard critical data, including operational data, control commands, and system settings, in CPS-enabled energy systems from being intercepted or altered by unauthorized parties. This helps to maintain the integrity and dependability of grid operations [73–75].

These theories and methods cover a variety of techniques, including as secure communication protocols, anomaly detection, and intrusion detection, to guarantee cybersecurity in CPS-enabled energy systems. Energy stakeholders may improve the security, dependability, and resilience of vital infrastructure by putting these strategies into practice. This will protect vital infrastructure from cyberattacks and guarantee the continuous operation of energy systems. Furthermore, continued research and development is necessary to keep up with the ever-evolving cyber threats and vulnerabilities. This allows cybersecurity solutions to be continuously improved and adjusted to meet new problems in CPS-enabled energy systems.

9.3.2.2 Resilience Concepts and Strategies for CPS

CPS must be resilient in order to survive and bounce back from cyberattacks, system failures, and other interruptions while preserving critical features and services. These tactics cover a range of actions meant to improve system resilience and maintain system continuity, including as redundancy, fault tolerance, and fast recovery techniques.

In order to provide redundancy of vital services and lessen the effects of failures or interruptions, redundancy in a CPS entails the duplication of crucial parts, subsystems, or operations. There are several ways to incorporate redundancy: data redundancy, software redundancy, and hardware redundancy. Deploying redundant systems or components that can take over operations in the case of a breakdown to ensure ongoing service delivery is known as hardware redundancy. Comparably, software redundancy refers to the replication of software elements or procedures to offer fallback capabilities in the event of cyberattacks or software malfunctions.

In order to minimize data loss and speed up data recovery in the case of a system failure or data corruption, data redundancy is the practice of keeping numerous copies of important data across dispersed places. Organizations may improve system availability, resilience, and dependability by adding redundancy to CPS architecture. This reduces downtime and service interruptions brought on by cyberattacks or malfunctions.

The capacity of a CPS to keep running and perform necessary tasks in the event that mistakes, malfunctions, or breakdowns occur inside the system is referred to as fault tolerance. In order to preserve the integrity and functionality of a system, fault tolerance methods identify, isolate, and recover from faults or mistakes. These mechanisms include methods for detecting problems in data transfer or processing, such as parity checks and checksums, which allow for quick error repair and recovery. Furthermore, error recovery techniques like rollback mechanisms and checkpointing are included in fault tolerance methods. These techniques allow the system to recover from faults or return to a known state without jeopardizing its integrity or dependability.

By putting fault tolerance techniques in place, CPS can lessen the effects of system failures, cyberattacks, and other interruptions, guaranteeing service delivery and continuing operation under difficult situations.

By limiting downtime and service interruptions, CPS's rapid recovery techniques allow it to bounce back from cyberattacks, system failures, and other disturbances swiftly and effectively. In the case of a cyberattack or system failure, these methods, which enable enterprises to restore operations and services, include backup and restoration processes, disaster recovery plans, and incident response protocols. In order to facilitate quick data recovery and system restoration in the case of data loss or corruption, backup and restoration methods entail routinely backing up important data, settings, and software components to secondary storage sites.

In order to provide a coordinated and efficient reaction to disruptive occurrences, disaster recovery plans include specified methods and practices for reacting to and recovering from cyberattacks, natural disasters, or other emergencies. By clearly defining roles, responsibilities, and escalation processes, incident response protocols help businesses respond to security events and cyberattacks in a way that successfully limits, mitigates, and recovers from system compromises and security breaches. Rapid recovery techniques can help CPS lessen the effects of disruptions, get back to business fast, and continue providing customers and stakeholders with critical services.

Enhancing the resilience of CPS against cyberattacks, system failures, and other disturbances requires the use of resilience ideas and tactics including redundancy, fault tolerance, and quick recovery mechanisms. Organizations can reduce downtime, lessen the effects of disruptions, and guarantee continuity of operations by implementing these strategies into the design and operation of CPS. This will allow CPS to withstand unfavorable events and recover while continuing to provide users and stakeholders with critical functions and services.

9.4 CHALLENGES AND BARRIERS ASSOCIATED WITH CPS INTEGRATION IN POWER AND ENERGY SYSTEMS, INCLUDING SECURITY, SCALABILITY, AND REGULATORY ISSUES

9.4.1 Threat Landscape Facing CPS-enabled Energy Infrastructure

CPS-enabled energy infrastructure faces a complex and dynamic threat landscape that is typified by a multitude of cyberattacks aimed against vital components including smart meters, SCADA (Supervisory Control and Data Acquisition) systems, and grid control devices. These dangers might have a serious negative impact on the economy and public safety in addition to posing serious hazards to the energy systems' resilience, security, and dependability. The foundation of contemporary energy metering and billing systems, smart meters are widely deployed and networked, making them easy targets for cyberattacks [76, 77].

Cyberattackers may alter meter readings, steal energy, or interfere with billing systems by taking advantage of flaws in smart meter firmware or communication protocols. This might result in revenue losses for utilities and possible interruptions to customer service. Similarly, because of their crucial role in grid operations, SCADA systems that monitor and manage vital infrastructure, including power plants, substations, and distribution networks, are very desirable targets for cyberattacks.

In order to get illegal access, change control settings, or interfere with grid operations, adversaries may take advantage of flaws in SCADA software or communication protocols. This might result in extensive power outages or actual physical harm to infrastructure. Furthermore, cyberattacks might jeopardize the integrity, availability, or confidentiality of grid control equipment, such as Intelligent Electronic equipment (IEDs), Remote Terminal Units (RTUs), and Programmable Logic Controllers (PLCs) [78–80].

Adversaries might jeopardize the stability and dependability of the grid by using flaws in communication protocols or control device firmware to carry out instructions, change settings, or retrieve private data. Furthermore, new attack vectors and propagation paths are introduced by the interconnectedness of CPS-enabled energy infrastructure, making it possible for cyberattacks to propagate quickly throughout the grid and intensify their impact. A cyberattack that targets a single SCADA system or smart meter, for instance, might cascade through the grid, creating extensive disruptions and making it more difficult for operators to bring things back to normal.

The growing amalgamation of information technology (IT) and operational technology (OT) systems in the energy infrastructure gives rise to supplementary susceptibilities and attack domains. This is because malevolent actors use shared vulnerabilities in both areas to get unapproved entry or carry out malevolent actions. The multiplicity of nation-state actors, organized cybercrime groups, and advanced persistent threats (APTs) that have the means, skills, and intent to carry out sophisticated cyberattacks against energy infrastructure for financial gain, geopolitical goals, or ideological reasons exacerbates these threats even more.

Energy stakeholders need to take a proactive and comprehensive strategy to cybersecurity in order to counter these ever-evolving threats. This approach should include rigorous risk management, threat intelligence, incident response, and resilience planning measures. To safeguard vital assets and infrastructure from cyberattacks, this entails putting security-by-design principles into practice, carrying out frequent security audits and assessments, and putting advanced security measures like IDS, firewalls, and encryption mechanisms into place. In order to detect new risks, improve situational awareness, and coordinate response activities in the case of a cyberattack or security incident, utilities, government agencies, industry partners, and cybersecurity specialists must collaborate and share information. Stakeholders may reduce risks, bolster defenses, and guarantee the ongoing dependability and security of energy systems in the face of changing cyberthreats and difficulties by giving cybersecurity and resilience top priority in CPS-enabled energy infrastructure.

9.4.2 Challenges Related to Safeguarding Sensitive Data Collected by CPS Devices

CPS devices gather sensitive data, and securing it from unwanted access and exploitation presents a number of challenges. Examples of these challenges include protecting customer energy usage patterns, personal information, and grid operating data. Large volumes of data are created, communicated, and stored as a result of the expansion of IoT devices, smart meters, sensors, and other connected devices in the energy infrastructure. This raises serious privacy and security issues for users, utilities, and grid operators [81, 82]. For instance, consumer energy use patterns offer insightful information about personal habits, preferences, and behavior. These insights can be used by bad actors for a variety of illicit activities, such as identity theft, targeted advertising, and even physical security risks. Customer names, addresses, and payment information are among the personal data collected by CPS devices. This data is a prime target for cyberattacks and data breaches because unauthorized access to it can result in financial losses, harm to a company's reputation, and legal ramifications for utilities and service providers. Similarly, to guarantee the integrity, availability, and dependability of the energy infrastructure, grid operational data such as system settings, network topology, and equipment status must be shielded from illegal access or alteration. Data security and privacy are further complicated and challenged by the interconnectedness of CPS devices and the integration of IT and OT systems inside energy infrastructure. Grid stability and dependability are under danger as a result of the convergence of IT and OT systems, which opens up new attack surfaces and vulnerabilities that cyber adversaries may use to get unauthorized access, alter data, or interfere with operations. Furthermore, utilities and service providers are subject to strict obligations under regulatory compliance requirements, such as the General Data Protection Regulation (GDPR) in Europe and the California Consumer Privacy Act (CCPA) in the United States, to protect consumer data privacy rights and guarantee the lawful collection, use, and disclosure of personal information. In order to comply with these requirements, utilities must have strong data protection measures in place to preserve sensitive data and reduce the risk of data breaches or fines for non-compliance. These methods include encryption, access limits, data anonymization, and privacy-enhancing technology [83, 84].

Data security and privacy face constant problems because of the quick speed of technological advancement, the growing complexity of CPS devices, and the emergence of new threats and vulnerabilities that render outdated protections. Utilities and other energy stakeholders need to take a proactive, multi-layered approach to data security and privacy in order to meet these issues. This strategy should include risk assessment, threat modeling, security controls, and incident response capabilities. This entails putting security-by-design into practice, carrying out frequent security audits and assessments, and encouraging an organization-wide culture of cybersecurity knowledge and accountability.

To improve situational awareness, identify new threats, and coordinate reaction actions in the case of a data breach or security incident, utilities, industry partners, governmental organizations, and cybersecurity specialists must collaborate and share information. In an increasingly digital and linked world, utilities can preserve sensitive information, foster customer confidence, and guarantee the continuous dependability and resilience of energy systems by placing a high priority on data security and privacy in CPS-enabled energy infrastructure.

9.4.3 Interoperability and Standardization Issues

In CPS, interoperability and standardization are crucial problems that affect how well different devices, systems, and technologies work together across different domains. Interoperability and standardization are essential for CPS, which integrate physical components with computational and communication capabilities, to guarantee compatibility, efficiency, and dependability in their functioning. The dynamic nature of CPS settings and applications, along with the multiplicity of devices, protocols, and interfaces, make it difficult to achieve interoperability and standardization in CPS. Interoperability, which allows interoperable CPS to efficiently communicate, collaborate, and coordinate their actions, is the capacity of various systems, devices, or components to share information and work together effortlessly. In contrast, standardization entails creating and implementing shared protocols, interfaces, and standards to guarantee dependability, consistency, and compatibility among various CPS implementations. Interoperability and standardization problems can take many different forms in the context of CPS, each with its own set of difficulties and complexity. These forms include communication protocols, data formats, interface requirements, and system architectures.

A major obstacle to attaining standardization and interoperability in CPS is the variety of communication protocols and technologies used by various systems and devices. Sensors, actuators, controllers, and networking devices are just a few of the many devices that fall under the umbrella of CPS. These devices use many protocols and technologies for communication, including Bluetooth, Wi-Fi, Zigbee, and MQTT. This variety makes interoperability efforts more difficult since it might lead to incompatible protocols or communication obstacles that prevent devices from easily exchanging data or interacting with one another. Interoperability is further hampered by the absence of standardized communication protocols and interfaces as devices made by various manufacturers may use proprietary protocols or interfaces that are incompatible with those of other systems or devices. To tackle these obstacles, standardized communication protocols and interfaces that facilitate smooth communication and interoperability across various CPS components must be developed and implemented.

The interoperability and data interchange can be severely hampered by the variety of data formats and semantics employed by various CPS devices and systems. Large volumes of data are produced by CPS from a variety of sensors, actuators, and control systems, each of which represents and interprets data in a unique way using a different set of units, formats, and semantics. This variability impedes interoperability and data interchange among CPS components by making it challenging to integrate and process data from various sources. The absence of uniform data models and ontologies intensifies these difficulties by creating an indistinguishable foundation for data representation and interpretation among various CPS implementations. In order to overcome these obstacles, work is being done to create ontologies, models, and data formats that are standardized and allow for easy data interchange and interoperability between CPS components. These standards make it easier to combine and interpret data from many sources, giving CPS the ability to get useful insights and decide on the basis of heterogeneous data.

Because various CPS deployments may use different architectures, platforms, and integration strategies, interoperability and standardization challenges in CPS extend to system architectures and integration frameworks. Sensing, communication, computing, and control are some of the layers that are included in CPS designs. Each layer has its own set of standards, interfaces, and protocols. Effective integration and orchestration of CPS components is difficult because of the absence of standardized designs and integration frameworks, which impedes communication and cooperation between various systems. Interoperability attempts are made more difficult by the dynamic nature of CPS settings and applications, which must instantly adjust to shifting conditions, restrictions, and needs. Standardized CPS architectures, integration frameworks, and reference models that offer common standards and best practices for planning, implementing, and overseeing interoperable CPS installations are being developed in an effort to overcome these issues. These standards facilitate smooth integration and operation across several domains by enabling cooperation and interoperability across various CPS systems and components.

The absence of industry consensus on best practices and regulatory frameworks exacerbates interoperability and standardization challenges in CPS. CPS installations operate in a relatively new and fast developing world, where standards and regulations lag behind technology improvements, in contrast to traditional IT systems, which are subject to established standards and regulatory constraints. Interoperability and standardization initiatives are hampered by the absence of industry consensus and regulatory direction because stakeholders may not have clear rules or incentives to embrace similar standards and practices.

The international scope of CPS installations complicates interoperability efforts since many jurisdictions may have unique laws, certification procedures, and standards, further fragmenting the environment and impeding interoperability. In order to establish and promote shared standards, laws, and best practices for CPS interoperability and standardization, stakeholders from academics, industry groups, governments, and standards bodies must work together to address these difficulties. Despite these obstacles, industry cooperation, research projects, and technology developments have resulted in a great deal of progress being made in resolving interoperability and standards concerns in CPS. The development and promotion of standards for CPS interoperability, which address topics like communication protocols, data formats, system architectures, and security, is greatly aided by organizations that set standards, including the Internet Engineering Task Force (IETF), the International Organization for Standardization (ISO), and the Institute of Electrical and Electronics Engineers (IEEE) [85, 86].

Collaboration and information exchange among stakeholders are facilitated by industry consortia, alliances, and forums including the Industrial Internet Consortium (IIC), the Open Connectivity Foundation (OCF), and the Thread Group. This promotes interoperability and innovation in CPS implementations. Furthermore, the advancement of cutting-edge technologies and approaches for attaining interoperability and standardization in CPS such as software-defined networking (SDN), edge computing, and semantic interoperability is facilitated by research projects and academic partnerships. The CPS community may solve interoperability and standardization problems and enable the seamless integration and operation of CPS across many domains and applications by utilizing these developments and promoting collaboration among stakeholders [87–89].

9.4.4 Insider Threats and Human Error

Insider attacks and human mistake continue to be major problems in CPS-enabled energy infrastructure, even with strong security measures in place. These threats can jeopardize system integrity, data security, and operational dependability. Insider threats pertain to the possibility of malevolent or careless acts by those who possess authorized access to CPS components, such as contractors, employees, or third-party service providers, jeopardizing the security and integrity of the system.

These risks might result in data breaches, service interruptions, or physical harm to vital infrastructure components. They could also take the form of sabotage, illegal access, data exfiltration, or manipulation of system settings. Conversely, human error describes inadvertent errors or misconfigurations committed by authorized users as a result of things like ignorance, insufficient training, or disregard for security procedures. Even if they are unintended, human mistakes can have serious repercussions for the energy infrastructure that supports CPS, such as device malfunctions, system outages, or unintentional disclosure of private data.

Within CPS installations, it is crucial to fortify user identification, access restrictions, and training programs to reduce the risks presented by insider threats and human error. To ensure that only authorized individuals have access to critical systems and data, this involves putting multi-factor authentication systems, role-based access restrictions, and least privilege principles into practice. Additionally, consistent security awareness education and training initiatives may aid in educating staff members and outside contractors on the value of cybersecurity best practices, including maintaining strong passwords, being alert of phishing scams, and following incident reporting guidelines.

In order to mitigate insider threats and human error, businesses should put in place monitoring and auditing procedures to identify and address suspicious activity or departures from standard behavior. CPS-enabled energy infrastructure can improve its resilience, integrity, and security posture by addressing these issues through a combination of technical controls, training programs, and organizational policies. This will protect against insider threats and human error while guaranteeing the dependable and continuous operation of critical energy systems.

9.5 ROLE OF EMERGING TECHNOLOGIES SUCH AS DIGITAL TWINS IN ADDRESSING CPS-RELATED CHALLENGES AND OPTIMIZING SYSTEM PERFORMANCE

9.5.1 Enhanced Monitoring and Simulation

9.5.1.1 Real-Time Monitoring and Simulation in CPS-Enabled Energy Infrastructure through Digital Twins

Within CPS-enabled energy infrastructure, the idea of digital twins offers a novel way to model and simulate real assets, processes, or systems. Digital twins are digital replicas of real-world objects, from single parts to whole systems that are identical to their analog counterparts in physical form. In order to mimic and assess the behavior, performance, and condition of physical assets or processes, these virtual replicas make use of real-time data, sensor measurements, and computational models. This allows for predictive maintenance, optimization, and decision-making in CPS-enabled energy infrastructure. Fundamentally, the idea of "digital twins" represents the coming together of the digital and physical domains, opening doors to new knowledge, possibilities, and energy system efficiency by bridging the gap between the two [90, 91]. Digital twins are essential to the real-time monitoring, analysis, and simulation of vital processes and assets in CPS-enabled energy infrastructure, including distribution networks, power plants, and transmission networks. Digital twins give stakeholders a thorough and integrated perspective of their energy infrastructure by generating virtual copies of actual assets and systems. This enables them to keep an eye on performance, spot irregularities, and anticipate problems before they happen. Digital twins, for instance, may mimic and model in real-time the behavior of turbines, generators, and other equipment in the context of power production, allowing operators to improve efficiency, minimize downtime, and optimize operations. Comparably, digital twins in the transmission and distribution networks can mimic the flow of power, forecast grid congestion, and enhance grid operations to guarantee customers receive dependable and resilient energy supply.

Utilizing real-time data and sensor readings to deliver actionable insights and streamline data-driven decision-making in CPS-enabled energy infrastructure is one of the main benefits of digital twins. Digital twins allow operators and engineers to see trends, patterns, and anomalies in real-time, allowing for proactive intervention and energy infrastructure optimization. Data from physical assets and systems is continually collected and analyzed. Digital twins, for instance, let operators to plan maintenance tasks, assign personnel, and proactively reduce risks by identifying maintenance requirements, detecting equipment problems, and forecasting future performance based on past data and predictive analytics.

Digital twins also make what-if analysis and scenario planning easier. This enables stakeholders to model various operational scenarios, evaluate the effects of proposed modifications or interventions, and optimize energy infrastructure appropriately.

For cooperative decision-making and stakeholder participation in CPS-enabled energy infrastructure, digital twins are effective instruments. Digital twins provide successful collaboration, information sharing, and real-time action coordination among stakeholders from many departments, disciplines, and organizations by offering a single platform and shared knowledge of energy systems. In order to improve grid operations, control demand, and guarantee grid stability, for instance, digital twins can help grid operators, utilities, regulators, and energy customers collaborate by allowing them to exchange data, insights, and projections. Digital twins offer insight into energy infrastructure, allowing stakeholders to follow progress, monitor performance, and evaluate compliance with industry standards or regulatory requirements. This can improve transparency, accountability, and confidence among stakeholders.

9.5.1.2 Predictive Maintenance in Power Systems Using Digital Twins and Advanced Analytics

Leveraging ML, digital twins based data analytics, and physics-based models is a potent way to forecast equipment failures, diagnose performance problems, and optimize maintenance schedules in the power production, transmission, and distribution networks. Digital twins record data in real time and mimic the behavior of physical assets, processes, or systems in a digital environment. They function as virtual copies of these entities. Through the incorporation of ML algorithms, physics-based models, and data analytics into digital twins, energy stakeholders may leverage predictive capabilities and actionable insights to improve resilience, efficiency, and dependability of power infrastructure [92, 93]. Large volumes of data from sensors, meters, and control systems placed across power generating, transmission, and distribution networks may be gathered and analyzed with the use of digital twins. This information gives a thorough picture of the whole energy system and includes operating characteristics, performance measures, environmental factors, and equipment health indicators. Energy stakeholders may get significant insights from this data by utilizing data analytics techniques like statistical analysis, time-series forecasting, and pattern recognition to find trends, anomalies, and patterns that may indicate equipment breakdowns or performance deterioration.

By automatically recognizing hidden patterns in complicated information, forecasting future events, and detecting connections, ML algorithms significantly improve the capabilities of digital twins. In order to help operators prevent downtime and optimize maintenance schedules, ML algorithms, for instance, can analyze historical data to identify early warning signs of equipment failures, such as changes in vibration patterns, temperature fluctuations, or energy consumption trends. Furthermore, in order to replicate the behavior of real assets and systems and get a better knowledge of their operational dynamics and performance traits, physics-based models are incorporated into digital twins. These models depict the fundamental physical processes that control the behavior of power production, transmission, and distribution networks by utilizing concepts from mathematics, physics, and engineering. Digital twins may reliably forecast equipment breakdowns, diagnose performance issues, and optimize maintenance schedules based on the underlying physics of the system by merging physics-based models with real-time data and ML techniques.

Physics-based models, for instance, can forecast the remaining usable life of vital assets, simulate the deterioration of equipment components over time, and suggest the best maintenance procedures to reduce hazards and extend asset lifespan. Furthermore, scenario analysis and what-if simulations are made possible by physics-based models, which let stakeholders evaluate how various operating scenarios, maintenance plans, and investment choices might affect the dependability and performance of a system.

The utilization of digital twins can aid condition-based maintenance techniques by offering ongoing monitoring and instantaneous evaluation of the health and functionality of equipment. Digital twins are able to identify signals of future malfunctions or performance decline by evaluating sensor data and comparing it to baseline levels. Utilizing these insights, condition-based maintenance techniques set priorities for maintenance tasks, plan interventions in advance, and allocate resources optimally to increase uptime and decrease downtime. When equipment metrics surpass certain thresholds, for instance, digital twins might initiate maintenance alerts that notify operators to do preventative maintenance or inspections to resolve any concerns before they worsen and become expensive failures or outages.

Utilizing ML, physics-based models, and data analytics based on digital twins is a potent way to identify performance problems, anticipate equipment breakdowns, and optimize maintenance schedules in power production, transmission, and distribution systems. Energy stakeholders may get actionable insights and predictive capabilities to improve power infrastructure resilience, efficiency, and reliability by combining real-time data, sophisticated analytics, and simulation capabilities into digital twins. Digital twins will be crucial for allowing proactive maintenance plans, maximizing asset performance, and guaranteeing the consistent and sustainable supply of energy to customers as they develop and become more sophisticated.

Energy stakeholders have the opportunity to revolutionize their operations, stimulate innovation, and mold the energy sector's future into a more intelligent and resilient energy ecosystem by utilizing digital twins.

9.5.2 Optimization and Decision Support

9.5.2.1 Real-Time Optimization of CPS-Enabled Energy Systems through Digital Twins

By offering insights into system behavior, spotting optimization possibilities, and assisting in real-time or almost real-time decision-making, digital twins play a revolutionary role in enabling the dynamic optimization of CPS-enabled energy systems. Digital twins are essentially digital copies of real-world systems, processes, or physical assets that are identical to their digital counterparts. Digital twins play a revolutionary role in allowing the dynamic optimization of CPS-enabled energy systems because they provide real-time or near-real-time decision-making support, insights into system behavior, and optimization options. Digital twins are essentially digital replicas of real-world physical assets, processes, or systems that are represented digitally [94]. These virtual representations allow stakeholders to monitor, control, and optimize system operations in a dynamic and responsive way by simulating and analyzing the behavior, performance, and condition of CPS-enabled energy systems using real-time data, sensor measurements, and computational models. Energy stakeholders may take use of actionable insights and predictive capabilities to improve resilience, efficiency, and dependability in energy infrastructure by incorporating data analytics, ML algorithms, and optimization approaches into digital twins. The capacity of digital twins to present a thorough and integrated picture of CPS-enabled energy systems allows stakeholders to track system behavior, spot optimization possibilities, and make deft choices instantly. This is one of the main benefits of using digital twins. Digital twins are able to identify trends, patterns, and anomalies that point to problems with system performance or operation by gathering and evaluating real-time data from sensors, meters, and control systems placed across the energy infrastructure.

Digital twins, for instance, can track trends in energy use, rates of equipment use, and environmental factors to spot inefficiencies, bottlenecks, or departures from ideal operating circumstances. Energy stakeholders may get important insights from this data to enhance system operations, increase energy efficiency, and save operating costs by incorporating data analytics methods like statistical analysis, time-series forecasting, and pattern recognition into digital twins. By utilizing past data and ML algorithms to predict future patterns, foresee possible problems, and suggest preemptive actions in CPS-enabled energy systems, digital twins allow predictive capabilities. Through the use of ML algorithms, stakeholders may foresee equipment failures, performance deterioration, or system disruptions before they happen by analyzing previous data to find connections, forecast future events, and uncover hidden patterns in large, complicated datasets. To estimate the remaining usable life of important assets, suggest the best maintenance plans, and reduce downtime, ML algorithms, for instance, can examine sensor data, operating parameters, and previous maintenance records. Digital twins also provide scenario analysis and what-if simulations, which enable interested parties to investigate various operational scenarios, try out other tactics, and evaluate the effects of prospective modifications on system dependability and performance. Through the integration of real-time data, predictive analytics, and optimization approaches into decision-making processes, digital twins facilitate closed-loop control and optimization of CPS-enabled energy systems.

Digital twins may dynamically modify control settings, improve resource allocation, and instantly adapt system operations to changing needs or restrictions by continually monitoring system behavior, performance indicators, and environmental circumstances. For instance, using real-time data and predictive analytics, digital twins can modify demand response plans, distribution routes, or power output levels to maximize energy delivery, reduce losses, and preserve grid stability. Furthermore, by spotting early warning indicators of equipment breakdown, suggesting prompt treatments, and streamlining maintenance plans to enhance asset uptime and dependability, digital twins can support predictive maintenance programs.

9.5.2.2 Assessing Resilience, Vulnerabilities, and Mitigating Risks

The overall dependability and security of energy infrastructure may be improved by using digital twins to evaluate system resilience, find vulnerabilities, and reduce risks related to cyberattacks, natural disasters, and other disruptive events.

A novel method for evaluating system resilience, spotting weaknesses, and reducing the risks connected to cyberattacks, natural catastrophes, and other disruptive events in energy infrastructure is the use of digital twins. By acting as virtual copies of actual resources, operations, or systems, digital twins allow interested parties to model and assess the functionality, state, and behavior of energy infrastructure in a virtual setting. Digital twins offer significant insights into the resilience of energy systems by combining real-time data, predictive analytics, and simulation capabilities. This helps stakeholders recognize possible threats, evaluate their impact, and take proactive steps to reduce risks and improve overall security and reliability.

The capacity of digital twins to model and evaluate how energy infrastructure reacts to different threats and interruptions is one of its main benefits when evaluating system resilience. Through the simulation of various scenarios, like cyberattacks, natural catastrophes, or equipment malfunctions, digital twins may assess the resilience of energy systems and detect any flaws or vulnerabilities that could jeopardize the integrity or operation of the system. Digital twins, for instance, can be used to model cyberattacks on transmission networks, control systems, or other vital infrastructure elements in order to determine how they might affect system dependability, safety, and operations.

Stakeholders can take proactive steps to improve system resilience and reduce risks by identifying possible sources of failure, crucial dependencies, or single points of failure within the energy infrastructure by assessing the outcomes of these simulations.

Stakeholders may dynamically monitor and adjust system resilience in response to changing conditions and emerging threats, thanks to digital twins. Digital twins are capable of identifying abnormalities, deviations, or odd patterns that might be signs of possible dangers or disruptions by continually monitoring real-time data from sensors, meters, and control systems installed across the energy infrastructure.

Digital twins have the capability to examine sensor data and identify anomalous operational circumstances, unexpected network activity, or unauthorized access attempts that might indicate a security compromise or cyberattack. Stakeholders can automatically recognize, categorize, and prioritize security events by incorporating ML algorithms and anomaly detection techniques into digital twins. This allows them to respond swiftly and efficiently to manage risks and limit the impact on system operations.

In order to improve system resilience and reduce the risks associated with cyberattacks, natural disasters, and other disruptive events, digital twins can also make it easier to adopt preemptive actions. Through the use of digital twins, stakeholders may assess how well various mitigation strategies—like redundancy, fault tolerance, or quick recovery mechanisms—achieve system resilience and risk reduction. To evaluate the effects of backup systems, emergency response plans, or disaster recovery methods on system performance, dependability, and safety, digital twins can be used as a simulation. In order to improve system resilience and lessen the impact of prospective threats or disruptions, stakeholders may select the most effective mitigation techniques and allocate resources appropriately by assessing the outcomes of these simulations.

9.6 CONCLUSION

The application of digital twins to evaluate system robustness, detect susceptibilities, and reduce risks in energy infrastructure is a significant development in the domain. Digital twins give stakeholders important insights into the behavior, performance, and condition of physical assets, processes, and systems through a comprehensive approach to modeling and simulating these elements. However, a number of significant gaps and difficulties need to be resolved in order to fully realize the promise of digital twins in improving the dependability and security of energy infrastructure. Interdisciplinary research collaboration is essential to fostering innovation in digital twin technology and bridging the gap across several fields. Standardization and interoperability are essential to provide compatibility, consistency, and dependability across diverse digital twin implementations. Resolving cybersecurity and privacy problems is necessary for securing sensitive data and preventing cyberattacks in energy systems enabled by digital twins. Lastly, there is a great deal of opportunity to enhance the capabilities and effectiveness of digital twins in energy infrastructure optimization through the integration of cutting-edge technologies such as AI, ML, and edge computing. By addressing these gaps and challenges, stakeholders may fully leverage digital twins to drive innovation, efficiency, and resilience in the energy sector and ultimately design a more secure and smarter energy future.

REFERENCES

1. Gunes, V., Peter, S., Givargis, T., & Vahid, F. (2014). A survey on concepts, applications, and challenges in cyber-physical systems, KSII Transactions on Internet and Information Systems (TIIS), 8(12), 4242–4268.
2. Oks, S. J., Jalowski, M., Lechner, M., Mirschberger, S., Merklein, M., Vogel-Heuser, B., & Möslein, K. M. (2022). Cyber-physical systems in the context of Industry 4.0: A review, categorization and outlook, Information Systems Frontiers, 26, 1–42.
3. Shi, X., Li, Y., Cao, Y., & Tan, Y. (2015). Cyber-physical electrical energy systems: Challenges and issues, CSEE Journal of Power and Energy Systems, 1(2), 36–42.
4. Delgado, J. M. D., & Oyedele, L. (2021). Digital twins for the built environment: Learning from conceptual and process models in manufacturing, Advanced Engineering Informatics, 49, 101332.

5. Segovia, M., & Garcia-Alfaro, J. (2022). Design, modeling and implementation of digital twins, Sensors, 22(14), 5396.
6. In power and energy systems, digital twins are useful instruments for monitoring, regulating, optimizing, and carrying out predictive maintenance.
7. Chaterji, S., Naghizadeh, P., Alam, M. A., Bagchi, S., Chiang, M., Corman, D., & Weller, J. (2019). Resilient cyberphysical systems and their application drivers: A technology roadmap. arXiv preprint arXiv:2001.00090.
8. Hasan, M. K., Habib, A. A., Shukur, Z., Ibrahim, F., Islam, S., & Razzaque, M. A. (2023). Review on cyber-physical and cyber-security system in smart grid: Standards, protocols, constraints, and recommendations, Journal of Network and Computer Applications, 209, 103540.
9. Qian, C., Liu, X., Ripley, C., Qian, M., Liang, F., & Yu, W. (2022). Digital twin—Cyber replica of physical things: Architecture, applications and future research directions, Future Internet, 14(2), 64.
10. Alam, K. M., & El Saddik, A. (2017). C2PS: A digital twin architecture reference model for the cloud-based cyber-physical systems, IEEE Access, 5, 2050–2062.
11. Guo, W., Zhang, Y., & Li, L. (2015). The integration of CPS, CPSS, and ITS: A focus on data, Tsinghua Science and Technology, 20(4), 327–335.
12. Wan, J., Chen, M., Xia, F., Di, L., & Zhou, K. (2013). From machine-to-machine communications towards cyber-physical systems, Computer Science and Information Systems, 10(3), 1105–1128.
13. Zhang, F. (2022). Cyber-Physical Robustness Enhancement Strategies for Demand Side Energy Systems (Doctoral dissertation, UNSW Sydney).
14. Paterakis, N. G., Erdinç, O., & Catalão, J. P. (2017). An overview of demand response: Key-elements and international experience, Renewable and Sustainable Energy Reviews, 69, 871–891.
15. Green, R. C., Wang, L., & Alam, M. (2013). Applications and trends of high performance computing for electric power systems: Focusing on smart grid, IEEE Transactions on Smart Grid, 4(2), 922–931.
16. Bhattacharya, S., Chengoden, R., Srivastava, G., Alazab, M., Javed, A. R., Victor, N., & Gadekallu, T. R. (2022). Incentive mechanisms for smart grid: State of the art, challenges, open issues, future directions, Big Data and Cognitive Computing, 6(2), 47.
17. Zhao, P., Simões, M. G., & Suryanarayanan, S. (2010, November). A conceptual scheme for cyber-physical systems based energy management in building structures. In 2010 9th IEEE/IAS International Conference on Industry Applications-INDUSCON 2010 (pp. 1–6).
18. Xu, G. (2015). Towards a Framework of Enabling Efficient and Secured Energy Based Cyber-Physical System (CPS), Towson University Institutional Repository.
19. Nasir, T., Bukhari, S. S. H., Raza, S., Munir, H. M., Abrar, M., Muqeet, H. A. U., & Masroor, R. (2021). Recent challenges and methodologies in smart grid demand side management: State-of-the-art literature review, Mathematical Problems in Engineering, 2021, 1–16.
20. Baheti, R., & Gill, H. (2011). Cyber-physical systems, The Impact of Control Technology, 12(1), 161–166.
21. Darwish, A., & Hassanien, A. E. (2018). Cyber physical systems design, methodology, and integration: The current status and future outlook, Journal of Ambient Intelligence and Humanized Computing, 9(5), 1541–1556.
22. Greer, C., Burns, M., Wollman, D., & Griffor, E. (2019). Cyber-physical systems and Internet of Things.
23. Kim, K. D., & Kumar, P. R. (2012). Cyber–physical systems: A perspective at the centennial. Proceedings of the IEEE, 100(Special Centennial Issue), 1287–1308.

24. Bordel, B., Alcarria, R., Robles, T., & Martín, D. (2017). Cyber–physical systems: Extending pervasive sensing from control theory to the Internet of Things, Pervasive and Mobile Computing, 40, 156–184.
25. Dafflon, B., Moalla, N., & Ouzrout, Y. (2021). The challenges, approaches, and used techniques of CPS for manufacturing in industry 4.0: A literature review, The International Journal of Advanced Manufacturing Technology, 113, 2395–2412.
26. Jakovljevic, Z., Majstorovic, V., Stojadinovic, S., Zivkovic, S., Gligorijevic, N., & Pajic, M. (2017). Cyber-physical manufacturing systems (CPMS). In Proceedings of 5th International Conference on Advanced Manufacturing Engineering and Technologies: NEWTECH 2017 5 (pp. 199–214). Springer International Publishing.
27. Pivoto, D. G., de Almeida, L. F., da Rosa Righi, R., Rodrigues, J. J., Lugli, A. B., & Alberti, A. M. (2021). Cyber-physical systems architectures for industrial Internet of Things applications in Industry 4.0: A literature review, Journal of Manufacturing Systems, 58, 176–192.
28. Pundir, A., Singh, S., Kumar, M., Bafila, A., & Saxena, G. J. (2022). Cyber-physical systems enabled transport networks in smart cities: Challenges and enabling technologies of the new mobility era, IEEE Access, 10, 16350–16364.
29. He, Y., Jolfaei, A., & Zheng, X. (2021). Cyber-Physical Systems in Transportation. In Intelligent Cyber-Physical Systems for Autonomous Transportation (pp. 129–149). Cham: Springer International Publishing.
30. Shah, T., Yavari, A., Mitra, K., Saguna, S., Jayaraman, P. P., Rabhi, F., & Ranjan, R. (2016). Remote health care cyber-physical system: Quality of service (QoS) challenges and opportunities, IET Cyber-Physical Systems: Theory & Applications, 1(1), 40–48.
31. Venkatasubramanian, K. K., Nabar, S., Gupta, S. K., & Poovendran, R. (2012). Cyber Physical Security Solutions for Pervasive Health Monitoring Systems. In E-Healthcare Systems and Wireless Communications: Current and Future Challenges (pp. 143–162). IGI Global.
32. Dey, N., Ashour, A. S., Shi, F., Fong, S. J., & Tavares, J. M. R. (2018). Medical cyber-physical systems: A survey, Journal of Medical Systems, 42, 1–13.
33. Moreno Escobar, J. J., Morales Matamoros, O., Tejeida Padilla, R., Lina Reyes, I., & Quintana Espinosa, H. (2021). A comprehensive review on smart grids: Challenges and opportunities, Sensors, 21(21), 6978.
34. Ghaemi, A. A. (2017, July). A cyber-physical system approach to smart city development. In 2017 IEEE International Conference on Smart Grid and Smart Cities (ICSGSC) (pp. 257–262). IEEE.
35. Llaria, A., Dos Santos, J., Terrasson, G., Boussaada, Z., Merlo, C., & Curea, O. (2021). Intelligent buildings in smart grids: A survey on security and privacy issues related to energy management, Energies, 14(9), 2733.
36. Xu, L., Guo, Q., Sheng, Y., Muyeen, S. M., & Sun, H. (2021). On the resilience of modern power systems: A comprehensive review from the cyber-physical perspective, Renewable and Sustainable Energy Reviews, 152, 111642.
37. Saxena, N., Chukwuka, V., Xiong, L., & Grijalva, S. (2017, November). CPSA: A cyber-physical security assessment tool for situational awareness in smart grid. In Proceedings of the 2017 Workshop on Cyber-Physical Systems Security and PrivaCy (pp. 69–79).
38. Guo, H., Zheng, C., Iu, H. H. C., & Fernando, T. (2017). A critical review of cascading failure analysis and modeling of power system, Renewable and Sustainable Energy Reviews, 80, 9–22.
39. Nguyen, T., Wang, S., Alhazmi, M., Nazemi, M., Estebsari, A., & Dehghanian, P. (2020). Electric power grid resilience to cyber adversaries: State of the art, IEEE Access, 8, 87592–87608.

40. Tyagi, A. K., & Sreenath, N. (2021). Cyber physical systems: Analyses, challenges and possible solutions, Internet of Things and Cyber-Physical Systems, 1, 22–33.
41. Amin, M., El-Sousy, F. F., Aziz, G. A. A., Gaber, K., & Mohammed, O. A. (2021). CPS attacks mitigation approaches on power electronic systems with security challenges for smart grid applications: A review, IEEE Access, 9, 38571–38601.
42. Voropai, N. I., Kolosok, I. N., Korkina, E. S., & Osak, A. B. (2020). Issues of cybersecurity in electric power systems, Energy Systems Research, 3(2 (10)), 19–28.
43. Rathi, R., Sharma, N., Manchanda, C., Bhushan, B., & Grover, M. (2020, April). Security challenges & controls in cyber physical system. In 2020 IEEE 9th International Conference on Communication Systems and Network Technologies (CSNT) (pp 242–247). IEEE.
44. Humayed, A., Lin, J., Li, F., & Luo, B. (2017). Cyber-physical systems security—A survey, IEEE Internet of Things Journal, 4(6), 1802–1831.
45. Törngren, M., & Sellgren, U. (2018). Complexity challenges in development of cyber-physical systems. Principles of modeling: Essays dedicated to Edward A. Lee on the occasion of his 60th birthday, 478–503.
46. Park, K. J., Zheng, R., & Liu, X. (2012). Cyber-physical systems: Milestones and research challenges, Computer Communications, 36(1), 1–7.
47. Haque, S. A., Aziz, S. M., & Rahman, M. (2014). Review of cyber-physical system in healthcare, International Journal of Distributed Sensor Networks, 10(4), 217415.
48. Tucker, D., Pezzini, P., & Bryden, K. M. (2018, June). Cyber-physical systems: A new paradigm for energy technology development. In ASME Power Conference (Vol. 51395, p. V001T04A001). American Society of Mechanical Engineers.
49. Alvarez-Alvarado, M. S., Apolo-Tinoco, C., Ramirez-Prado, M. J., Alban-Chacón, F. E., Pico, N., Aviles-Cedeno, J., & Rengifo, J. (2024). Cyber-physical power systems: A comprehensive review about technologies drivers, standards, and future perspectives, Computers and Electrical Engineering, 116, 109149.
50. Dowdeswell, B., Sinha, R., & MacDonell, S. G. (2020). Finding faults: A scoping study of fault diagnostics for industrial cyber–physical systems, Journal of Systems and Software, 168, 110638.
51. He, Y., Nie, B., Zhang, J., Kumar, P. M., & Muthu, B. (2022). Fault detection and diagnosis of cyber-physical system using the computer vision and image processing. Wireless Personal Communications, 127, 1–20.
52. Alippi, C., Ntalampiras, S., & Roveri, M. (2016). Model-free fault detection and isolation in large-scale cyber-physical systems, IEEE Transactions on Emerging Topics in Computational Intelligence, 1(1), 61–71.
53. Wan, J., Yan, H., Suo, H., & Li, F. (2011). Advances in cyber-physical systems research, KSII Transactions on Internet and Information Systems (TIIS), 5(11), 1891–1908.
54. El-Kady, A. H., Halim, S., El-Halwagi, M. M., & Khan, F. (2023). Analysis of safety and security challenges and opportunities related to cyber-physical systems. Process Safety and Environmental Protection, 173, 384–413.
55. Devenish, A. (2023). Institutional and contextual drivers of and barriers to incentive-based demand response: A comparative case study in the Pacific Northwest, Utilities Policy, 84, 101645.
56. Sajadi, A., Strezoski, L., Strezoski, V., Prica, M., & Loparo, K. A. (2019). Integration of renewable energy systems and challenges for dynamics, control, and automation of electrical power systems, Wiley Interdisciplinary Reviews: Energy and Environment, 8(1), e321.
57. Hussain, H. M., Narayanan, A., Nardelli, P. H., & Yang, Y. (2020). What is energy internet? Concepts, technologies, and future directions, IEEE Access, 8, 183127–183145.
58. Inderwildi, O., Zhang, C., Wang, X., & Kraft, M. (2020). The impact of intelligent cyber-physical systems on the decarbonization of energy, Energy & Environmental Science, 13(3), 744–771.

59. Khaitan, S. K., & McCalley, J. D. (2014). Design techniques and applications of cyber-physical systems: A survey, IEEE Systems Journal, 9(2), 350–365.
60. Zheng, P., Wang, H., Sang, Z., Zhong, R. Y., Liu, Y., Liu, C., & Xu, X. (2018). Smart manufacturing systems for industry 4.0: Conceptual framework, scenarios, and future perspectives, Frontiers of Mechanical Engineering, 13, 137–150.
61. Mishra, A., Jha, A. V., Appasani, B., Ray, A. K., Gupta, D. K., & Ghazali, A. N. (2023). Emerging technologies and design aspects of next generation cyber physical system with a smart city application perspective, International Journal of System Assurance Engineering and Management, 14(Suppl 3), 699–721.
62. Jóźwiak, L. (2015). Embedded computing technology for highly-demanding cyber-physical systems, IFAC-PapersOnLine, 48(4), 19–30.
63. Chae, J., Lee, S., Jang, J., Hong, S., & Park, K. J. (2023). A survey and perspective on industrial cyber-physical systems (ICPS): From ICPS to AI-augmented ICPS. IEEE Transactions on Industrial Cyber-Physical Systems 1, 247–272.
64. Nighoskar, A., Gautam, S., & Lamba, K. (2024). Role of Artificial Intelligence and Machine Learning in Power Systems with Fault Detection and Diagnosis. In Artificial Intelligence Techniques in Power Systems Operations and Analysis (pp. 143–162). Auerbach Publications.
65. Ahmad, A. S., Hassan, M. Y., Abdullah, M. P., Rahman, H. A., Hussin, F., Abdullah, H., & Saidur, R. (2014). A review on applications of ANN and SVM for building electrical energy consumption forecasting, Renewable and Sustainable Energy Reviews, 33, 102–109.
66. Odyurt, U., Sapra, D., & Pimentel, A. D. (2021). The Choice of AI Matters: Alternative Machine Learning Approaches for CPS Anomalies. In Advances and Trends in Artificial Intelligence. From Theory to Practice: 34th International Conference on Industrial, Engineering and Other Applications of Applied Intelligent Systems, IEA/AIE 2021, Kuala Lumpur, Malaysia, July 26–29, 2021, Proceedings, Part II 34 (pp. 474–484). Springer International Publishing.
67. Wang, H. (2020). Research on real-time reliability evaluation of CPS system based on machine learning, Computer Communications, 157, 336–342.
68. Konstantopoulos, G. C., Alexandridis, A. T., & Papageorgiou, P. C. (2020). Towards the integration of modern power systems into a cyber–physical framework, Energies, 13(9), 2169.
69. Voropai, N. I., Stennikov, V. A., & Senderov, S. M. (2020). Infrastructural cyber-physical energy systems: Transformations, challenges, future appearance, Energy Systems Research, 3(3 (11)), 18–29.
70. Khraisat, A., Gondal, I., Vamplew, P., & Kamruzzaman, J. (2019). Survey of intrusion detection systems: Techniques, datasets and challenges, Cybersecurity, 2(1), 1–22.
71. Biermann, E., Cloete, E., & Venter, L. M. (2001). A comparison of intrusion detection systems, Computers & Security, 20(8), 676–683.
72. Jyothsna, V. V. R. P. V., Prasad, R., & Prasad, K. M. (2011). A review of anomaly based intrusion detection systems, International Journal of Computer Applications, 28(7), 26–35.
73. Parmar, H., & Gosai, A. (2015). Analysis and study of network security at transport layer, International Journal of Computer Applications, 121(13), 21604–4716.
74. Garimella, D., & Kumar, R. (2015). Secure shell—Its significance in networking (SSH), Int. J. of Application or Innovation in Engineering & Management, 4(3/12), 187–196.
75. Khanvilkar, S., & Khokhar, A. (2004). Virtual private networks: An overview with performance evaluation, IEEE Communications Magazine, 42(10), 146–154.
76. Pliatsios, D., Sarigiannidis, P., Lagkas, T., & Sarigiannidis, A. G. (2020). A survey on SCADA systems: Secure protocols, incidents, threats and tactics, IEEE Communications Surveys & Tutorials, 22(3), 1942–1976.

77. Lamnabhi-Lagarrigue, F., Annaswamy, A., Engell, S., Isaksson, A., Khargonekar, P., Murray, R. M., & Van den Hof, P. (2017). Systems & control for the future of humanity, research agenda: Current and future roles, impact and grand challenges, Annual Reviews in Control, 43, 1–64.
78. Zavoda, F. (2008, December). The key role of intelligent electronic devices (IED) in advanced Distribution Automation (ADA). In 2008 China International Conference on Electricity Distribution (pp. 1–7). IEEE.
79. Ab Ghani, M. R., Jusoh, W. N. S. E. W., Hanafiah, M. A. M., Raman, S. H., & Jano, Z. (2013). A review of communication protocols for intelligent remote terminal unit development, TELKOMNIKA (Telecommunication Computing Electronics and Control), 11(4), 819–826.
80. Sehr, M. A., Lohstroh, M., Weber, M., Ugalde, I., Witte, M., Neidig, J., & Lee, E. A. (2020). Programmable logic controllers in the context of industry 4.0, IEEE Transactions on Industrial Informatics, 17(5), 3523–3533.
81. Ashibani, Y., & Mahmoud, Q. H. (2017). Cyber physical systems security: Analysis, challenges and solutions, Computers & Security, 68, 81–97.
82. Han, S., Xie, M., Chen, H. H., & Ling, Y. (2014). Intrusion detection in cyber-physical systems: Techniques and challenges, IEEE Systems Journal, 8(4), 1052–1062.
83. Hoofnagle, C. J., Van Der Sloot, B., & Borgesius, F. Z. (2019). The European Union general data protection regulation: What it is and what it means, Information & Communications Technology Law, 28(1), 65–98.
84. Harding, E. L., Vanto, J. J., Clark, R., Hannah Ji, L., & Ainsworth, S. C. (2019). Understanding the scope and impact of the California Consumer Privacy Act of 2018, Journal of Data Protection & Privacy, 2(3), 234–253.
85. Cooper, D., Santesson, S., Farrell, S., Boeyen, S., Housley, R., & Polk, W. (2008). Internet Engineering Task Force. Fremont, CA, May.
86. Ali, S., Al Balushi, T., Nadir, Z., Hussain, O. K., Ali, S., Al Balushi, T., & Hussain, O. K. (2018). Standards for CPS. Cyber Security for Cyber Physical Systems, 161–174.
87. Serpanos, D., Wolf, M., Serpanos, D., & Wolf, M. (2018). Industrial Internet of Things. Internet-of-Things (IoT) Systems: Architectures, Algorithms, Methodologies, 37–54.
88. Park, S. (2017, March). OCF: A new open IoT consortium. In 2017 31st international conference on advanced information networking and applications workshops (WAINA) (pp. 356–359). IEEE.
89. Benzekki, K., El Fergougui, A., & Elbelrhiti Elalaoui, A. (2016). Software-defined networking (SDN): A survey, Security and Communication Networks, 9(18), 5803–5833.
90. Eckhart, M., & Ekelhart, A. (2019). Digital twins for cyber-physical systems security: State of the art and outlook. Security and Quality in Cyber-Physical Systems Engineering: With Forewords by Robert M. Lee and Tom Gilb, 383–412.
91. Radanliev, P., De Roure, D., Nicolescu, R., Huth, M., & Santos, O. (2022). Digital twins: Artificial intelligence and the IoT cyber-physical systems in Industry 4.0, International Journal of Intelligent Robotics and Applications, 6(1), 171–185.
92. van Dinter, R., Tekinerdogan, B., & Catal, C. (2022). Predictive maintenance using digital twins: A systematic literature review, Information and Software Technology, 151, 107008.
93. Falekas, G., & Karlis, A. (2021). Digital twin in electrical machine control and predictive maintenance: State-of-the-art and future prospects, Energies, 14(18), 5933.
94. Lampropoulos, G., & Siakas, K. (2023). Enhancing and securing cyber-physical systems and industry 4.0 through digital twins: A critical review, Journal of Software: Evolution and Process, 35(7), e2494.

10 Digitally Enhanced Adaptive Security

A Solution for Power Theft in Smart Cities

Prateek Pandey and Deepak Sharma

10.1 INTRODUCTION AND BACKGROUND

The transmission and reception of electricity, ensuring both security and reliability, has remained a persistent challenge for researchers. Recent surveys by various socioeconomic organizations have highlighted electricity theft as a significant obstacle to India's economic growth, amounting to approximately 1.5% of the nation's GDP annually. This issue isn't confined to India alone; it affects countries worldwide. For instance, the United States faces losses of about $6 billion annually due to power theft, while in developing countries like those in Africa, the theft rate can soar as high as 30%, significantly impacting social power supply and consumption (Jokar et al., 2016).

Theft not only hampers financial growth but also erodes moral and ethical values. Major causes include high tariff rates, unaffordability for socially and economically vulnerable users, and production inadequacies. Hence, addressing and curbing electricity theft is crucial for any country's socioeconomic development. Despite advancements like smart grids, theft rates persist, indicating that technological solutions alone are insufficient. There's a need for behavioral changes alongside technological innovations to combat this menace effectively (Ahmad et al., 2014).

Conventional detection methods rely heavily on manual operations, such as meter reading and analysis by operators. While some hardware enhancements like specialized watt-hour metering boxes and anti-theft functions have been implemented, there's a lack of robust algorithms to analyze historical consumption data effectively. Thus, the power industry must focus on developing new blockchain-based technologies to ensure secure transmission and distribution of electricity (Al-Dahidi et al., 2019).

Presently, smart grid detection frameworks are popular, where consumption data is collected through smart meters and analyzed centrally using algorithms to detect theft. Various data mining algorithms such as clustering and neural networks are employed for this purpose. However, the wide-scale deployment of smart meters consumes considerable resources, necessitating sustainable approaches, including renewable energy systems, for the long-term viability of smart cities (Li et al., 2019).

Novel detection methods, like the one-class SVM algorithm, have shown promise. These methods rely on analyzing historical data to establish models and detect inconsistencies indicative of theft. To address this issue comprehensively, a proposed

DOI: 10.1201/9781003537823-13

solution involves a secure blockchain-based encrypted electricity distribution system for smart cities. This approach integrates technological advancements with robust security measures to combat electricity theft effectively (Dou et al., 2018).

Blockchain technology is a decentralized ledger system that securely records transactions across a network of computers. It operates on the principles of transparency, immutability, and decentralization (Pandey, 2023; Soner et al., 2021b, 2022). In essence, it creates a digital record of transactions or data that cannot be altered or tampered with once recorded, providing a high level of security and trust.

In the realm of electricity distribution, blockchain offers several advantages. One prominent application is in managing and optimizing energy distribution networks. Traditional centralized systems face challenges such as inefficiencies, lack of transparency, and susceptibility to cyberattacks (Pandey & Litoriya, 2020a, 2020b; Soner et al., 2021a; Srivastava et al., 2021). By leveraging blockchain technology, electricity distribution can become more efficient, transparent, and secure.

One use case involves the creation of a decentralized energy marketplace. Through blockchain, consumers can directly trade excess energy generated from renewable sources with other users on the network. Smart contracts, self-executing contracts with the terms of the agreement directly written into code, facilitate these transactions autonomously, eliminating the need for intermediaries and reducing transaction costs.

Another significant application is in certificate generation for renewable energy sources. With the increasing emphasis on sustainability and reducing carbon footprints, there is a growing demand for renewable energy certificates (RECs) to track and verify the production of clean energy. Blockchain can streamline the process of generating, tracking, and trading RECs by providing a transparent and immutable record of energy production and consumption (Pandey & Litoriya, 2020c).

Blockchain technology ensures the authenticity and integrity of certificates by securely recording each transaction on the distributed ledger. This eliminates the risk of fraud or manipulation associated with centralized certificate issuance systems (Dutta et al., 2020; Kim & Huh, 2018; Luo, 2022; Ren et al., 2019; Thakker et al., 2021; Wong et al., 2020). Additionally, blockchain enables real-time verification of certificates, reducing administrative burdens and improving the overall efficiency of renewable energy markets.

Overall, blockchain holds immense potential to revolutionize the electricity distribution sector by enhancing efficiency, transparency, and security. By leveraging this technology, stakeholders can create a more sustainable and resilient energy infrastructure that meets the demands of a rapidly evolving energy landscape (Andreadou et al., 2016; Celik et al., 2023; Mohammadi & Taylor, 2017; Pal & Kant, 2019).

A typical block structure is shown in Figure 10.1, where each block is shown to be composed of a block header and the transactions. The block header contains previous hash, i.e., the hash of the previous block, current block hash (which is computed from the previous hash value and the Merkle root of the transactions), and the timestamp values. If the contents of any block are tampered with, all the subsequent blocks' hashes got changed and the intrusion is detected. In this way, a blockchain architecture provides a tamperproof, secure, and transparent solution.

The novelty of the proposed scheme lies in the use of phase modulation to encrypt electrical signals is a novel approach to securing electricity distribution networks.

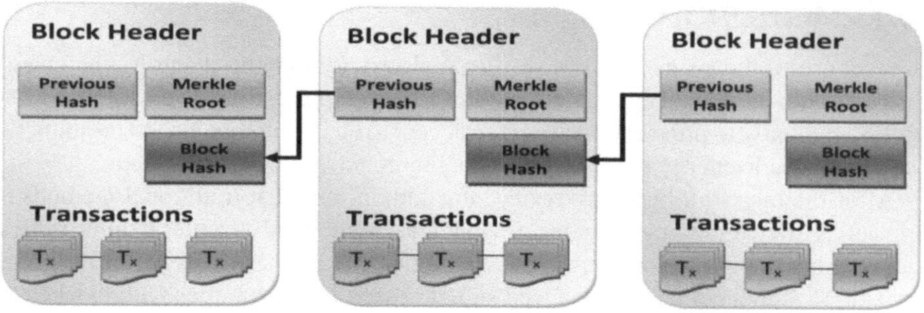

FIGURE 10.1 Typical block structure in a blockchain.

This method makes it difficult for unauthorized users to tap into the power lines without detection, addressing the power theft issue at a fundamental level. By encrypting the signals, the proposed system ensures that only authorized users with the correct decryption keys can access the electricity, effectively creating a theft-proof distribution system.

The potential research questions that this chapter is trying to address are as follows:

- How can a signal modulation technique be effectively implemented to encrypt electrical signals in a distribution network?
- What are the challenges and solutions associated with using modulation to prevent unauthorized access to electricity?

The main contributions of this chapter can be summarized as follows:

- Innovation in Signal Encryption: The chapter introduces the use of phase modulation as a novel technique for encrypting electrical signals in the distribution network. This method enhances the security of electricity distribution by making it significantly more difficult for unauthorized users to intercept and steal power.
- Real-World Application: The chapter outlines a plan for conducting a full-load test in a real smart city setting. This contribution is critical for moving from theoretical research to practical implementation.
- Holistic Approach: The chapter presents a comprehensive solution that not only addresses power theft but also promotes the adoption of renewable energy by promising that only intended receivers will make use of the supplied energy. This dual focus on security and sustainability is a significant contribution to the field.

The rest of this paper is structured as follows: Section 10.2 entails methodology, Section 10.3 is describing the proposed system, Section 10.4 presents implementation and simulation results, Section 10.5 is about implications of the proposed system, and the Section 10.6 concludes this chapter.

10.2 METHODOLOGY

In response to the rising concern of illegal electricity tapping plaguing the energy distribution networks, the authors propose to develop a cutting-edge solution. The vision is not only to prevent unauthorized access to electricity but also to promote the use of renewable energy sources through a secure and transparent system.

After the thorough literature review, the authors find that traditional methods of preventing illegal tapping were insufficient and often reactive rather than proactive. Drawing inspiration from emerging technologies, the authors conceive a revolutionary approach that combines advanced signal processing techniques with blockchain technology.

The heart of this methodology lies in the transmission of phase-shifted AC signals to households, a seemingly unconventional approach with significant potential. By transmitting electricity with a phase shift, the authors aim to deter unauthorized access while maintaining safe and reliable operation for legitimate consumers.

To complement this innovative transmission method in the context of smart city infrastrucutre, the authors propose the integration of blockchain technology for generating RECs. These certificates would serve as irrefutable proof of the origin and sustainability of the electricity supplied to households, incentivizing the adoption of renewable energy sources. In order to prove the concept, the authors resort to the simulation using python and three libraries namely:

1. SciPy – https://www.scipy.org/
2. NumPy – https://numpy.org/
3. Matplotlib – https://matplotlib.org/

To realize the proposed method, specialized equipment is needed to be installed at distribution units (Double Pole Structures) to transform the received masking signals into phase-shifted AC signals, incorporating random bit sequences (Figure 10.2) transmitted through the cloud.

Simultaneously, submeters equipped with internet connectivity are to be installed at households to receive the coded bit sequences, ensuring synchronization with the distribution units. At the household level, additional equipment is to be deployed to generate the original signal from the transformed one, utilizing the unique coded sequences transmitted through the cloud.

Parallel to the implementation of the signal processing infrastructure, the authors propose to develop smart contracts and decentralized applications (DApps) to leverage blockchain technology for generating RECs. Each unit of renewable energy produced by authorized sources is to be assigned a digital certificate, recorded on a secure and immutable blockchain ledger.

The certificates, encoded with relevant data such as energy production metrics and environmental attributes, provided transparent and tamper-proof evidence of renewable energy generation. Smart contracts automatically generated and managed certificates based on predefined criteria, ensuring accuracy and reliability in certificate issuance.

Figure 10.2 shows an algorithm for generating 8-bit random bit sequence to be fed further to the cloud from where the transmitter and receiver modules will fetch the

Algorithm for Random Bit-Sequence Generation

```
Function GenerateRandomBitSequence():
  // Initialize a Random object
  random = new Random()

  // Generate a random integer between 0 and 255 (inclusive)
  randomNumber = random.nextInt(256)

  // Convert the random integer to a binary string
  binaryString = toBinaryString(randomNumber)

  // Pad the binary string with leading zeros if necessary to ensure it's 8 bits long
  while length(binaryString) < 8:
    binaryString = "0" + binaryString

  // Return the resulting 8-bit sequence
  return binaryString
```

FIGURE 10.2 Algorithm to generate an 8-bit random sequence.

bit-sequence. It is to be noted here that the 8-bit random sequence will be generated after random intervals and the new sequences will be updated in the cloud, which will make an attempt of breaking the code sequence difficult.

10.3 PROPOSED SYSTEM

The proposed system offers a robust method to thwart electricity theft by masking the input energy signal, depicted in Figure 10.3, with transmitter and receiver sections. The transmitter comprises a code receiver, bipolar NRZ-level encoder,

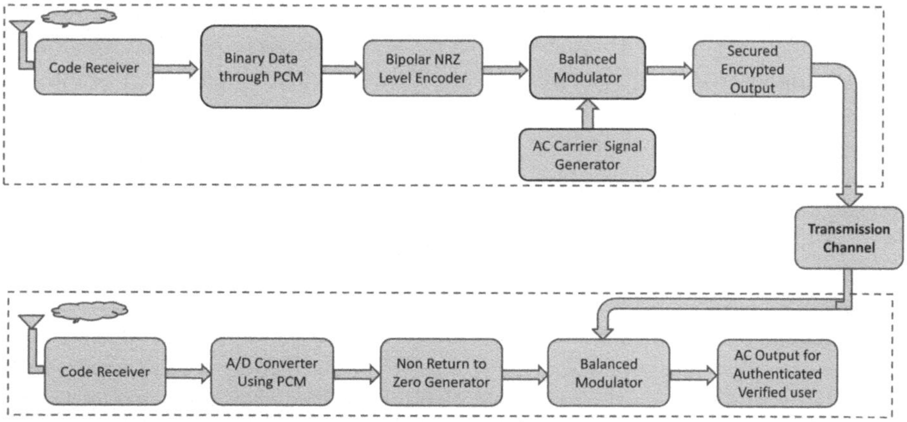

FIGURE 10.3 The proposed transceiver communication system.

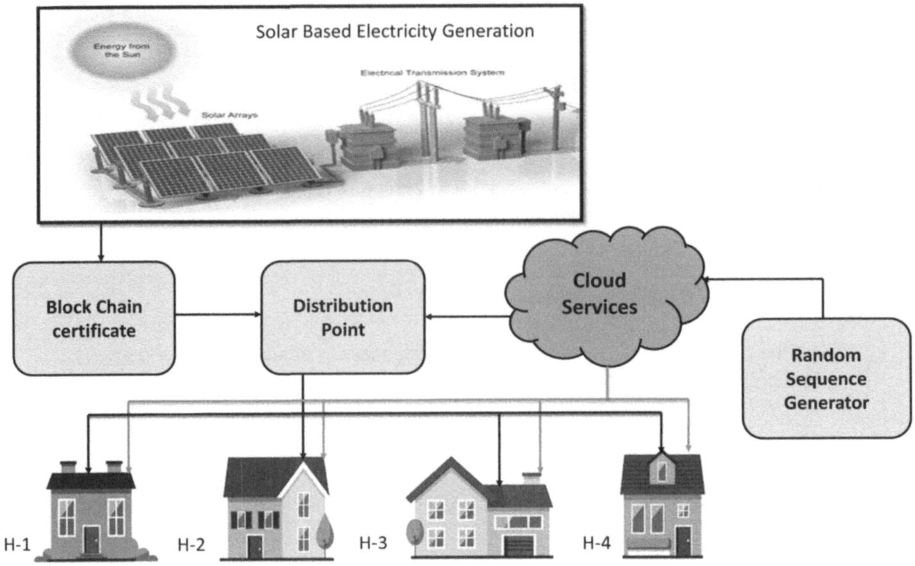

FIGURE 10.4 Detailed process showing distribution of electricity to the households.

balanced modulator, AC carrier signal, secured encrypted output, and PCM system. Conversely, the receiver includes an AC output, NRZ-level encoder, balanced modulator, PCM system, and code receiver. Figure 10.4 shows the complete picture detailing the generation, distribution, and reception of the electricity through the proposed anti-theft distribution system for smart city development.

In the transmission component, a PCM system digitizes a masking signal received from the cloud. The digitized signal is then processed through a NRZ-level encoder represented in Figure 10.5, converting it into a non-return-to-zero (NRZ) form, where logic one (+1) and zero (–1) are denoted. The balanced modulator multiplies this encoded signal with the input AC signal (220V/50Hz), generating a secure output that deters electricity theft. This encrypted output is transmitted to the customer's premises via transmission lines, potentially causing malfunctions or damage to the home appliances if used in an unauthorized manner. However, as current household equipment may not accept this form, the secured output needs transformation before use, requiring a receiver system to extract the desired output from the secured signal. In addition, Figure 10.5 provides an insight into the generation process of BPSK signal required for the secured transmission of electricity. The generated BPSK signal produces 180° phase shifted output with the masked signal generated from the random bit-sequence. In essence, the reception component receives the same masking signal from the cloud, undergoing PCM processing to digitize it. The NRZ signal is reobtained using an NRZ-level encoder and forwarded to the balanced modulator. Here, it is multiplied by the secured signal to produce the intended outcome.

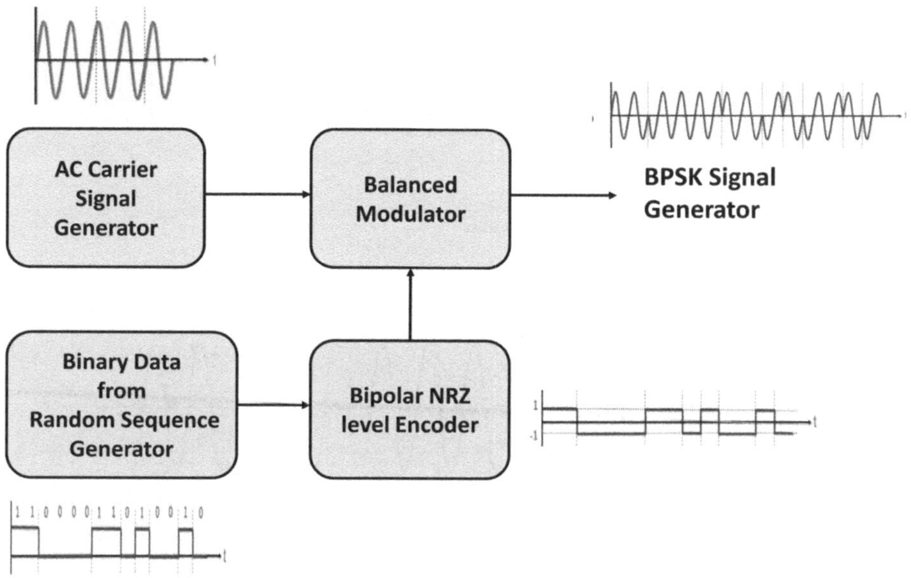

FIGURE 10.5 BPSK signal generation process.

10.4 EXPERIMENTAL SETUP AND SIMULATION RESULTS

The simulated model comprises two crucial components: the transmitter and the receiver. While the transmitter creates a secure signal, the receiver is responsible for accurately restoring the original signal. With the aid of the 64-bit Core i7 processor and ample 16GB RAM, computational limitations are minimized, enabling thorough exploration of the proposed approach's performance and reliability. Conducting qualitative and quantitative analyses within the Python environment provides valuable insights into the method's efficacy, enhancing understanding of its applications and security implications. The simulation aims to validate the transmitter's capability to generate a secure signal and the receiver's proficiency in restoring the original signal accurately. This systematic evaluation is essential for establishing the viability of the proposed methodology and its potential contributions to secure signal processing in diverse applications. Figure 10.6 shows the random sequence

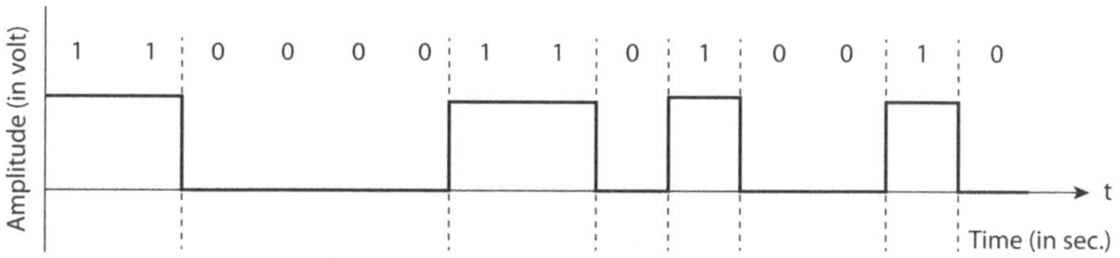

FIGURE 10.6 Output from the random sequence generator.

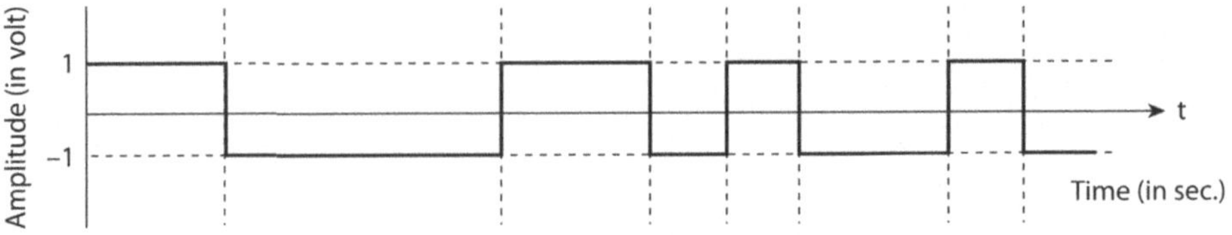

FIGURE 10.7 Bipolar encoded NRZ signal.

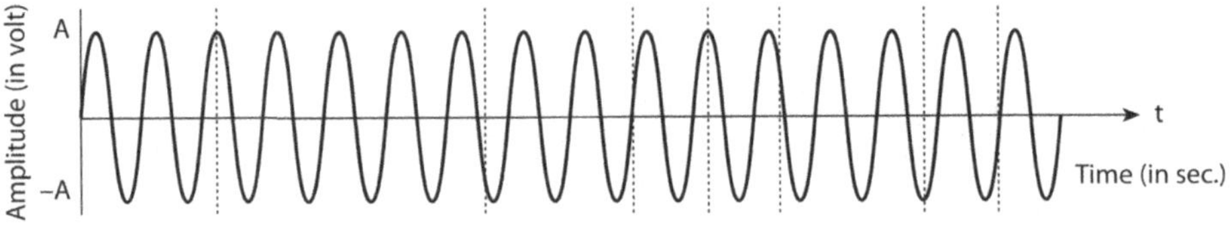

FIGURE 10.8 AC carrier signal.

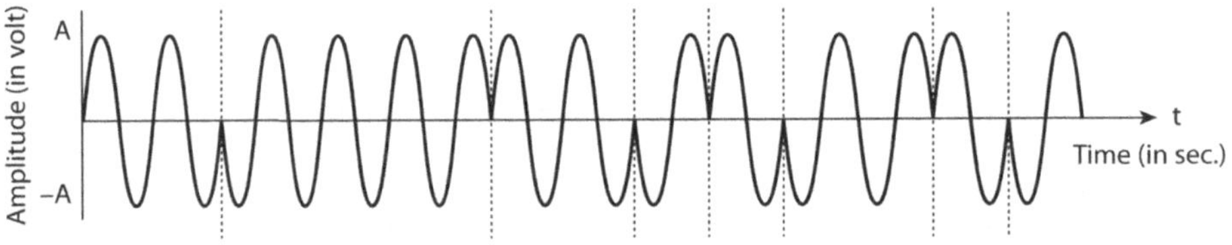

FIGURE 10.9 Binary phase-shifted modulated signal.

generator output whereas Figure 10.7 display the NRZ waveform generated based on the code received from the cloud. Figure 10.8 illustrates the unencrypted sinusoidal wave input to the transmitter and the results of the simulated transmitter section are depicted in Figure 10.9. Signal portrayed in Figure 10.9 is actually the output encrypted secured waveform produced by the transmitter, intended for use in power transmission processes.

10.5 IMPLICATIONS

Implementing such a system would deter the theft of electricity and its implications will be multifaceted and extend to various domains, including technological, social, and economical. Some of these implications are listed as follows:

a. The proposed method, relying on phase modulation for signal encryption within the framework of smart cities, introduces innovative technological solutions to combat power theft.

b. Implementation of this approach could lead to advancements in secure electricity distribution systems, potentially inspiring further research and development in the field of smart grid technologies.
c. Reduction in power theft has direct economic implications, positively affecting revenue streams for utility companies. This, in turn, may contribute to the overall economic development of regions grappling with the consequences of electricity theft.
d. A more reliable and secure electricity distribution system can attract investments and encourage businesses to operate in areas where power theft was previously a significant concern.
e. Reliable access to electricity enhances the quality of life for citizens, promoting the use of electrical appliances, supporting education, and facilitating business activities.
f. Reduction in power theft can lead to a more equitable distribution of electricity, ensuring that those who are willing to pay for power receive uninterrupted service, contributing to social welfare.
g. Efficient electricity distribution systems reduce the need for emergency measures to compensate for power theft losses. This can lead to a more sustainable use of resources and a decrease in the environmental impact associated with power generation.
h. The research contributes to the realization of smart city initiatives by addressing a critical challenge in urban environments. Implementing theft-proof electricity distribution aligns with the broader goals of creating efficient, sustainable, and technologically advanced cities.

10.6 CONCLUSIONS

In conclusion, the transformation of power signals into secured formats and their subsequent delivery to consumers form the bedrock of a system designed to effectively thwart unauthorized tapping from distribution lines. This strategic approach represents a comprehensive solution to the persistent threat posed by unlawful tapping in our modern society. These illicit activities not only have adverse effects on the local economy but also impact the financial stability of transmission companies, both of which can be mitigated through the implementation of this innovative system.

One notable advantage of this method is its ability to autonomously detect and respond to incidents of theft, eliminating the necessity of reporting to vigilance teams. Any illegal tapping automatically triggers malfunctions in the appliances of thieves, streamlining the process of addressing and rectifying such occurrences.

Importantly, the integration of blockchain certificates enhances the credibility and transparency of the system. By utilizing blockchain technology for certification of renewable energy sources, the proposed system ensures the authenticity and sustainability of the supplied electricity. This not only strengthens consumer trust but also incentivizes the adoption of renewable energy, further contributing to environmental sustainability.

Moreover, the development of a working model for this system can be achieved at a reasonable cost without necessitating significant capital expenditure. This investment proves minimal when compared to the substantial capital losses incurred by governments or utility companies due to electricity theft.

Future research will focus on optimizing the encryption algorithm for minimal impact on transmission efficiency, exploring advanced blockchain frameworks for better scalability, and conducting extensive field tests to refine the system. Additionally, the integration of smart meters and IoT devices will be explored to enhance the monitoring and management capabilities of the electricity distribution network.

REFERENCES

Ahmad, S., Mathai, M. V., & Parayil, G. (2014). Household electricity access, availability and human well-being: Evidence from India. *Energy Policy*, *69*(C), 308–315.

Al-Dahidi, S., Ayadi, O., Adeeb, J., & Louzazni, M. (2019). Assessment of artificial neural networks learning algorithms and training datasets for solar photovoltaic power production prediction. *Frontiers in Energy Research*, 7, 1–18. https://doi.org/10.3389/fenrg.2019.00130

Andreadou, N., Guardiola, M., & Fulli, G. (2016). Telecommunication technologies for smart grid projects with focus on smart metering applications. *Energies*, *9*(5), 375. https://doi.org/10.3390/en9050375

Celik, Y., Petri, I., & Barati, M. (2023). Blockchain supported BIM data provenance for construction projects. *Computers in Industry*, *144*, 103768. https://doi.org/10.1016/j.compind.2022.103768

Dou, J., Liu, X., Lu, J., Wu, D., & Wang, X. (2018). Research on electricity anti-stealing method based on power consumption information acquisition and big data. *Electronic Measuring Instruments*, *55*(21), 43–49.

Dutta, P., Choi, T.-M., Somani, S., & Butala, R. (2020). Blockchain technology in supply chain operations: Applications, challenges and research opportunities. *Transportation Research Part E: Logistics and Transportation Review*, *142*, 102067. https://doi.org/10.1016/j.tre.2020.102067

Jokar, P., Arianpoo, N., & Leung, V. (2016). Electricity theft detection in AMI using Customers' consumption patterns. *IEEE Transactions on Smart Grid*, *7*, 216–226.

Kim, S.-K., & Huh, J.-H. (2018). A study on the improvement of smart grid security performance and blockchain smart grid perspective. *Energies*, *11*(8), 1973. https://doi.org/10.3390/en11081973

Li, S., Han, Y., Yao, X., Yingchen, S., Wang, J., & Zhao, Q. (2019). Electricity theft detection in power grids with deep learning and random forests. *Journal of Electrical and Computer Engineering* 2019, 4136874. https://doi.org/10.1155/2019/4136874

Luo, L. (2022). Application of blockchain technology in intellectual property protection. *Mathematical Problems in Engineering*, *2022*, 1–12. https://doi.org/10.1155/2022/4641559

Mohammadi, N., & Taylor, J. E. (2017). Smart city digital twins. *2017 IEEE Symposium Series on Computational Intelligence (SSCI)*, 1–5. https://doi.org/10.1109/SSCI.2017.8285439

Pal, A., & Kant, K. (2019). Using blockchain for provenance and traceability in Internet of Things integrated food logistics. *Computer*, *52*(12), 94–98. https://doi.org/10.1109/MC.2019.2942111

Pandey, P. (2023). Empowering Hindu women by way of smart marriages. *Wireless Personal Communications*. https://doi.org/10.1007/s11277-023-10805-6

Pandey, P., & Litoriya, R. (2020a). Implementing healthcare services on a large scale: Challenges and remedies based on blockchain technology. *Health Policy and Technology*, *9*(1), 69–78. https://doi.org/10.1016/j.hlpt.2020.01.004

Pandey, P., & Litoriya, R. (2020b). Promoting trustless computation through blockchain technology. *National Academy Science Letters*, *44*, 225–231. https://doi.org/10.1007/s40009-020-00978-0

Pandey, P., & Litoriya, R. (2020c). Securing and authenticating healthcare records through blockchain technology. *Cryptologia*, *44*(4), 341–356. https://doi.org/10.1080/01611194.2019.1706060

Ren, Q., Man, K. L., Li, M., & Gao, B. (2019). Using blockchain to enhance and optimize IoT-based intelligent traffic system. *2019 International Conference on Platform Technology and Service (PlatCon)*, 1–4. https://doi.org/10.1109/PlatCon.2019.8669412

Soner, S., Litoriya, R., & Pandey, P. (2021a). Exploring blockchain and smart contract technology for reliable and secure land registration and record management. *Wireless Personal Communications*, *121*(4), 2495–2509. https://doi.org/10.1007/s11277-021-08833-1

Soner, S., Litoriya, R., & Pandey, P. (2021b). Making Toll Charges Collection Efficient and Trustless: A Blockchain-Based Approach. *2021 3rd International Conference on Advances in Computing, Communication Control and Networking (ICAC3N)*, 1533–1538. https://doi.org/10.1109/ICAC3N53548.2021.9725447

Soner, S., Litoriya, R., & Pandey, P. (2022). Combining blockchain and machine learning in healthcare and health informatics: An exploratory study. In *Blockchain Applications for Healthcare Informatics* (pp. 117–135). Elsevier. https://doi.org/10.1016/B978-0-323-90615-9.00014-1

Srivastava, S., Gupta, R., Pandey, P., & Litoriya, R. (2021). Ethical and legal aspects of using blockchain technology for 5G-based health-care systems. In *Blockchain for 5G Healthcare Applications: Security and Privacy Solutions* (pp. 35–51). Institution of Engineering and Technology. https://doi.org/10.1049/PBHE035E_ch2

Thakker, U., Patel, R., Tanwar, S., Kumar, N., & Song, H. (2021). Blockchain for diamond industry: Opportunities and challenges. *IEEE Internet of Things Journal*, *8*(11), 8747–8773. https://doi.org/10.1109/JIOT.2020.3047550

Wong, P. F., Chia, F. C., Kiu, M. S., & Lou, E. C. W. (2020). The potential of integrating blockchain technology into smart sustainable city development. *IOP Conference Series: Earth and Environmental Science*, *463*(1), 012020. https://doi.org/10.1088/1755-1315/463/1/012020

Section IV

Distribution Systems and Sustainable Energy Infrastructure

11 An Integrated Approach to Enhance Distribution System Reliability

Analysis of Load Growth Dynamics

Sivasankari G.S. and Narayanan K.

11.1 INTRODUCTION

The efficient management of distribution networks (DNs) becomes challenging in areas with high load density due to fluctuating load demand. With a fixed network topology and changing load demands, radial DNs are prone to non-minimal power losses and improved load curtailment. To mitigate this, optimal sizing and siting of distributed generators (DGs) and network reconfiguration (NR) are commonly employed methods. NR involves altering the topological structure of feeders by adjusting the open/closed state of tie lines. DGs, which are small power-generating units connected directly to the existing DN, play a crucial role in this process. By strategically placing DGs and reconfiguring the network, overloaded feeders can be relieved, leading to a reduction in system loss and reliability. Some literature provides NR as the measure to deal with the objective of loss minimization [1]. In [2], a technique termed the sequential switch opening and exchange method for distribution network reconfiguration (DNR) was introduced. In [3], the focus was on reducing energy losses by re-configuring distribution systems, considering various load and distributed generation (DG) scenarios. The study in [4] explored an ideal solution aiming to maximize the system's load-carrying capacity while minimizing power losses. Employing the improved harmony search algorithm, Tyagi et al. [5] investigated DNR in balanced IEEE 33 and unbalanced 25-bus distribution systems.

Reference [6] proposed a DNR with simultaneous allocation of DGs using the Whale Optimization Algorithm (WOA). Reference [7] suggested to proceed with the Equilibrium Optimization Algorithm (EOA) and an improved EOA (IEOA) for DNR. This method was tested on 69-node and 33-node systems under varied load conditions. Reference [8] proposed the DNR with the simultaneous allocation of DGs using the improved Salp Swarm Algorithm (ISSA) and reference [9] proposed the same using the Firefly (FF) algorithm. Reference [10] explored the combination of levy fights with a Sine-Cosine Algorithm (ISCA) for DNR.

Additionally, reference [11] utilized a genetic algorithm-based feeder reconfiguration for loss minimization, incorporating electric vehicles (EVs) in IEEE 69 and

DOI: 10.1201/9781003537823-15

118 bus systems. Considering the curtailment of renewable energy, real power loss, voltage deviation, and generation cost, reference [12] provided insights into various aspects of distribution system optimization.

Optimally allocating DGs followed by DNR contributes to reduced system losses, as discussed in [13]. The study suggested using the Harris Hawks Optimizer to strategically position DGs, aiming to minimize power loss and enhance system voltage profiles. Furthermore, [14] proposed the use of the Plant Propagation Algorithm to optimize the sizing and placement of DGs within load flow analysis.

Addressing the concurrent optimization challenge of DG allocation and DNR, [15] presented a method targeting multiple objectives, including minimizing voltage deviation and loss while enhancing voltage stability indices. This approach was evaluated on small- and medium-sized 33 and 69-bus networks. In [16], a strategy for DNR alongside the placement and sizing of various DG types was proposed using the ant colony optimization algorithm and evaluated in balanced IEEE 33 and 69 bus systems.

EVs play a vital role in complementing intermittent renewable energy sources (RES). Integrating EVs with DG sources can enhance system's effectiveness by analyzing EV charging/discharging and travel patterns [17]. However, studies focusing solely on reducing EV charging costs may overlook grid performance [18, 19]. On the contrary, grid operator-controlled charging emphasizes grid performance, aiming to improve voltage stability and reduce energy loss [20, 21]. For instance, the average rate charging method proposed in [22] suggests charging EVs at a consistent rate throughout the planned duration to mitigate peaks during off-peak periods and enhance grid performance. The comprehensive review on DS allocation is presented in [23]. The probabilistic DS allocation is presented in [24]. The modeling of DS using Differential Evolution Algorithm (DEA) has been explored in [25]. Reference [26] used the Cuckoo Search Algorithm (CSA) for optimal allocation of DG and DSTATCOM in radial DN and [27] used Gravitational Search Algorithm (GSA) for the siting and sizing of DS.

The initial phase involves assessing the impact of load growth (G) on the system, prompting the system operator to evaluate the maximum loading capacity of the lines. Before enhancing the conductors/cables to accommodate the LG, it is imperative to ascertain the system's capacity to bear additional load, thereby enabling the utilization of the lines' maximum loading capacity. This study is dedicated to this objective, with a specific emphasis on investigating the impact on system reliability. The process of radial reconfiguration for maximizing loadability is executed, analyzing the effect of LG on the line's capacity to withstand it, both initially and after reconfiguration. The analysis adopts the standard IEEE system. The system's reliability is thoroughly examined for its efficacy, with all reliability indices referenced comprehensively in [28].

The major contributions of this work include:

- Identification of the potential annual load growth that the system can accommodate with minimized losses.
- Determination of the optimal radial reconfiguration to enhance the system's loadability.

- Examination of the system's load growth in conjunction with the integration of EVs and analysis conducted with DSs to achieve optimal outcomes.
- Furthermore, the analysis delves into the impact of system reliability on both the base case and the reconfigured case, aimed at addressing observed reliability issues in the pursuit of loadability maximization.

The chapter is organized as follows: The proposed methodology is explained in Section 11.2. Section 11.3 provides the result and discussion of the test system and the conclusion is presented in Section 11.4. The load data and line data pertaining to the IEEE 33 bus systems are documented in a separate appendix at the end of the document.

11.2 METHODOLOGY

11.2.1 Problem Formulation

The primary objective is the maximization of loadability by considering load growth. This objective is analyzed in both the initial configuration and after reconfiguration. To achieve loadability maximization, the PSO algorithm is applied to identify the optimal configuration for attaining the highest loading capacity. The objective function is maximizing the loadability.

$$F_f = max\left[\omega_{ij}, min\left(RP_{Loss}\right)\right] \quad (11.1)$$

'ω_{ij}' represents the line loading capacity in between '*i*th' node and '*j*th' node.
'RP_{loss}' represents amount of real power loss.

'P_{base}' and 'Q_{base}' are the initial real and reactive power loads, connected with *i*th node. The real and reactive load is increased on all buses with equal loading factor, and 'ω' is the loading factor. 'P_{new}', and 'Q_{new}' are the new real and reactive power values of the system. It should be identified in order to obtain the maximum allowable load growth in the system.

$$P_{new} = P_{base} + \left(P_{base} * \omega\right) \quad (11.2)$$

$$Q_{new} = Q_{base} + \left(Q_{base} * \omega\right) \quad (11.3)$$

The factor 'ω' is identified subject to the following constraints:

Voltage Constraint: The value of 'U_i^{limit}' represents the boundary for the voltage magnitude at the '*i*th' node. It signifies the occurrence of either upper or lower limit violations when reached. Here, the value of 'U_i^{min}' and 'U_i^{max}' is 0.95 and 1.05 pu, respectively.

$$U_i^{min} < U_i^{limit} < U_i^{max} \quad (11.4)$$

Current Constraint: 'I_i' and '$I_i^{k\,max}$' denote the current and the maximum current for the 'kth' branch, respectively. The thermal limits (i.e., ampacity of the lines) are considered in the analysis.

$$I_i^k \le I_i^{k\,max} \tag{11.5}$$

Loadability Constraint: 'I_{ij}' and 'I_{ij}^{max}' denote the current loadability and the maximum allowable loadability between the 'ith' node and the 'jth' node, respectively.

$$I_{ij} \le I_{ij}^{max} \tag{11.6}$$

Radiality Constraint: The distribution system's topology must adhere to a radial configuration, and no isolated nodes are allowed within the system.

11.2.1.1 Load Growth Calculation

$$P_{load}(n) = P_{load}(0) * (1+g)^n \tag{11.7}$$

'$P_{load}(0)$' is the initial year's real power load of the system at the initial year. 'n' is the number of years, and 'g' is the growth rate in the system's annual load. For the considered analysis, it is fixed at 7.5%.

11.2.1.2 Loadability Calculation

$$P_{Ly}(n) = \frac{P_{load}(n) - P_{load}(n-1)}{P_{load}(n-1)} * 100 \tag{11.8}$$

where '$P_{load}(n)$' represents the amount of real power load at 'nth' year and '$P_{load}(n-1)$' represents the amount of real power load at '$(n-1)$th' year (i.e., previous year).

11.2.1.3 Percentage Change in Real Power Loss Calculation

$$RP_{Loss}(n) = \frac{RP_{loss}(n) - RP_{loss}(n-1)}{RP_{loss}(n-1)} * 100 \tag{11.9}$$

where '$RP_{loss}(n)$' represents the amount of real power loss at 'nth' year and '$RP_{loss}(n-1)$' represents the amount of real power loss at '$(n-1)$th' year (i.e., previous year). The flowchart of the proposed work is shown in Figure 11.1.

11.2.2 Importance of Loadability

In the domain of distribution system management, the importance of loadability cannot be overstressed. It acts as a crucial metric for guaranteeing the seamless functioning and dependability of power distribution to end-users. At its essence, loadability analysis delves deeply into comprehending and addressing the challenges

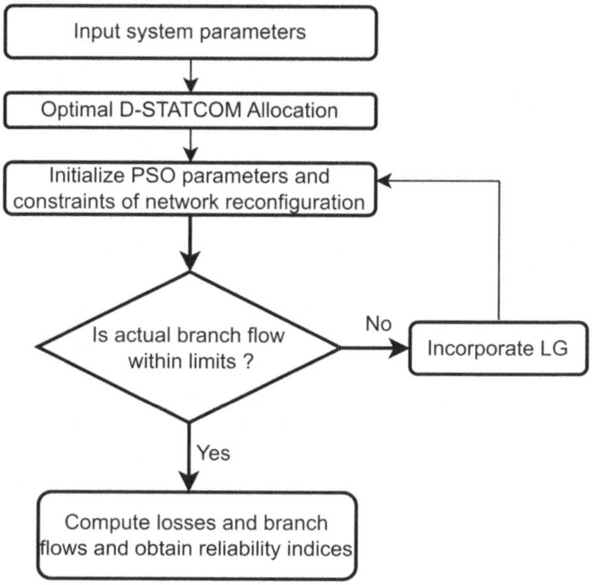

FIGURE 11.1 Flowchart for the proposed work.

brought about by the rising LG. By meticulously aligning this upsurge in energy consumption with the existing line flow capacity, the analysis aims not just to meet but surpass expectations. It underscores a proactive approach—one that foresees and adjusts to the evolving LG, whether in terms of system demand or the integration of EV load into the system. The flowchart for the loadability maximization of the proposed work is given in Figure 11.2.

11.2.3 Significance of Load Growth

The system operator's attention to LG within the distribution system is of utmost importance and necessity. Initially, as the electricity demand progressively rises, comprehending and accommodating LG becomes vital to uphold the system's dependability and operational effectiveness. Neglecting to foresee and tackle this expansion may result in strained infrastructure, leading to system breakdowns, voltage fluctuations, and increased downtime, thus jeopardizing service reliability. Furthermore, proactive management of LG empowers the system operator to make well-informed decisions regarding system reinforcement or the optimal integration of DS sources. By accurately predicting trends in LG, operators can strategize and execute infrastructure enhancements, such as expanding transmission and DNs or upgrading existing equipment, to efficiently meet growing demand. Additionally, the incorporation of D-STATCOM (DS) sources provides a sustainable approach to bolstering system capacity and mitigating the impacts of LG. Deliberately situating

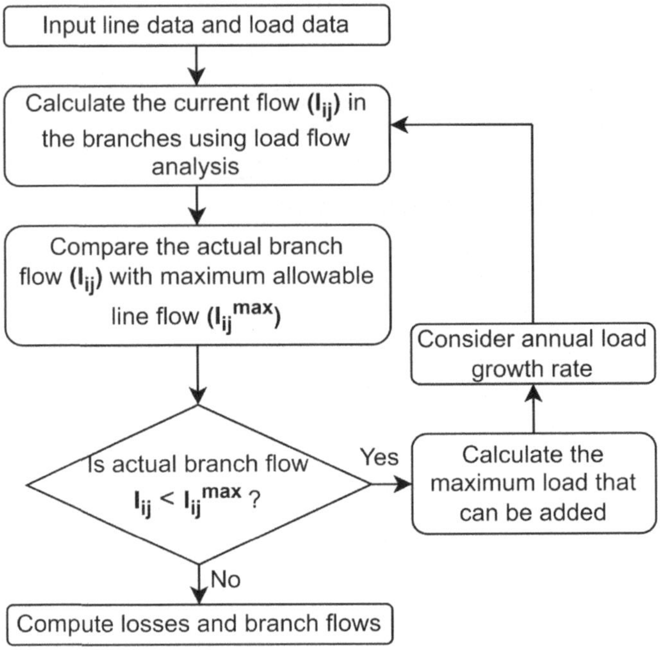

FIGURE 11.2 Flowchart for loadability maximization.

DS units at strategic points within the DN can alleviate congestion/overloading on centralized generation facilities, enhance system flexibility, and fortify resilience against disruptions.

Acknowledging the significance of LG empowers system operators to proactively navigate and adjust to changing demand trends. Through prioritizing investments in infrastructure and embracing innovative approaches such as integrating DS, operators can uphold the continuous reliability, resilience, and sustainability of the distribution system.

11.2.4 Power Flow Calculation

The power flow of a distribution system is computed by the backward forward sweep algorithm [29]. The flowchart is given in Figure 11.3.

Backward Sweep: The backward sweep serves as a solution for current or power flow, with the potential for voltage updates. It initiates from branches in the terminal layer and advances toward those linked to the root node. During backward propagation computation, the updated effective power flows in each branch are ascertained by considering node voltages from the preceding iteration. Consequently, voltage values obtained in the forward path remain constant during the backward sweep, while updated power flows in each branch are retrogressively transmitted along the feeder using the backward path. Thus, the backward sweep commences at the farthest end node and progresses towards the source node.

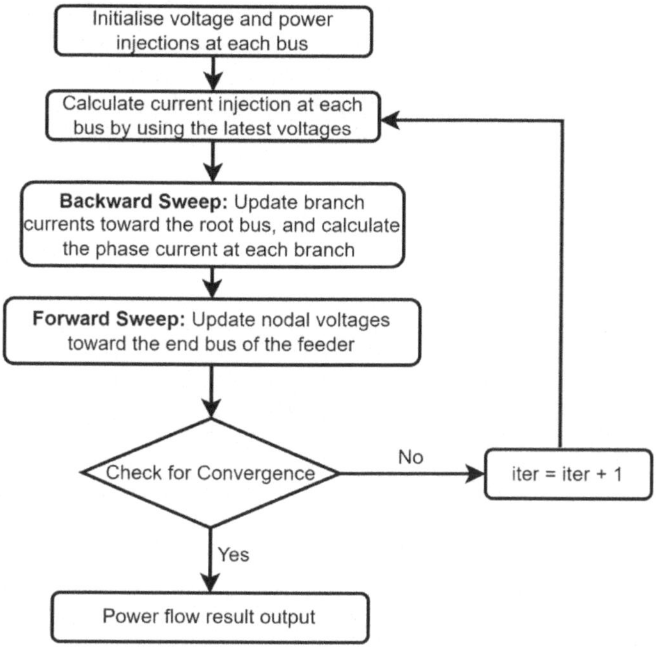

FIGURE 11.3 Flowchart of forward-backward sweep algorithm.

The node current calculation in the backward sweep is calculated using Eqn. 11.10.

$$I_i^{(k)} = \left(\frac{S_i}{V_i^{(k-1)}} \right)^* \tag{11.10}$$

$I_i^{(k)}$ Current injection at ith node on kth iteration
S_i Power injection at ith node
$V_i^{(k-1)}$ Voltage at ith node on $(k-1)$th iteration

The branch current calculation in the backward sweep is calculated using Eqn. 11.11.

$$I_{\mathrm{Br}((i-1),i)}^{(k)} = I_i^{(k)} + \sum I_{\mathrm{Br}(i,(i+1))}^{(k)} \tag{11.11}$$

$I_{\mathrm{Br}((i-1),i)}^{(k)}$ is the current flowing through the branch that connects (ith) node to its upstream node which is $(i-1)$th

$\sum I_{\mathrm{Br}(i,(i+1))}^{(k)}$ is the total sum of all currents of branches initiating from (ith) node

Forward Sweep: The forward sweep entails calculating voltage drops, with the option to update current or power flow. Nodal voltages are iteratively updated in a forward sweep process, commencing from branches in the initial node and progressing toward those in the last. The primary objective of forward sweep is to determine voltages at each node, originating from the feeder source node, while maintaining the feeder substation voltage at its actual value. Throughout the forward sweep, the effective power in each branch remains constant, a value determined during the backward process.

The voltage calculation of the forward sweep is calculated using Eqn. 11.12.

$$V_i^{(k)} = V_{i-1}^{(k)} - I_{\mathrm{Br}(i-1)}^{(k)} Z_{(i-1),i} \tag{11.12}$$

$V_i^{(k)}$ Voltage of ith node at kth iteration

$V_{i-1}^{(k)}$ Voltage of $(i-1)$th node (upstream node) at kth iteration

$I_{\mathrm{Br}(i-1)}^{(k)}$ Branch current between ith node and the upstream node which is $(i-1)$th node

$Z_{(i-1),i}$ Impedance between ith node and the upstream node which is $(i-1)$th node

The terminating criteria for LFA are the voltage magnitude convergence computed by calculating the differences in voltage magnitude between two successive iterations. It is given in Eqn. 11.13.

$$max(\Delta V) \leq 0.00001 \tag{11.13}$$

ΔV Difference in voltage magnitude and $max(\Delta V)$ element with the higher magnitude

11.2.5 Significance of Reconfiguration

The importance of reconfiguration in adapting to LG lies in its capacity to optimize the performance and resilience of power distribution systems as demand patterns evolve. Reconfiguration strategies enable DNs to dynamically adjust their layout, effectively addressing challenges associated with rising load levels.

Through strategic adjustments to network layout, system operators can efficiently manage load growth by redistributing power flows, mitigating congestion in overloaded segments, and optimizing voltage profiles. This proactive approach enhances system reliability and efficiency, ensuring consistent electricity supply to end-users despite increasing demand.

Moreover, reconfiguration empowers utilities to maximize the utilization of existing infrastructure without requiring substantial investments in system expansion or reinforcement. By tapping into the latent capacity within the distribution system, reconfiguration minimizes operational constraints and reduces the risk of system overloads, ultimately enhancing overall network performance.

In the analyzed scenario, reconfiguration is executed to identify a configuration with reduced losses and improved line loading capacity using Particle Swarm Optimization (PSO).

11.2.6 PSO for Reconfiguration

Within the context of optimizing loadability through reconfiguration, PSO is utilized for efficient decision-making [30]. Under the PSO framework, the problem is explored within a defined search space, where the current position of the objective function acts as the reference point. In this analysis, the objective is to maximize line loading while minimizing system losses. Random variables dictate particle movement within the confined search space, influenced by the particle's current position, previous velocity, and the best-known location, with random perturbations introduced. Throughout the iterative process, particle velocities are adjusted to approach the optimal solution. The position of each particle is updated in each iteration. This swarm-based technique converges toward the optimal solution through particle interactions [4] and [31].

Eqns. 11.14 and 11.15 are used when initializing particle velocities and positions (open switches) respectively.

$$\vartheta_k = \vartheta_{max} \times rand(m) - \frac{\vartheta_{max}}{2} \tag{11.14}$$

$$\chi_k = round\left(\chi_{min} + \left[\chi_{max} - \chi_{min}\right] \times rand(m)\right) \tag{11.15}$$

ϑ_k	Velocity of particle k
ϑ_{max}	Maximum permissible velocity for particles
m	Number of particles considered
$\text{rand}(m)$	Function that returns j random numbers between 0 and 1
χ_k	Position of particle k (Open switches)
χ_{min}	Position with open switches of highest possible numbers
χ_{max}	Position with open switches of the lowest possible numbers

Eqns. 11.16 to 11.18 are used to update the inertial weights, velocities, and positions of particles respectively.

$$\omega = \omega_{max} - \left(\frac{\omega_{max} - \omega_{min}}{I_{max}}\right) \times iter \tag{11.16}$$

$$\vartheta_k^{new} = \omega\vartheta_k + c_1 rand(1) \times \left(P_{best,k} - \chi_k\right) + c_2 rand(1) \times \left(G_{best} - \chi_k\right) \tag{11.17}$$

$$\chi_k^{new} = round\left(\chi_k + \vartheta_k^{new}\right) \tag{11.18}$$

$Iter_{\text{max}}$	Number of iterations to be performed
Iter	Current iteration number
ϑ_k^{new}	Updated velocity of particle k
$P_{\text{best},k}$	Particle best position
G_{best}	Global best position
χ_k^{new}	Updated position of particle k

The parameters considered for PSO are shown in Table 11.1. Figure 11.4 explains the process in detail.

TABLE 11.1
PSO Parameters

Parameters		Value
Inertial weight constants	ω_{min}	0.25
	ω_{max}	0.75
Cognitive acceleration constant	c_1	1.5
Social accelerating constant	c_2	1.5
Number of particles	m	50
Number of iterations	$Iter_{max}$	100

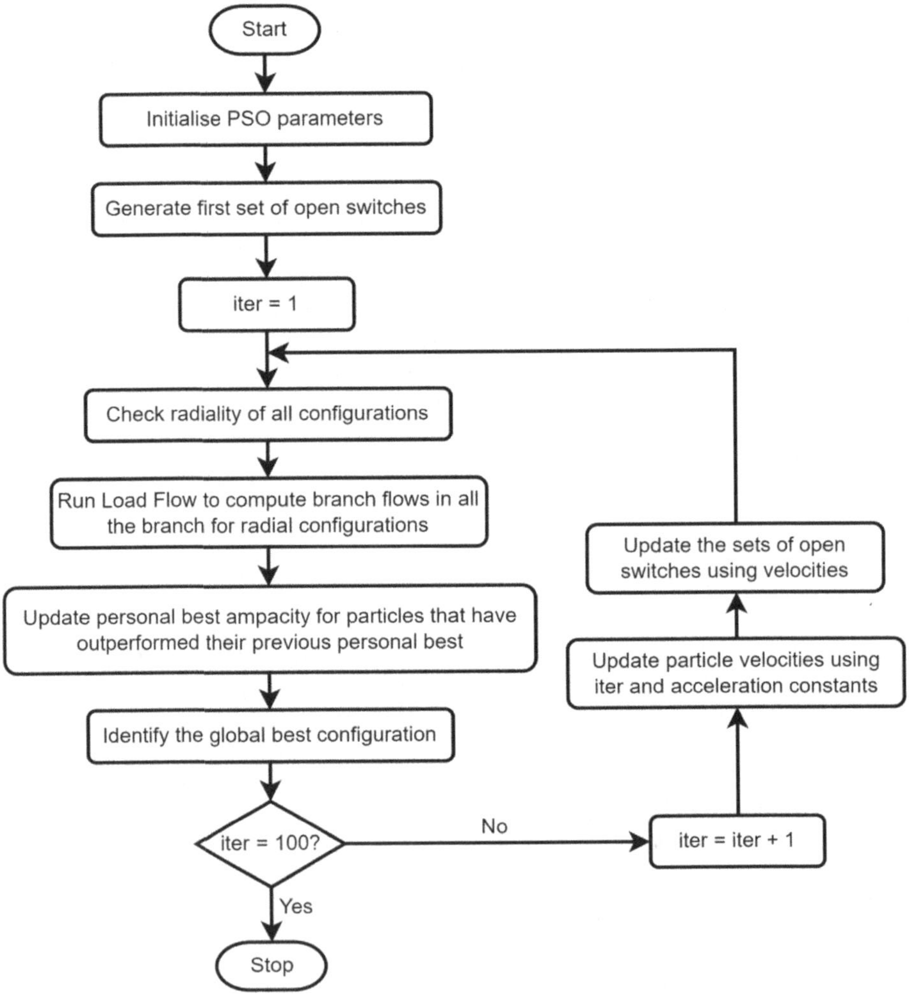

FIGURE 11.4 Flowchart of PSO.

11.2.7 D-STATCOM (DS) Allocation

The process of integrating DS involves two distinct stages: identifying suitable locations and determining optimal sizes. This two-step sizing approach effectively reduces computational burdens while ensuring robust optimization of DS integration within the distribution system. To identify suitable locations for DS placement, all buses within the DN are ranked based on their loading capacities, with priority given to those with the lowest loading capacities. The "To buses" of the branches connected to these selected buses are then earmarked for DS allocation, adhering to the following criteria:

1. Only one DS is permitted per selected feeder.
2. Selected locations must not overlap within the same feeder.
3. DS placement is not allowed at the starting node of lateral lines.

This process is described in detail in Figure 11.5.

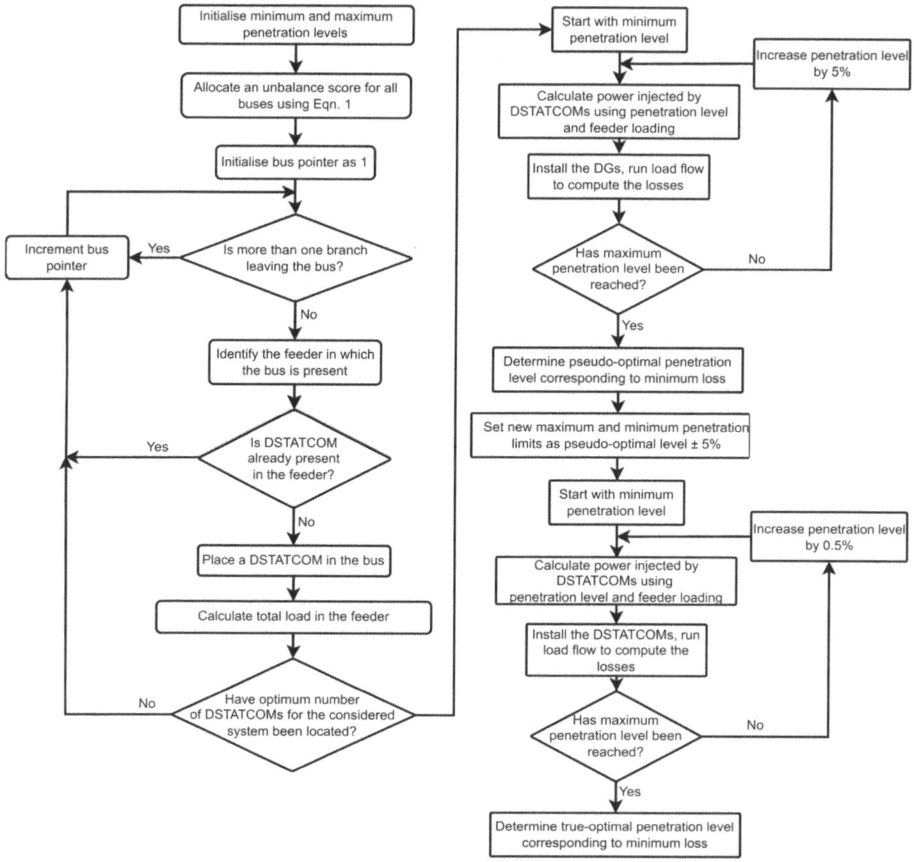

FIGURE 11.5 Flowchart of DS allocation.

11.2.7.1 Penetration Level Calculation for DS Allocation

The penetration level (Φ) of DS unit is calculated using Eqn. 11.19 taken from [32].

$$\Phi = \left(\frac{Q_{DS}}{Q_{DS} + Q_{grid}} \right) * 100\% \tag{11.19}$$

The limits are:

$$0 \leq Q_{DS} \leq Q_{DS,m} \tag{11.20}$$

where Q_{DS} is the power from each DS unit (here, considered as 2000 kVAr) and Q_{grid} is the power drawn from the grid. $Q_{DS,m}$ is considered to vary from 20% to 90% of the feeder loads. It can be rewritten to find Q_{DS} as:

$$Q_{DS} = \left(\frac{\Phi \times \left(Q_{DS} + Q_{grid} \right)}{100\%} \right) \tag{11.21}$$

From Eqn. 11.20, the constraints for the size of *DS* are as follows:

$$Q_{DS_{min}} \leq Q_{DS} \leq Q_{DS_{max}} \tag{11.22}$$

The penetration level increases gradually by 5% until reaching 90%. Load flow analyses are performed at each penetration level, and the level where the lowest losses occur is noted as the near-optimal level. Following this, new penetration limits are established as the near-optimal level plus or minus 5%. This adjusted range undergoes iterative refinements, with each iteration increasing by 0.5%. The true optimal penetration level, identified as the point where losses reach a minimum during subsequent iterations, is then determined.

11.2.7.2 Cost Calculation of DS:

The Total Annual Cost Savings (TACS) of DS is calculated using the following equations [24] and [33]:

$$DSC_{year} = \frac{C_{cost} \times \left(1 + B\right)^{n} \times B}{\left(1 + B\right)^{n} - B} \tag{11.23}$$

$$TACS = K_p * T_i \sum_{i=1}^{N_{DS}} \left(\left(RP_{Loss-without\ DS} - RP_{Loss-with\ DS} \right) \right) - DSC_{year} \tag{11.24}$$

DSC_{year} is the annual D-STATCOM cost, C_{cost} is the cost of DS (50$/kVAr), B is the asset rate (0.1), N_{DS} is the number of DS, n is the year (30 years), K_p is 0.06 ($/kWh), T_i is 8760, $RP_{Loss-without\ DS}$ denotes the real power loss without DS, $RP_{Loss-with\ DS}$ denotes the real power loss DS.

11.2.8 Significance of Reliability

The ability of a power system to supply energy to the customer without interruption with the specified standard of power quality is referred to as the reliability of the system. A system can be termed as more efficient with increased system reliability. Assessment of the reliability indices also helps enhance the system's reliability when required [34]. Each branch within a DN is characterized by a failure rate denoted as 'λ' (in failures per hour) and a repair time denoted as 'r' (in hours). Equations 11.25–11.29 facilitate the computation of equivalent system indices. The dataset used for reliability indices computations is illustrated in separate appendix at the end of the document and the number of consumers at each load point has been referred from [35].

The indices calculated are Energy Not Supplied (ENS), and average ENS (AENS). AENS is the index that provides an idea of how much amount of energy is not served during a particular time interval.

$$U_i = \frac{\sum_i \lambda_i r_i}{\sum \lambda_i} \tag{11.25}$$

$$ENS = Total\ energy\ not\ supplied\ by\ the\ system \tag{11.26}$$

$$ENS = \sum L_{a(i)} U_i \tag{11.27}$$

AENS is the average energy not supplied index and is expressed in terms of kilowatt hour per customer.

$$AENS = \frac{Total\ energy\ not\ supplied}{Total\ number\ of\ customers\ served} \tag{11.28}$$

$$AENS = \frac{\sum L_{a(i)} U_i}{N_i} \tag{11.29}$$

where '$L_{a(i)}$' is the average load connected to load point 'ith', 'U_i' is the annual outage time of the 'ith' load point, 'λ_i' is the repair time of the component in 'ith' load point, 'r_i' is the repair time of the component in 'ith' load point, and 'N_i' is the total number of customers connected at the 'ith' load point.

The proposed analysis aims to determine how long the existing system can withstand the addition of new loads. This has been conducted with the objective of maximizing loadability while reducing real power loss. The two objectives have been evaluated using a well-known common metric reliability index. Ideally, the value of ENS should be zero for an exceptionally reliable system, but practically this is not possible. Therefore, lower indices indicate higher system reliability. In the base case, we have the standard topology of the system, whereas in the reconfigured network, power is re-routed with the help of existing tie lines. All the components in the system have unique failure rates and repair times. Consequently, the value of ENS needs to be significantly lower in the proposed or reconfigured case compared to the base case. Thus, using the reliability indices is appropriate for evaluating the proposed objective.

11.3 RESULTS AND DISCUSSION

With the objective of maximizing loading capacity while minimizing losses, as detailed in Section 11.2, the proposed methodology is validated within the IEEE 33 bus systems. The objective of maximizing load capacity entails determining the maximum load that the system can handle while adhering to operational constraints.

- **LG with EV:** LG may also entail the integration of EVs into the system. EVs serve as dynamic energy sources, affecting system dynamics through their power demand during charging and supply during discharging. To address this, a dataset containing EV load profiles from [36] is integrated into the analysis.

The analysis is in two primary conditions:

- **Base Case:** This denotes the typical system setup with existing loads.
- **Proposed Case:** In this scenario, the base case undergoes reconfiguration with the objective of maximizing load capacity.

For the proposed case, reconfiguration is achieved using the PSO algorithm [4] and [30]. The outcomes of the test systems are sequentially presented for both the base case and the proposed case, accompanied by a comparative case drawn from previous literature. Additionally, reliability indices are computed for both the base and proposed cases.

11.3.1 IEEE-33 Bus System

In this section, the proposed approach aimed at maximizing load capacity, with a specific focus on the analysis carried out on the 33 bus system. Figure 11.6 illustrates the single-line diagram of the standard IEEE 33 bus distribution system. In the base case scenario, switches 33, 34, 35, 36, and 37 remain open. The line data, load data, and maximum allowable line limits are obtained from [37].

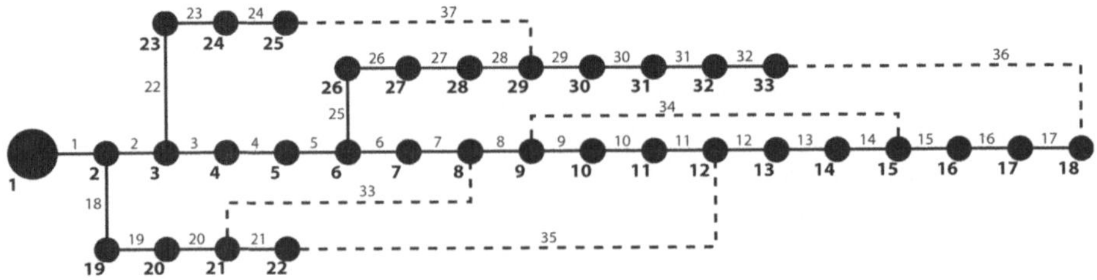

FIGURE 11.6 Single line diagram of IEEE 33 bus distribution system—base case.

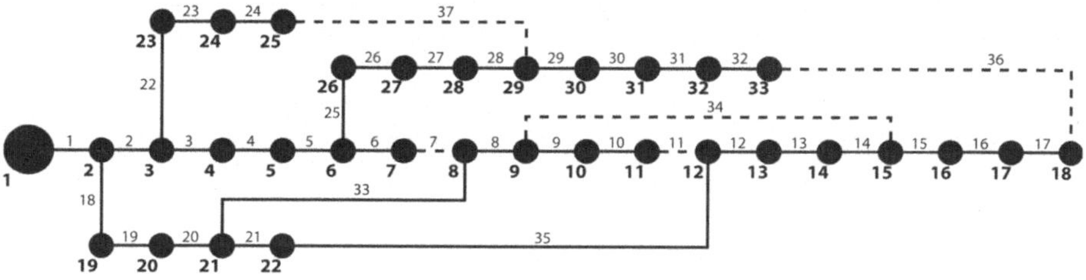

FIGURE 11.7 Single line diagram of IEEE 33 bus distribution system—proposed case.

Utilizing the PSO algorithm, the reconfiguration process leads to the opening of switches 7, 36, 34, 11, and 37 to achieve the defined objective. The single-line diagram of the proposed case is depicted in Figure 11.7.

The sizes and locations of DS inclusion are determined using the penetration level calculation Eqns. (11.19)–(11.22), with the optimal penetration level set at 80%. Further details regarding the sizes and locations are provided in Table 11.2. The impact of reconfiguration and DS allocation on the system's voltage magnitude has been analyzed. The values of the voltage magnitude of the base and the proposed case are given in Table. 11.2.

TABLE 11.2

2 Result Analysis of the Proposed Reconfiguration in Comparison with Base Case

Parameters	Base Case			Proposed Case				
Condition of DS	Without DS	With DS		Without DS	With DS			
Year	Year 0	Year 0	Year I	Year 0	Year I	Year II	Year III	Year IV
D-STATCOM Size (kVAr) Location	***	450 (19) 900 (26) 450 (23)		***	450 (19) 900 (26) 450 (23)			
Switches Open	33, 34, 35,36, 37			7, 36, 34, 11, 37				
P- Loss (kW)	202.677	159.704	188.514	144.537	115.906	135.796	159.362	187.29
% Loss Reduction	***	21.203	6.988	28.686	42.813	32.999	21.371	7.592
Sum KVA	33913	30890	33400	30028	26893	29064	31434	34017
Line Flow (A)	364.361	309.812	336.343	359.216	306.141	331.878	360.091	390.976
V min (pu)	0.9131	0.9230	0.9060	0.9336	0.9438	0.9385	0.9329	0.9267
TACS (USD/ Year)	***	22586.609	7444.336	30558.331	45606.943	35152.654	22766.364	8087.197

*** denotes that no value is computed.

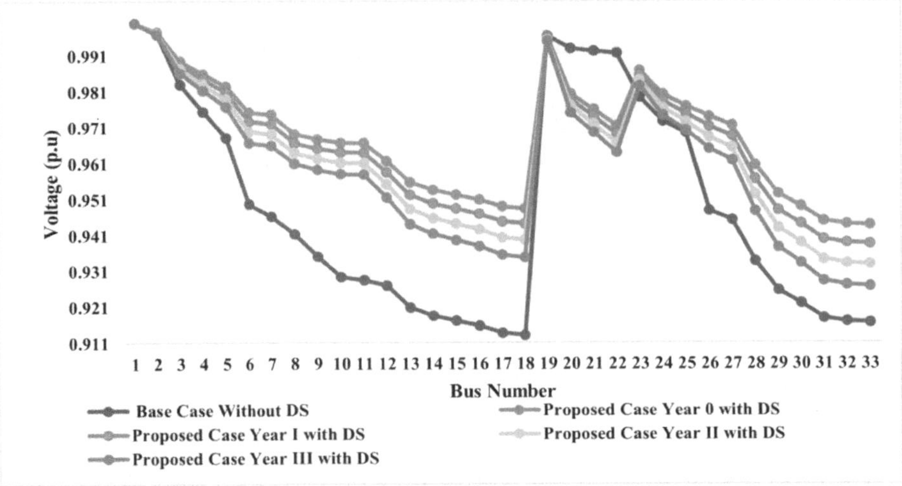

FIGURE 11.8 Voltage magnitude—33 bus system.

Upon analyzing, a comparison between the base case and the proposed case reveals significant improvements, particularly concerning line flow and losses. Notably, the proposed case showcases a considerable decrease in loss (kW) compared to the base case. First, the active power losses, branch flow, voltage magnitude, and the TACS of 'Year 0' of the base case (i.e., standard condition without any annual LG) are computed and considered as a reference case for all the following comparisons. The considered annual load growth rate is 7.5%. The voltage magnitudes of both the base case and the proposed case are shown in Figure 11.8.

After incorporating the DS to the base case, the considered parameters are evaluated. The loss is reduced to 21.203% and 6.988% in Years 0 and I, respectively. But, there is no evident improvement in line flow. Meanwhile, the TACS in Year 0 and Year I with DS are $22586.609 and $7444.336, respectively. The proposed case is a reconfigured case with the objective of line flow maximization. Initially, the proposed reconfigured is analyzed without allocating the DS. Without DS allocation, the proposed case shows loss reduction of 28.686% compared to the base case. The TACS is $30558.331.

Specifically after including DS, in Year 0, there is a loss reduction of approximately 42.813%, while in Year I, the decrease amounts to about 32.99%. The voltage magnitude is also improved from 0.9131 to 0.9438 p.u and 0.9385 p.u in Year 0 and I respectively. Meanwhile, the TACS in Year 0 and Year I with DS are $45606 and $35152.654 respectively. The line flow increased by 8.4% in Year I compared with Year 0. In Year II, the loss reduction is about 21.371% and the improvement in line flow is 8.5% with the TACS of $22766.364. In Year III, the loss reduction is about 7.592% and the improvement in line flow is 8.57% with the TACS of $8087.197.

The reliability analysis of the proposed work is shown in Table 11.3. The percentage decrease in ENS between the base case and the proposed case is approximately 28.50%. The percentage change in AENS between the base case and the proposed

TABLE 11.3
Reliability Analysis of 33 Bus System

Case/Index	Base Case	Proposed Case
ENS	7097.95	5075.15
AENS	2.1112	1.5654

case is approximately 25.84%. From the reliability analysis, it is clearly evident that there is a notable reduction in the average equivalent number of service interruptions per customer in the proposed case compared to the base case. These results emphasize the improvement in reliability achieved in the proposed case compared to the base case, with fewer service interruptions experienced by customers on average.

Table 11.4 depicts the comparison of existing reconfiguration with the proposed DS size and location. From Table 11.4, it is notable that the methods WOA [6] and

TABLE 11.4
Comparison of Existing Reconfiguration with Proposed DS Size and Location

Method	Parameters	Year 0	Year I	Year II	Year III
WOA [6]	Line flow (A)	308.387	334.41	**	**
(7, 8, 9, 27, 36)	Loss (kW)	141.113	163.8949	**	**
	TACS (USD/Year)	32358.0384	20383.8718	**	**
EOA [7]	Line flow (A)	307.98	333.96	**	**
(8, 27, 33, 34, 36)	Loss (kW)	137.051	159.3496	**	**
	TACS (USD/Year)	34493.0256	22772.8814	**	**
IHS [5] **(7, 9, 14, 28, 32)**	Line flow (A)	307.32	333.15	361.47	**
	Loss (kW)	129.2726	149.957	174.3263	**
	TACS (USD/Year)	38581.3526	27709.6320	14901.1279	**
ISSA [8] **(9, 7, 14, 28, 36)**	Line flow (A)	307.44	333.3	361.65	**
	Loss (kW)	130.7021	151.6879	176.4219	**
	TACS (USD/Year)	37830.0074	26799.8710	13799.6806	**
FF [9] **(8, 9, 28, 32, 33)**	Line flow (A)	307.74	333.69	362.13	**
	Loss (kW)	134.1267	155.8995	181.582	**
	TACS (USD/Year)	36030.0377	24586.2540	11087.5320	**
IEOA [7] **(7, 10, 13, 27, 31)**	Line flow (A)	308.91	335.1	363.81	395.28
	Loss (kW)	146.5461	170.2721	198.2868	231.4018
	TACS (USD/Year)	29502.4010	17032.0154	2307.4891	−15097.7549
ISCA [10] **(7, 9, 14, 28, 31)**	Line flow (A)	307.86	333.81	362.28	393.45
	Loss (kW)	134.7078	156.3163	181.7953	211.8618
	TACS (USD/Year)	35724.6115	24367.1839	10975.4215	−4827.5309
Proposed PSO (7, 36, 34, 11, 37)	Line flow (A)	306.15	331.89	360.09	390.99
	Loss (kW)	115.9058	135.796	159.362	187.2904
	TACS (USD/Year)	45606.9427	35152.6536	22766.3640	8087.1970

** denotes that no value is computed.

EOA [7] withstand the LG only for one year. After that, the operating limits get violated. While in the case of IHS [5], ISSA [8], and FF [9] methods, the system withstands for an LG of up to Year II. In the methods IEOA [7] and ISCA [10], the systems withstands LG for III years like the proposed PSO. But it is noteworthy that there are no savings in Year III. The negative value of TACS represents the same. The loss is also 19.06% and 11.59% higher in IEOA and ISCA methods, respectively compared with the loss achieved in the proposed PSO with a loss of 187.2904 kW. Hence, the proposed PSO is promising in the aspect of loadability maximization with reduced system loss.

Table 11.5 shows the comparison of the proposed reconfiguration with the existing DS size and location. From Table 11.5, it is notable that the methods DEA [6], CSA [26], and GSA [27] withstand the LG only for two years. After that, the operating limits get violated. The proposed PSO is promising in the aspect of loadability maximization with reduced system loss which goes up to Year III.

Similarly, the analysis of loadability through line flow (A) indicates notable enhancements in the proposed case relative to the base case. The cessation of LG by Year III is notable, as it breached the prescribed ampacity limits by Year IV.

This highlights a significant improvement in system performance, indicating more efficient management of power distribution. The proposed reconfiguration emerges as the more optimal choice, primarily due to the decrease in losses along

TABLE 11.5
Comparison of Proposed Reconfiguration with Existing DS Size and Location

Method	DS Size (kVAr) (Location)	Parameters	Year 0	Year I	Year II	Year III
DEA [25]	1252.7 (30)	Line flow (A)	314.55	341.07	369.93	**
		Loss (kW)	104.5086	120.4949	139.7012	**
		TACS (USD/Year)	51597.3110	43194.9117	33100.0804	**
CSA [26]	1370.5 (23)	Line flow (A)	314.73	341.61	370.92	**
		Loss (kW)	134.0663	156.0225	181.8751	**
		TACS (USD/Year)	36061.7839	24521.6052	10933.4786	**
GSA [27]	927 (27)	Line flow (A)	324.03	351.39	381.12	**
		Loss (kW)	117.6895	137.9476	161.9098	**
		TACS (USD/Year)	44669.4300	34021.7726	21427.2403	**
Proposed PSO	450 (19)	Line flow (A)	306.15	331.89	360.09	390.99
	900 (26)	Loss (kW)	115.9058	135.796	159.362	187.2904
	450 (23)	TACS (USD/Year)	45606.9427	35152.6536	22766.3640	8087.1970

** denotes that no value is computed.

with the improvement in line loading. This has been supported by evaluating the reliability of the base case and proposed case. In the proposed case, the value of the ENS and AENS index is lesser which, in turn, shows the improved system reliability.

11.4 CONCLUSION

In the analysis of enhancing the loading capacity (loadability) of distribution systems, the study meticulously examines scenarios with and without the integration of DSs and EVs. Leveraging the standardized IEEE 33 bus test system, the focus lies on load growth, and the maximum line loading capacity of branches. Additionally, a thorough reconfiguration analysis is conducted to measure system response to load growth dynamics.

The findings underscore the significance of reconfiguration in optimizing the siting and sizing of DS, particularly evident in the substantial improvements observed within the IEEE 33 bus systems. Here, the proposed configurations exhibit remarkable stability, maintaining operational limits up to Year IV without violating loading constraints. Beyond these technical nuances, the analysis also sheds light on the broader implications for real-world DN planning and management. By significantly reducing losses and optimizing line flow, the proposed case offers tangible improvements in operational efficiency. This indicates that the proposed configuration potentially offers higher system reliability, as evidenced by the lower values of reliability indices.

Looking ahead, the ongoing work aims to expand the scope to encompass concurrent load growth across diverse consumer segments, alongside reliability enhancement and loss minimization strategies. Additionally, exploring energy management strategies post-implementation is anticipated to further refine system performance.

ACKNOWLEDGMENT

The authors like to acknowledge the University Grants Commission (UGC) India, for the financial support to Sivasankari GS under the scheme Savitribai Jyotirao Phule Fellowship For Single Girl Child (SJSGC). Reference ID: 202223 - UGCES-22-OB-TAM-F-SJSGC-9020.

APPENDIX

This test system and its data are referred from [25]. The single line diagram of the standard IEEE 33 bus system is shown in Figure 11.6. The total real and reactive power loads on the system are 3715 kW and 2300 kVAr, respectively. The voltage magnitude of the system is $1\angle 0$ p.u.

TABLE 11.6
System and Load Parameters for 33-bus Distribution System

Branch Parameters				Load on - To Bus		Reliability Parameters	
From Bus	To Bus	Resistance (Ω)	Reactance (Ω)	P (kW)	Q (kVAr)	Failure Rate (f/yr)	Repair Time (hours)
1	2	0.0922	0.0470	100.00	60.00	0.0500	1.0000
2	3	0.4930	0.2511	90.00	40.00	0.3000	1.0000
3	4	0.3660	0.1864	120.00	80.00	0.2200	1.0000
4	5	0.3811	0.1941	60.00	30.00	0.2300	1.0000
5	6	0.8190	0.7070	60.00	20.00	0.5100	1.0000
6	7	0.1872	0.6188	200.00	100.00	0.1100	1.0000
7	8	0.7114	0.2351	200.00	100.00	0.4400	1.0000
8	9	1.0300	0.7400	60.00	20.00	0.6400	1.0000
9	10	1.0440	0.7400	60.00	20.00	0.6500	1.0000
10	11	0.1966	0.0650	45.00	30.00	0.1200	1.0000
11	12	0.3744	0.1238	60.00	35.00	0.2300	1.0000
12	13	1.4680	1.1550	60.00	35.00	0.9100	1.0000
13	14	0.5416	0.7129	120.00	80.00	0.3300	1.0000
14	15	0.5910	0.5260	60.00	10.00	0.3600	1.0000
15	16	0.7463	0.5450	60.00	20.00	0.4600	1.0000
16	17	1.2890	1.7210	60.00	20.00	0.8000	1.0000
17	18	0.7320	0.5740	90.00	40.00	0.4500	1.0000
2	19	0.1640	0.1565	90.00	40.00	0.1000	0.5000
19	20	1.5042	1.3554	90.00	40.00	0.9300	0.5000
20	21	0.4095	0.4784	90.00	40.00	0.2500	0.5000
21	22	0.7089	0.9373	90.00	40.00	0.4400	0.5000
3	23	0.4512	0.3083	90.00	50.00	0.2800	0.5000
23	24	0.8980	0.7091	420.00	200.00	0.5600	0.5000
24	25	0.8960	0.7011	420.00	200.00	0.5500	0.5000
6	26	0.2030	0.1034	60.00	25.00	0.1200	0.5000
26	27	0.2842	0.1447	60.00	25.00	0.1700	0.5000
27	28	1.0590	0.9337	60.00	20.00	0.6600	0.5000
28	29	0.8042	0.7006	120.00	70.00	0.5000	0.5000
29	30	0.5075	0.2585	200.00	600.00	0.3100	0.5000
30	31	0.9744	0.9630	150.00	70.00	0.6000	0.5000
31	32	0.3105	0.3619	210.00	100.00	0.1900	0.5000
32	33	0.3410	0.5302	60.00	40.00	0.2100	0.5000
8	21	2.0000	2.0000	0.00	0.00	1.2400	0.5000
9	15	2.0000	2.0000	0.00	0.00	1.2400	0.5000
12	22	2.0000	2.0000	0.00	0.00	1.2400	0.5000
18	33	0.5000	0.5000	0.00	0.00	0.3100	0.5000
25	29	0.5000	0.5000	0.00	0.00	0.3100	0.5000

Other Data: Current carrying capacity of branches 1-5 is 400 A,
6-7 is 300 A, 25-27 is 300 A and for all other branches including tie lines is 200 A

REFERENCES

1. T. T. Nguyen, and A. V. Truong, "Distribution network reconfiguration for power loss minimization and voltage profile improvement using cuckoo search algorithm," *International Journal of Electrical Power & Energy Systems*, vol. 68, pp. 233–242, 2015.
2. J. Zhan, W. Liu, C. Chung, and J. Yang, "Switch opening and exchange method for stochastic distribution network reconfiguration," *IEEE Transactions on Smart Grid*, vol. 11, no. 4, pp. 2995–3007, 2020.
3. A. M. Tahboub, V. R. Pandi, and H. Zeineldin, "Distribution system reconfiguration for annual energy loss reduction considering variable distributed generation profiles," *IEEE Transactions on Power Delivery*, vol. 30, no. 4, pp. 1677–1685, 2015.
4. M. Aman, G. Jasmon, A. H. A. Bakar, and H. Mokhlis, "A new approach for optimum simultaneous multi-DG distributed generation units placement and sizing based on maximization of system loadability using HPSO (hybrid particle swarm optimization) algorithm," *Energy*, vol. 66, pp. 202–215, 2014.
5. A. Tyagi, A. Verma, and P. Bijwe, "Reconfiguration for loadability limit enhancement of distribution systems," *IET Generation, Transmission & Distribution*, vol. 12, no. 1, pp. 88–93, 2018.
6. E. Mahdavi, S. Asadpour, L. H. Macedo, and R. Romero, "Reconfiguration of distribution networks with simultaneous allocation of distributed generation using the whale optimization algorithm," *Energies*, vol. 16, no. 12, p. 4560, 2023.
7. A. M. Shaheen, A. M. Elsayed, R. A. El-Sehiemy, and A. Y. Abdelaziz, "Equilibrium optimization algorithm for network reconfiguration and distributed generation allocation in power systems," *Applied Soft Computing*, vol. 98, p. 106867, 2021.
8. R. Fathi, B. Tousi, and S. Galvani, "Allocation of renewable resources with radial distribution network reconfiguration using improved Salp swarm algorithm," *Applied Soft Computing*, vol. 132, p. 109828, 2023.
9. O. Badran, H. Mokhlis, S. Mekhilef, W. Dahalan, and J. Jallad, "Minimum switching losses for solving distribution NR problem with distributed generation," *IET Generation, Transmission & Distribution*, vol. 12, no. 8, pp. 1790–1801, 2018.
10. U. Raut, and S. Mishra, "An improved sine–cosine algorithm for simultaneous network reconfiguration and dg allocation in power distribution systems," *Applied Soft Computing*, vol. 92, p. 106293, 2020.
11. J. Singh, and R. Tiwari, "Real power loss minimisation of smart grid with electric vehicles using distribution feeder reconfiguration," *IET Generation, Transmission & Distribution*, vol. 13, no. 18, pp. 4249–4261, 2019.
12. Y. Li, G. Hao, Y. Liu, Y. Yu, Z. Ni, and Y. Zhao, "Many-objective distribution network reconfiguration via deep reinforcement learning assisted optimization algorithm," *IEEE Transactions on Power Delivery*, vol. 37, no. 3, pp. 2230–2244, 2021.
13. H. U. R. Habib, A. Waqar, S. Sohail, A. K. Junejo, M. F. Elmorshedy, S. Khan, Y.-S. Kim, and M. M. Ismail, "Optimal placement and sizing problem for power loss minimization and voltage profile improvement of distribution networks under seasonal loads using Harris Hawks optimizer," *International Transactions on Electrical Energy Systems*, vol. 2022, 2022.
14. A. Waqar, U. Subramaniam, K. Farzana, R. M. Elavarasan, H. U. R. Habib, M. Zahid, and E. Hossain, "Analysis of optimal deployment of several DGS in distribution networks using plant propagation algorithm," *IEEE Access*, vol. 8, pp. 175546–175562, 2020.
15. S. Kamel, M. Khasanov, F. Jurado, A. Kurbanov, H. M. Zawbaa, and M. A. Alathbah, "Simultaneously distributed generation allocation and network reconfiguration in distribution network considering different loading levels," *IEEE Access*, vol. 11, pp. 105916–105934, 2023, http://dx.doi.org/10.1109/ACCESS.2023.3319456.

16. A. Raza, M. Zahid, J. Chen, S. M. Qaisar, T. Ilahi, A. Waqar, and A. Alzahrani, "A novel integration technique for optimal location & sizing of dg units with reconfiguration in radial distribution networks considering reliability," *IEEE Access*, vol. 11, pp. 123610–123624, 2023, http://dx.doi.org/10.1109/ACCESS.2023.3329704.
17. M. Honarmand, A. Zakariazadeh, and S. Jadid, "Integrated scheduling of renewable generation and electric vehicles parking lot in a smart microgrid," *Energy Conversion and Management*, vol. 86, pp. 745–755, 2014.
18. L. Jian, Y. Zheng, and Z. Shao, "High efficient valley-filling strategy for centralized coordinated charging of large-scale electric vehicles," *Applied Energy*, vol. 186, pp. 46–55, 2017.
19. Y. Cao, S. Tang, C. Li, P. Zhang, Y. Tan, Z. Zhang, and J. Li, "An optimized EV charging model considering TOU price and SOC curve," *IEEE Transactions on Smart Grid*, vol. 3, no. 1, pp. 388–393, 2011.
20. J. E. Cardona, J. C. López, and M. J. Rider, "Decentralized electric vehicles charging coordination using only local voltage magnitude measurements," *Electric Power Systems Research*, vol. 161, pp. 139–151, 2018.
21. F. Un-Noor, S. Padmanaban, L. Mihet-Popa, M. N. Mollah, and E. Hossain, "A comprehensive study of key electric vehicle (EV) components, technologies, challenges, impacts, and future direction of development," *Energies*, vol. 10, no. 8, p. 1217, 2017.
22. G. Mills, and I. MacGill, "Assessing electric vehicle storage, flexibility, and distributed energy resource potential," *Journal of Energy Storage*, vol. 17, pp. 357–366, 2018.
23. R. Sirjani, and A. R. Jordehi, "Optimal placement and sizing of distribution static compensator (D-STATCOM) in electric distribution networks: A review," *Renewable and Sustainable Energy Reviews*, vol. 77, pp. 688–694, 2017.
24. E. Shahryari, H. Shayeghi, and M. Moradzadeh, "Probabilistic and multi-objective placement of D-STATCOM in distribution systems considering load uncertainty," *Electric Power Components and Systems*, vol. 46, no. 1, pp. 27–42, 2018.
25. J. Sanam, S. Ganguly, A. Panda, and C. Hemanth, "Optimization of energy loss cost of distribution networks with the optimal placement and sizing of DSTATCOM using differential evolution algorithm," *Arabian Journal for Science and Engineering*, vol. 42, pp. 2851–2865, 2017.
26. T. Yuvaraj, K. Ravi, and K. Devabalaji *et al.*, "Optimal allocation of DG and dstatcom in radial distribution system using cuckoo search optimization algorithm," *Modelling and Simulation in Engineering*, vol. 2017, 2017.
27. A. K. Arya, A. Kumar, and S. Chanana, "Analysis of distribution system with D-STATCOM by gravitational search algorithm (GSA)," *Journal of The Institution of Engineers (India): Series B*, vol. 100, pp. 207–215, 2019.
28. "IEEE guide for electric power distribution reliability indices," in *IEEE Std 1366-2022 (Revision of IEEE Std 1366-2012)*, pp. 1–44, 2022, http://dx.doi.org/10.1109/IEEESTD.2022.9955492.
29. S. Kawambwa, R. Mwifunyi, D. Mnyanghwalo, N. Hamisi, E. Kalinga, and N. Mvungi, "An improved backward/forward sweep power flow method based on network tree depth for radial distribution systems," *Journal of Electrical Systems and Information Technology*, vol. 8, pp. 1–18, 2021.
30. N. Gupta, A. Swarnkar, and K. R. Niazi, "Reconfiguration of distribution systems for real power loss minimization using adaptive particle swarm optimization," *Electric Power Components and Systems*, vol. 39, no. 4, p. 317–330, Feb. 2011. [Online]. Available: http://dx.doi.org/10.1080/15325008.2010.528532
31. M. R. Bonyadi, and Z. Michalewicz, "Impacts of coefficients on movement patterns in the particle swarm optimization algorithm," *IEEE Transactions on Evolutionary Computation*, vol. 21, no. 3, pp. 378–390, 2016.

32. S. Shwetha, S. B. RG, K. Narayanan, A. Sharma, A. Aguila, S. Shwetha, and K. Narayanan, "Analysis of distribution systems in the presence of electric vehicles and optimal allocation of DGS considering power loss and VSI," 2023.
33. R. S. Rao, S. Narasimham, and M. Ramalingaraju, "Optimal capacitor placement in a radial distribution system using plant growth simulation algorithm," *International Journal of Electrical Power & Energy Systems*, vol. 33, no. 5, pp. 1133–1139, 2011.
34. N. Gupta, A. Swarnkar, and K. Niazi, "Distribution network reconfiguration for power quality and reliability improvement using genetic algorithms," *International Journal of Electrical Power & Energy Systems*, vol. 54, pp. 664–671, 2014.
35. D. Kumar, S. Samantaray, I. Kamwa, and N. Sahoo, "Reliability-constrained based optimal placement and sizing of multiple distributed generators in power distribution network using cat swarm optimization," *Electric Power Components and Systems*, vol. 42, no. 2, pp. 149–164, 2014.
36. P. Senthilkumar, S. Samidurai, V. Nedunchezian, N. Krishnan, G. Sharma, and T. Senjyu, "Profit maximization in hybrid microgrid system incorporating demand response and plug in EV," *International Transactions on Electrical Energy Systems*, vol. 31, no. 12, p. e13260, 2021.
37. W. Zhou, Y. Wang, F. Peng, Y. Liu, H. Sun, and Y. Cong, "Distribution network congestion management considering time sequence of peer-to-peer energy trading," *International Journal of Electrical Power & Energy Systems*, vol. 136, p. 107646, 2022.

12 A Review of Power Trading Models between Distribution Systems

The South African Context

Zipho Khumalo, Pitshou N. Bokoro, and Gulshan Sharma

12.1 INTRODUCTION

Prior to 1994, a significant portion of the South African population did not have access to electricity [1]. This led to a rapid implementation of electrification projects, which gave access to both on and off-grid electricity for over 6.7 million citizens between 1994 and 2016 [2]. These connections were predominantly implemented at the distribution level across the country, while little attention was afforded to generation and transmission capability. The provision of new electricity connections to communities at the current rate of 5% to 10% per annum is insufficient to achieve the targeted universal grid access to citizens by 2030 [3]. The little attention paid to electricity generation and transmission has contributed to electricity supply constraints, leading to planned rotational power cuts problem experienced in South Africa. Rotational power cuts, or load shedding, are last-resort measures initiated to prevent system collapse as soon as the electricity demand exceeds available supply capacity. In South Africa, this is achieved in collaboration with municipal electricity utilities [1]. According to the Council for Scientific and Industrial Research [4], a total of 2455 GWh of energy has been load curtailed by Eskom between 2007 and 2021. Under the current South African electricity sector structure, the following challenges were encountered during the 2022/2023 financial year alone: R58.5 billion in arrear debt owed to Eskom by municipalities, Eskom received a total of 80 environmental legislation contraventions and South Africa experienced total of 280 days of power cuts. The impact of power cuts has been directly related to the gross domestic product decline of the country. The cost to the South African economy of one full day of stage 1 load shedding is approximated to be R235.5 million [5]. The increase in reliability requirements and network operating costs has increased the burden of raising prices of electricity distribution tariffs. This poses a challenge for electricity distribution utilities [6]. The price escalation to consumers has encouraged electricity utility customers to adopt energy optimization measures, curtail electricity utilization or to go off the grid completely. This has negatively impacted the revenue generated by South African municipal utilities from the sale of electricity. South African municipalities are subjected to the challenge of buying electricity

DOI: 10.1201/9781003537823-16

from the national electricity utility, Eskom, and re-selling it to consumers at additional costs [7].

The trading of power between municipal electricity distribution systems, currently taking place in South Africa, is reliant on the services of the Eskom national grid and related tariffs. Municipalities have tariffs that are specifically tailored for their customers that are municipalities. These tariffs are based on Eskom's Megaflex tariff with a surcharge incorporated and are applicable where the inter-municipal power trading is taking place [8, 9]. The rapid population expansion in Southern Africa has made it challenging for utilities in the region to supply peak load demands of customers. This has necessitated the upgrade of power system infrastructure to enhance electricity production. Unfortunately, Eskom's coal-based power stations are old and require prolonged unplanned maintenance [10]. Mbanda et al. [11] suggest that the rapid energy demand increase in Southern Africa is a result of industrialization partnered with political and financial stability in the region. Energy trading was the solution alluded to in [11] to solve the electricity supply shortage. A negative relationship between electricity supply and the population growth rate in South Africa has been highlighted by Hlongwane and Daw [12], who further recommended a revision of government and industry policies to mitigate the power cut problem in South Africa. Population growth and industrialization in South Africa are further highlighted by Madoda et al. [13] as the challenges contributing to the electricity supply shortage experienced by the country, leading to power cuts to citizens. The coal-fired power stations providing 85% of the country's electricity are aged and fast approaching the culmination of their use life cycle [13]. As early as the year 2010 [14] expressed the strain on the national power grid imposed by industrialization and population growth. Most problems experienced in coal-fired power stations can be attributed to boilers, which are part of the energy conversion process in this category of power station [13]. At Eskom's Majuba power station alone, a total of R933 million was credited to rectify the 69 unplanned cycling operations start in 2016. Sixty percent of those outages were attributed to boiler-related failures. The resultant damage and repairs have augmented the rate of aging of the infrastructure in the power station [15]. According to Davidson et al. [16], the national power utility in South Africa has struggled to supply the electricity demands of the increasing population since 2010. This has resulted in the utility implementing power cuts to citizens. Aging infrastructure and growing electricity costs are the contributors to the challenges faced by South Africa's national utility Eskom, which supplies 95% of the power generation, 60% of electricity distribution and owns the country's transmission grid [16]. Mushavhanamadi et al. [17] described South Africa's electricity sector as being in a state of crisis. With South Africa's economy in the post-COVID-19 recovery process, Eskom's electricity supply shortage has damagingly been impacted by deteriorating generations ascribed to aging power station infrastructure. This has led to an estimated weekly occurrence of power cut by the national utility [17].

Over 90% of the electricity generated in South Africa is from Eskom, the national power utility [18]. Being a regulated industry, South African electricity distribution utilities are subjected to licensing requirements prescribed by the national electricity regulator [19]. The electricity industry practices and orientation have resulted in constraints to the electricity grid, resulting in the power cut problem experienced by the country [1, 5, 11, 13, 14, 16, 17]. Based on the orthodox industry trends, The National Energy Regulator of South Africa has raised concerns termed over the unsustainability of the electricity systems model of operation within the current market [17, 20]. The situation has a particularly negative impact on the South African government given that of the 2574 municipalities existing in South Africa, 70% earned over a quarter of their total revenue income from electricity sales [21]. The deterioration in electricity supply infrastructure and electricity pricing that was historically below economic viability have resulted in tariff increases at rates that consumers are incapable of absorbing. The performance of the electricity sector has resulted in South Africa falling in ranking on international benchmarks [22]. With a baseline of 1.285 million new customer connections to the national electricity grid and 85,993 non-grid electrification connections between 2014 and 2019, the South African government aims to achieve a 97% population target of access to basic electricity by 2030 [23]. Davidson et al. [11] specify that the African progressions in attaining the global targets of universal access to electricity are an active illustration of hope. Therefore, regional power amalgamation offers an optimal option for enabling synchronized and rapid access to electricity [11]. The concept of peer-to-peer power-sharing enables power system users to trade within a localized competitive market and contributes significantly to eliminating electricity supply and demand mismatching of neighboring power system users [24]. Zhou et al. [25] suggest that peer-to-peer energy trading is the process of enabling electricity consumers with electricity generation capacity to share and/or trade electrical energy. The process enables energy balance and improved management of distributed energy resources, which have become a common feature of modern power systems. However, it is essential to consider peer-to-peer energy trading in the context of public power systems because most distributed energy resources are connected to public networks. There has been extraordinarily little bulk commercial utilization of these methods throughout the world markets [25]. The upscaling of power trade can in principle reduce the cost of electricity supply and contribute to improved levels of access to electricity within communities [26]. According to Xie et al. [27], the optimized incorporation of clean energy within the power trading market may contribute to improved power grid security, electricity generation efficiency and electricity supply and demand balancing. The transition to a green energy-based electricity market requires a thorough understanding of demand side management practices, as it introduces a degree of uncertainty in the power system user's consumption and inducement. This brings challenges to the decision-making process for electricity sellers. A financial incentive-focused decision-making strategy was developed by Wang et al. [28] to solve this problem within an unregulated electricity market. Demand side management strategies have also been employed in [29] to match electricity supply and demand in constrained power systems and in power systems with a high injection of distributed energy resources. The method was efficient in a flexible load customer base. Davidson et al. [30] identified the burden on South Africa's electrical infrastructure triggered by the electrification projects described in [1]. These new connections resulted in infringements to technical performance parameters such as voltage limits, thermal loading, and technical losses. South African electricity utilities' remedy to these challenges require enormous amounts in capital expenditure and extended periods of time for implementation. Distributed generation incorporation at distribution level of power systems is one of the feasible short-term, less capital-intensive solutions available [30]. Khan et al. [31] established that the privatization of Eskom Generation and introducing wholesale retail competition to the South African power trade market would not be a feasible short to medium-term solution for the markets' problem and may lead to increased electricity costs to the consumers [31]. As stated in [22], South African electricity users are burdened by and inept to absorb the electricity tariff increases already imposed on them. The electricity supply shortage mitigation strategies prescribed in the literature predominantly encompass power trading and variations of renewable energy resource dispatchment. Power-sharing is one of the fundamental approaches in the evolving electricity market. It is imperative to explore power trading in power systems with distributed energy resources given the increasing number of consumers installing solar panels and battery storage schemes onto their internal reticulation systems. The presence of distributed generation changed the nature of consumers within the power system ushering in prosumers. Prosumers are electricity customers who can consume as well as generate electricity in the power system [11]. The purpose of this chapter is to review the different technics that enable power trade and sharing between distribution power systems considering distributed generation resource availability. The chapter aims to review the different power trading strategies that have been and are being deployed in various parts of the world to alleviate the demand vs supply shortages in power trading.

This chapter is structured as follows: Section two provides a critical review of the relevant power trading models reported in the literature and in practical use across electricity markets. Section three introduces major power trading frameworks adopted in transmission and distribution systems. Section four deals with the state-of-the-art pertaining to currently applied models and algorithms currently to power trading problems. The last section of this chapter consists of a discussion on the advantages and disadvantages associated with the existing power trading models and algorithms.

12.2 RELATED WORK

This section consists of a systematic review of power trading models between electricity distribution power systems and markets.

Ji et al. [32] set a model that utilizes smart contracts based on the highest offer principle in Tianjin municipality, China. This principle determines the lowest combined operation cost and then prescribes an electricity price based on average profit sharing among the prosumers connected to the distribution network. Each generator or prosumer submits an offer that specifies all the associated power settings and related costs. The smart contract that is programmed at each of the soft open points on the network evaluates the physical feasibility of each offer received during the trading period, before implementing the power trade configuration on the distribution network. Power is regulated at individual soft open points within the distribution network. This regulation is used to automatically implement control instructions by the network operator. The regulation is dictated to by trade contracts that are based on power volumes and varying tariff charges. The proposed method was evaluated on an existing flexible distribution network with four soft operating points over a 24-hour period. The result is that the distribution system operating costs were reduced. The need for distributed power generation curtailment is significantly reduced and the voltage profile of the distribution network was improved (more aligned to prescribed limits). The overall power demand from the main power grid was reduced. The impact and regulation of reactive power were not considered. The system was evaluated on flexible distribution networks with soft open points, which are not common in industry. The performance of regular distribution networks remains to be assessed [32]. A blockchain-based strategy is discussed in [33], which was concentrated on transaction security. Chen et al. [34] proposed and evaluated an auction theory method that reduced physical distribution network power losses and promoted increased distributed generation power utilization among the market participants in Guangdong province. This increase may be undesirable in some markets. The model input is derived from the consumer demand requirements, the available energy to sell by the prosumer and the main grid conditions. The market participant assesses if selling power and fulfilling grid conditions is beneficial to themselves. The power objective function at each participant is given as follows:

$$P_{i,t} = sP_{i,t} + fP_{i,t} \tag{12.1}$$

where $P_{i,t}$ is the total load consumption of the *i*th soft open point at time *t*, $sP_{i,t}$ is the prosumer's flexible load at time *t* and $fP_{i,t}$ represents prosumer i's fixed load at time *t*. Equation (12.2) is used to establish if the prosumer will be participating as seller or buyer in the market.

$$nP_{i,t} = P_{i,t} - E_{i,t} \tag{12.2}$$

The market participants' role is determined by $nP_{i,h}$. If $nP_{i,t} > 0$, the market participant is a buyer and when $nP_{i,t} < 0$, their role is that of a seller. $E_{i,t}$ represents the power generated by prosumer i at time t. Should the conditions be favorable, then a matching between buyer and seller was determined to find the optimal transaction (buyer/seller market participants). The seller and buyer matching are given by Equation (12.3):

$$M(i_s, i_b) = \left(P_{P2P,i_s,i_b,t} \middle| P_{S,i_s,i_b,t} = P_{B,i_s,i_b,t}, i_s \in N_s, i_B \in N_B \right) \tag{12.3}$$

The matching pair $M(i_s, i_b)$ is a function of seller i_s and buyer i_b. Where $P_{P2P,i_s,i_b,t}$ is the power trading matrix composed of power trading between seller i_s and buyer i_b. $P_{S,i_s,i_b,t}$ and $P_{B,i_s,i_b,t}$ are the energy sold by seller i_s to buyer i_b and the energy bought by seller i_s to buyer i_b, respectively. The model framework consists of the cyber layer and the physical layer. The distribution system operator maintains control of the network operations. Wireless channels are used to facilitate communication. The power seller's available capacity is matched with a purchaser's power shortage. The facilitated power-sharing is constrained by the physical capacity of the distribution grid therefore the power-sharing schedules are readjusted based on the grid operating conditions. To attain a degree of alignment with the grid operating parameters, the objective function Equation (12.4) is employed.

$$\min \Delta E_{\Sigma}\left(\phi, sP_{i,t}, P_{P2P,t}\right)$$
$$= \sum_{t=1}^{T}\sum_{i\in N}\sum_{k=1}^{N-1}\left(\eta_{k,t}R_{k}\frac{\left(\Gamma_{k}\left(nP_{i,t}-\left(P_{s,i,t},P_{b,i,t}\right)\right)+\eta_{i_s,i_b,t}^{H}\left(P_{P2P,t}A_{k}\right)\right)^{2}+Q_{k,t}^{2}}{U_{k,t}^{2}}+\gamma_{i_s,i_b}\left(P_{s,i,t},P_{b,i,t}\right)\right) \tag{12.4}$$

where H is the maximum number of iterations, η represented the state of the electrical switch between the buyer and seller considering multiple branches within the network, γ_{i_s,i_b} is the loss factor between the electrical nodes separating the buyer and seller access to the network. R_k equates to the resistance in branch k, Γ_k is the power flow calculation function within the kth branch, ϕ is the *kth* branch electrical switch state at a specified time. Main grid factors emphasized are the node voltages, current carrying capacity, and the radial structure of the network (which must accommodate bidirectional power flow). The grid quality of supply parameters is adjusted by sectionalizing the network. The power losses associated with the energy trade are compensated for by the prosumer as part of the transaction costs. This model is excessively reliant on the predicted generation capacity even for prices. The switch-based controlling strategy requires a smart distribution grid and frequent equipment operation that has proven detrimental in countries like South Africa, from the implementation of rotational power cuts [34]. From the power system context, there is close to no control over the current being used by the system therefore power quality predominantly implies controlling voltage within prescribed limits [35]. A voltage sampling at periodic intervals controller-unit model was projected in [36] for power-sharing among microgrids in Shanghai, China. The method is based on load sharing between two direct current bus supplied microgrids, one of which is connected to the main power grid and the other is not. Bidirectional DC-to-DC converters are utilized to vary the system voltage to stabilize and reduce the voltage drop on the DC buses. Due to the low voltages and low transmission distances, the distribution line power losses are considered negligible. Power outage durations due to no available capacity are decreased. The model does not show evidence of adaptability to distribution networks with numerous customers and higher distribution voltages [36]. In the Henan province of China, a three-layer energy sharing model is prescribed by Lui et al. [37]. The grid connected to the top layer consisted of energy hubs that generated electrical and heat energy. The middle layer is composed of the hub energy management system (of interest to this study), microturbines, and waste heat boilers. The bottom layer comprised the power-sharing market prosumers and consumers of energy. Prosumers can only buy thermal energy but are able to buy and sell electrical energy. Constrained optimization contractual agreements were implemented between power-sharing market participants. The energy hubs determined the contracts for each prosumer by calculating the derivative of the utility function, Equation (12.5):

$$u^{e}_{i,t} = \begin{cases} \theta^{e}_{i,t} \ln\left(1 + p^{e}_{i,t} + R_{i}\right) - r^{e}_{i,t} - d_{t} \times R_{i}, p^{e}_{i,t} \geq 0 \\ -\left(\theta^{e}_{i,t}\right) \ln\left(1 + p^{e}_{i,t} + R_{i}\right) - r^{e}_{i,t} - d_{t} \times R_{i}, p^{e}_{i,t} < 0 \end{cases} \tag{12.5}$$

The function $\theta^{e}_{i,t} \ln\left(1 + p^{e}_{i,t} + R_{i}\right)$ is applied to determine the energy consumption "$p^{e}_{i,t} + R_{i}$" by prosumer i and r represents the reward payment for market participation reward. Where d_{t} is the unit cost of energy production, θ is the prosumer's type and willingness-to-pay factor, p represents the energy traded between market participants, and R indicates the photovoltaic production. This considers the prosumers' willingness to buy factor and the prices offered by the prosumer. Contracts are drawn up an hour before the actual transaction takes place. The heat energy utility function is separate from the electrical and incorporates a heat transfer delay time that is incurred in the heat transmission pipes. For model feasibility, each prosumer's utility function is required to be greater than zero and the prosumer only achieved maximum benefits if it chose the contract aligned with its prosumer type classification. The energy hubs had their own operating contract utility function which considered the energy trade with prosumers, trade with other energy hubs, and self-operating costs. In the upper level of the power trading, consisting of energy hubs and the grid, a social welfare factor is included in the market operating equation. This is done to promote fairness and efficiency among the market participants. Voltage violations that occur during operation are rectified by optimizing the system's reactive power. This model is not adaptable to various system reconfigurations and abnormalities [37]. An energy cost reduction from 17.12% up to 28.91% is achieved by Chen et al. [38] in Chengdu, China. A cooperative Nash Equilibrium strategy was utilized to achieve welfare within the market using the cost-based objective function Equation (12.6) expressed as follows:

$$min\left[C_{i,user\_bat} + C_{i,user\_grid} + C_{i,load\_adjust} + \sum\nolimits_{t=1}^{T} \nu_{i,t} + \mu_{i}\tau_{i}\right] \tag{12.6}$$

where $C_{i,user\_bat}$ represents the cost to the prosumer for trading with embedded battery storage schemes (if applicable), $C_{i,user\_grid}$ is the market participants cost of energy from the distribution system operator grid, $C_{i,load\_adjust}$ denotes the cost of any load adjustments implemented by the prosumer, μ_{i} accounts for the random error in distributed generation output, τ_{i} was the uncertainty factor, and $\nu_{i,t}$ is the integer inequality variable governed by Equation (12.7):

$$0 \leq Z^{t}_{i} \leq 1, \; \forall t \in T, \forall i \in N : vi,t \tag{12.7}$$

where Z is an auxiliary variable introduced to mitigate the penalties of price deviations, t is the sampling point of time, and T is the maximum time. The model was not dynamic to real-time power quality factors within the power-sharing scheme and relied on market condition predictions [38]. On Bangladesh networks, Khan et al. [39] used cost objective-based methodology to encourage power-sharing participation among grid-connected prosumers and to reduce the demand on the distribution grid during periods of peak demand. The distribution grid operator governed the power-sharing using the mathematical strategy summarized in Equation (12.8):

$$P_{s,g} = 2a\left(E_p - E_{thresh}\right) + b \tag{12.8}$$

where $P_{s,g}$ is the distribution grid operator's price for selling power, E_{thresh} is the anticipated peak demand on the distribution grid for the period in consideration, a is variable of preference by the individual consumer based on their level of willingness to purchase or sell energy, E_p is the available capacity on the distribution grid, and b is a natural logarithm-based function of a. The results indicated a 28% reduction in the cost of electricity to consumers. The model is not dynamic to large-scale power system operations due to its reliance on the lambda-iteration method applied to the auction strategy to achieve social welfare. This is due to the time required for iterations to establish economic dispatchment calculations [39].

Eskom's time of use tariff costs was used in benchmarking the cost of electricity from the national grid in the model proposed by Kusakana [40] for the South African market. Power flow within the localized grid between the prosumers was considered free. Prosumers are primarily required to use their own generation, secondary to sharing the excess power generated. The load balance is determined individually for each prosumer. The sum of the generated power is required to be less than or equal to the maximum generation power of each prosumer. The model autonomously ensured that power flow from distributed generation and battery systems is always within the manufacturer's prescribed design specification, using nonlinear programming methods. Given the equations used to govern the model limits, the model is presented as a nonlinear programming problem and any tool able to solve nonlinear problems can be used for the model implementation. The model intends to optimize operation and reduce the associated costs of prosumers in a localized grid that are participating in electrical power-sharing through optimization of the primary objective equations: Equations (12.9) and (12.10):

$$f_1 = min\sum_{j=1}^{N}\left[\rho_j\left(P_{4(j)} + P_{7(j)}\right)\right]\Delta t \tag{12.9}$$

$$f_2 = \max\sum_{j=1}^{N}\left[\left(P_{1(j)} + P_{3(j)}\right)\right]\Delta t \tag{12.10}$$

where f_1 is a cost function resultant from the grid power flow, ρ_j is the grid power usage expenditure, j is the selected sampling interval, and Δt is simulation sampling time. P_4 is a representation of the grid power supplied to commercial customers variable, P_7 is the power supply to residential customers from the grid, P_1 is the self-consumed power generated by commercial customers, and P_3 is the power used to charge the battery storage scheme where applicable. Optimization is achieved by managing the electrical power flow between the consumers, prosumers, and power storage facilities while reducing dependence on the national power grid. Assessment was done using Eskom's time of use tariff. Dependence on electricity from the national grid is reduced by using the energy management algorithm to increase local energy generation and storage. The cost of operation for the prosumers is also reduced [40]. The transaction cost of the power-sharing was not considered. The model does not interrogate the performance impact on the local power distribution grid by power-sharing.

A model incorporating distributed generation and power-sharing model is recommended by Shukla et al. [41]. The method is applied in a microgrid; however, it can be explored for application in medium voltage networks. Distributed generation resources are controlled using a combination of a modified droop control to match demand and supply in the power-sharing scheme. The modified droop controller employed in the model facilitated power-sharing between prosumers based on their power ratings to restore frequency and voltage to their nominal values irrespective of the load variation in the distribution power system. This is achieved through modifications to the droop voltage and frequency equations. When more than one distribution power system with distributed generation is connected, then the modified power control technique is used. In the suggested modified power control technique, the distributed generation source in each distribution network is controlled using a reference power from the main power grid distribution system, irrespective of the load variation. Equations (12.11) and (12.12) are the electrical equations employed in the model.

$$I_{ref\_DER} = \frac{Q_{ref\_DER}}{V_{DER}} \tag{12.11}$$

$$I_{ref\_DER} = \frac{P_{ref\_DER}}{V_{DER}} \tag{12.12}$$

where I_{ref_DER} is the reference currents for any distributed generation resource that may be incorporated in the system, Q_{ref_DER} and P_{ref_DER} are the reactive power and active power of the distributed energy resources, respectively, V_{DER} is the voltage and the distributed energy resource network connection point. Load control is also utilized to compensate harmonic currents, current imbalances and reactive power demand in the distribution system. The average power demand in each connected power system is used as the reference power value for the distributed generation in each distribution network. The effects of system disturbances such as current imbalance, reactive power compensation, and harmonics on the load-sharing scheme are reduced. The modified droop controller provided sufficient load sharing, and the modified power controller ensured the system was free of transients. Equations (12.13) and (12.14) are utilized to govern the voltage and frequency variations within the model.

$$\omega_k = \omega_0 + m_{DER-k}\left(s_k \frac{\sum_{j=L1+1}^{L2} P_{EPS-j}}{\sum_{i=1}^{L1} S_i H_i} - P_{DER-k} \right) \tag{12.13}$$

$$V_{DER-k} = V_0 + n_{DER-k}\left(s_k \frac{\sum_{j=L1+1}^{L2} Q_{EPS-j}}{\sum_{i=1}^{L1} S_i H_i} - Q_{DER-k} \right) \tag{12.14}$$

where L1 is an integer value representing the total number of distributed energy resources connected to the power-sharing system, L2 is the number of distributed energy sources at local level, S_i is the distributed energy resource rating, H_i is the operating status of the distributed generation power source, $\omega_0 = 2\pi f_0$, with f_0 being the nominal frequency and V_0 is power-sharing scheme operating voltage. The model is dependent on the availability of droop control capability [41]. The model does not take into account the communication methods between the distribution power system operators and the power-sharing market. The model is heavily reliant on this communication as the power controller sets the nominal values based on values obtained from the communication. Two or more electricity distribution utilities participating in power-sharing can be considered as microgrids connected to the main transmission grid participating in power-sharing. A mathematics-intensive algorithm is proposed by Han et al. [42], which may possibly produce fault-finding challenges when put in the practical context. The results showed that the Gain scheduling scheme and predictive control strategies are better equipped to counter the unsought reactive power-sharing and offer improved operability of power-sharing systems. A primary Proportional Integral (PI) controller is infused with a secondary dual PI controller thus ensuring optimized regulation of reactive and active power being supplied to the main distribution grid. The methods used are not appropriate for complex loads and require intricate algorithms for implementation. However, they have a high reliability and redundancy factor and generally do not require communication lines. Active power-sharing control approaches discussed included the following:

1. equivalent active power-sharing with inductive feeder impedance: this method does not consider the active power-sharing under evolving operating conditions and is based on Equations (12.15), (12.16), and (12.17):

$$\omega_i = \omega_i^* - m_i P_i - m'_d \frac{dP_i}{dt} \tag{12.15}$$

$$E_i = E_i^* - n_i Q_i - n'_d \frac{dQ_i}{dt} \tag{12.16}$$

$$J = \sum_{i=1}^{n} \sum_{k=1}^{l} \left[kTE_{\omega i}(k) \right]^2 \tag{12.17}$$

where ω_i is the angular frequency, ω_i^* is the initial frequency of the distributed generation input to the system, m_i and n_i are the active power and reactive droop coefficients, m'_d and n'_d are the adaptive gains, J is the adaptive index aimed at stabilizing the system during operation, $E_{\omega i}(k)$ is the error in frequency at a specific time k, E_i is the nominal voltage, E_i^* was the distributed generation injection voltage amplitude, T is the sampling time, l is the number of samples, n is the total number of prosumers connected to the system.

2. The power/voltage and reactive power/frequency droop control, Equation (12.18):

$$E_i = E_i^* - (m_i + m_D s)P_i = \omega_i^* + \left(n_p + \frac{n_i}{s} + n_D s\right)Q_i \tag{12.18}$$

In Equation (12.18), m_i is the active power coefficient, m_D is the reactive power coefficient and n_p is the reactive power gain. This segment of the model was incorporated to remove the illogicality between voltage and performance in power-sharing. Proportional active power-sharing is regulated through a strategy focused on the distributed generation input point. A cost-based objective function is applied. The active power-sharing under unknown impedance conditions is accounted for using Equation (12.19) and applied through correction of the command angle at the distributed generation injection level.

$$\delta_{DG_i}(t) = \omega_{nl} t + (\Upsilon_P . \acute{P}_{tot} - P_i) \times \left(\frac{m_i}{S} + K_{P_i}\right) \tag{12.19}$$

The command angle is δ_{DG_i}, ω_{nl} was frequency when the distributed generation is operating at no load, Υ_P is the desired active power-sharing, $\acute{P}_{tot}$ is the total active power of the power-sharing scheme, K_{P_i} is the gain for the active power-sharing controller, and S is the apparent power. These methods eradicated voltage and frequency deviations within the power-sharing scheme and are not easily affected by system disturbances, having superior transient response. The dynamic active power-sharing is also more stable. Nevertheless, these control methods are not suitable for complex loads and systems consisting of multiple market participants [42]. The key virtual layer, the energy management system aspect of power-sharing proposed in [43] adopted an energy-sharing operational strategy based on cooperative alliance relationships to obtain financial benefits given the vagueness of the renewable energy source considered. A centralized system management center was used to forecast the network loading and generation capacity for the following day. The management center used weather forecasts and historical load data to perform the forecast. The forecasted load profile for each utility connected to the main grid system was considered and an optimal power-sharing operating strategy was derived prioritizing financial benefits and the optimized output from the renewable energy sources. Each utility buys and sells electricity based on the optimization strategy and associated prices. In deriving the optimal operating point, the central management center considered the extreme scenarios (minimum/maximum) for the distributed generation (each market participant) output. The result was an increased renewable distributed generation utilization reducing pressure on the main power grid [43]. Another main grid connected scheme is recommended in [44] that used reinforced learning techniques. Although not very dynamic, the model considered three types of customers: loads, load with generation and storage capacity, and load with generation. Power at a specific time (t) was summed with the product of photovoltaic power at time (t) and the Wiener Process co-efficient to determine the photovoltaic power output to the grid. The net load consumption was determined by summating the consumption at a specific time (t) with the product of power consumption at (t) and the load Wiener Process co-efficient. The net power traded results by subtracting the load power from the power transmitted. Mathematically the power-sharing model can be modeled using Equations (12.20), (12.21), and (12.22). These equations are applied at the power-sharing market participants' point of connection to the power-sharing scheme.

$$P^{i}_{gen,trans}\left(t\right) \triangleq P^{i}_{gen,his}\left(\mathrm{t}\right)\left(1+\omega P^{i}_{gen}\left(t\right)\right) \tag{12.20}$$

$$P^{i}_{load,trans}\left(t\right) \triangleq P^{i}_{load,his}\left(t\right)\left(1+\omega P^{i}_{load}\left(t\right)\right) \tag{12.21}$$

$$P^{i}_{net,trans}\left(t\right) = P^{i}_{gen,trans}\left(t\right) - P^{i}_{load,trans}\left(t\right) \tag{12.22}$$

The distributed generation power output was represented by $P^{i}_{gen,trans}$, the load consumption of prosumer i is $P^{i}_{load,trans}$, $P^{i}_{gen,his,}$ and $P^{i}_{load,his}$ are the historical power output and load consumption, respectively, ωP^{i}_{gen} and ωP^{i}_{load} are the time domain scalar Wiener Process coefficients. $P^{i}_{net,trans}$ is the total power of the market participant at a time t. A positive value implies that the prosumer/s are selling power to the grid and a negative value indicates that the prosumer is buying power from the system. A supply and demand ratio (SDR) is utilized to establish the transaction charges. The SDR is a ratio of the sum of the distributed generation and the grid power, divided by the sum of the localized load and the distribution grid load. The power selling and buying price is a function of the SDR. For the application of the model, the Markov Decision process (MDP) is segmented into state space, action space, reward function, state transition (probabilistic model), and finally the MDP. Reinforcement learning is used to solve the MDP, make power-sharing decisions, and to optimize expenditure. Using the proposed reinforcement learning, prosumers (where applicable) can select their energy storage system arrangements optimally and autonomously for peer-to-peer power-sharing. The income for electricity sales has increased for prosumers and electricity costs are reduced for consumers [44].

A European continent used a method of implementing anti-wind-up logic via a PI controller at points of control embedded in the power-sharing system was used to achieve regulatory compliance. Using this method, imbalances in power-sharing are proportionally reallocated within the power system [45]. The Piclo project was established with government aid in the United Kingdom (UK) to facilitate energy sharing through software development solutions. The project has provided support to various transmission and distribution system operators across the UK [46]. A model from Barcelona, Spain is discussed in [47]. The model produces results that show increased independence from the main power grid. The power-sharing scheme consists of a cluster of mobile network operators as market participants. The energy is generated from base stations where the mobile network operators can control the percentage of power that is yielded. The algorithm initially determined the power required for transmission. Then the signal-to-noise ratio is established along with the bit rate and the power needs for each mobile network operator base station. The information is sent to a charge control system that facilitates power-sharing. The charge control system controls the procurements by measuring and allocating the available energy to where it is required. The energy sharing is conducted through an aggregator. A double action strategy is employed by the market aggregator to incorporate the infrastructure provider into the transaction benefits and compensation. The iterative nature of the algorithm makes it difficult to assure convergence speed [47]. di Lorenzoa et al. [48] submitted a system where only the market aggregator controlling the point of connection to the main grid had bidirectional functionality. This ensures compliance with the main grid regulatory constraints in the event of feeding power into the main distribution grid. The power available from the peer-to-peer energy-sharing scheme entities is supplied to a direct current busbar termed the power-sharing link. From this bus, each cluster of consumers can tap into the renewable energy available via alternating current to direct current converters. The algorithm amplifies power output using Equation (12.23):

$$\frac{dP}{dV} = \frac{d(VI)}{dV} = I + V\frac{dI}{dV} \cong I + V\frac{\Delta I}{\Delta V} \tag{12.23}$$

$$\frac{\Delta I}{\Delta V} = -\frac{I}{V} \tag{12.24}$$

$$e = \frac{I}{V} + \frac{dI}{dV} \tag{12.25}$$

where $\frac{dP}{dV}$ is the derivative of the power active power equation, V represents the system voltage, I represents the current, and Equation (12.25) represents the error in voltage signal. The optimal power point is identified when Equation (12.24) becomes true. Power is distributed among the customer clusters by the aggregator according to their consumption and not any restriction imposed by the main grid system operator. The power users' inverter ratings are the only restriction to power supplied to each user. The aggregator is the custodian for the implementation of smart contracts within the power-sharing scheme. The main grid system operator is still able to monitor each consumers' consumption through meters installed at the inverter points. If the aggregator system is faulty or offline, then the energy-sharing scheme power is fed to the distribution grid [48]. UK consumers predominantly preferred to procure hydrogen-generated power in the peer-to-peer power-sharing model prescribed in [49]. Market participants are required to interact through a digital platform every 30 minutes to decide which buyer or seller they wish to transact with. Transaction decisions are made in relation to energy source type, physical distance from the generator or consumer, and the energy quantity required [49]. Gomez et al. [50] implemented a power-sharing platform that is discretionary price auction based. Bidders pay the price they bid for, and the system does not attempt social welfare interventions. Bids are executed on an hourly forecast basis. An auction synchronizer gives each seller the opportunity to broadcast the energy available for sale including the bid starting price. Potential buyers submit their bids based on a linear decreasing price strategy that was determined by adding the purchaser's lowest price offer to the product of the difference between the highest price offer and the lowest price off, and the division of the available energy by the bidder's energy requirement subtracted from the number one. The peer-to-peer power-sharing forecast is based using historical data from the past 10 days [50]. The energy management system for peer-to-peer power-sharing tried in Monte Carlo by Mirzaei et al. [51] produced an 8% reduction in reliance on the main electricity grid. The literature does not solve the electrical power flow aspect and associated parameter limits. The primary objective function focuses on operational costs and is defined by the difference in the costs to import and export power to the main power grid added to the distributed generation operating cost function. System constraints factored in the model were the distributed generation dispatchment costs, the energy storage system, the demand response program (participant load profiles), the plug-in enabled electric vehicles, and the supply/demand parameters. The Nash equilibrium method was used to solve the objective function. To reduce the impact on the market emanating from system disturbances, market uncertainty, and budget constraints robust optimization factor is incorporated in the solution [51]. In Germany, the Sonnen project was launched in 2015 and utilized virtual power plant control mechanisms to incorporate battery storage technology into community-based electricity-sharing schemes [52]. German municipal electricity utilities can sustain operationality using the following strategies: contracting with other electricity distribution utilities as an external service provider, municipal utility forms a joint venture with private companies, and the municipal utility being a shareholder in an electricity distribution utility. The power-sharing strategy employed smart energy trading contracts that change at 15-minute intervals. Power traders publish their trade prices and available capacities up front, and purchasers submit their power demand needs and the monetary value they request to pay for the supply. Fewer than 10% of municipalities were direct participants in the power trading market. This was attributed to the specialized knowledge required operation of the power trading markets and the related mechanisms. Larger municipal utilities offer services such as procurement and portfolio management, as service providers. The municipal model may be viewed as a profit-sharing scheme. The scheme consists of a fixed service fee and a portion of the revenue received from the power trade. The bulk of revenue generated is from procurement and portfolio management activities. Municipal distribution utilities generate and sell electrical power via third-party service providers. The model aimed at controlling customer demand-side flexible loads. However, the strategy is resisted by large power users who do not desire distribution utilities having control of their plant and production facilities [53]. In Italy, Moretti et al. [54] demonstrated that using the physical power available and consumption to govern the power-sharing scheme is more beneficial compared to predictive model control methods founded on historical data. In this model, a DC network with all the generators connected to it has the power-sharing being aggregated from the DC network. A single point of connection to the national distribution grid via a DC/AC converter to sell power to the grid is used. This is beneficial in reducing the power system technical losses by 1.5%, therefore availing more electricity to be utilized in the power-sharing scheme [54]. A federation of energy-sharing clusters is utilized by [55] in Milford Haven, South Wales, in the UK. The developed model is very dependent on the market participants' ability to communicate precise information among themselves and at regular intervals. It does not cater to misinformation or communication failure scenarios. Equations (12.26)–(12.29) are the primary set of governing equations utilized.

$$Total_{exchanges} = \sum_{i=1}^{n} \left(e_{if} \right) \tag{12.26}$$

$$Total_{cost} = \sum_{j=1}^{m} \left(\frac{quantity(t)}{energy.network(t)} \times cost(t) \right) \tag{12.27}$$

$$Total_{consumtion} = \sum_{k=1}^{r} \left(s_k \right) \tag{12.28}$$

$$C = \sum t = 1T \left(El_t^{Buy} \times P_t^{Sell} \right) - \left(El_t^{Sell} \times P_t^{Fit} \right) \tag{12.29}$$

where $Total_{exchanges}$ is the total energy sharing taking place, e_{if} is the amount of energy shared between energy nodes i and f, and n is the quantity of energy exchanges within the energy sharing scheme. The total energy cost $Total_{cost}$ is determined using Equation (12.27) where the total energy in the system energy. Network (t) is a function of time, $quantity(t)$ represents the magnitude of the energy being shared between the nodes, and $cost(t)$ accounts for the price of trading energy and factored in the trade participation incentive. The expression s_k is the energy consumption of node k and r is the total number of consumers within a cluster. The cost objective function is optimized using Equation (12.29), which represents the cost of electricity to the energy sharing scheme. T is the total number of time steps in the period of consideration, t is the time step, P_t^{Sell} is the electricity sold and measured in kWh, P_t^{Fit} was the feed-in tariff price of electricity in cost/kWh, El_t^{Buy} is the price of buying electricity during a time step, El_t^{Sell} is the price of selling electricity during a time step. Prosumers are arranged into clusters. Power-sharing takes place internally within the cluster and/or with the external power grid where each cluster is represented by a cluster manager and electrically as an energy node. The energy exchange transactions are constructed by the following parameters: energy type (wind, solar, gas), energy quantity, price, and payoff (Incentive for participating in the energy exchange) [55]. In the city of Riga in Latvia a model that increases the social welfare by 14.25% for the market participants is employed by Zidane et al. [56]. The model consists of a collective of microgrids that perform energy sharing locally among themselves. Each microgrid collective is connected to the main distribution grid via a single point of common coupling and thus acts as a prosumer relative to the main grid. Local power-sharing within the microgrids is controlled by a market aggregator referred to as the community manager in the article. The community manager is the entity that communicates with the main power grid and is responsible for the implementation of bilateral agreements between the main distribution grid and the represented microgrid. The SDR is applied to determine market participation. Equation (12.30) is the SDR applied in the study. The SDR in combination with the real-time energy price from the distribution grid is further utilized to determine buying and selling prices within the power-sharing scheme.

$$SDR = \frac{\sum_{g=1}^{G} P_g^n}{\sum_{d=1}^{D} C_g^n} \tag{12.30}$$

where P_g^n is the local energy resources and C_g^n their consumption, in each microgrid n, indexed by g and d. A separate objective function is applied for sellers, buyers, and the power-sharing scheme. The method of Alternating Direction Method of Multipliers is used to solve the equations to establish social welfare within the market. According to [56], the use of Alternating Direction Method of Multipliers provides information privacy for the market participants. The model is not dynamic and requires numerous iterations to produce the final market operating results. Moncecchi et al. [57] found that incentivizing shared energy quantities resulted in optimal power-sharing among market participants. In contrast, energy generation-focused incentives yield oversized energy capacities. Equation (12.31)–(12.32) are the underpinning evaluation criteria utilized in the model.

$$CF(y) = Inv(y) + Cost(y) + Sav(y) + Rev(u) + Fin_{inc}(y) \tag{12.31}$$

$$NPV = \sum_{y=0}^{EL} \frac{CF(y)}{(1+DR)^y} \tag{12.32}$$

where all the input functions in Equation (12.31) are financial parameters dependent on the year y. The net present value of the power-sharing scheme investment is determined using Equation (12.32), where EL is the asset life span, DR is the discount rate achieved, and CF is the cash flow projection. It was established that an increased capital investment such as that required for higher generation capacity does not yield a higher financial return within an energy sharing scheme. The model is not dynamic to common network abnormality events that occur on distribution level networks and to the unpredictable impetuous nature of renewable energy resources [57].

A non-cooperative game theory approach was utilized in [58] where every prosumer participant had the option of buying or selling power across different various buses within the distribution network, with the intention of maximizing personal interest. Linear programming is used to establish market equilibrium. The method assumes that all market participants are independent and will not form an alliance but present true information to the market collector. The sampling period is hourly, and sellers have the option of selling to the main grid if there is surplus energy after the power-sharing scheme obligations have been fulfilled. The energy transaction data is collected by a market collector referred to as an aggregator. The aggregator solves the equilibrium problem and submits the results to the distribution system operator. The distribution system operator solves the power flow problem using the branch flow model with the aim of minimizing the energy cost at optimized power flow under the distribution system constraints. The relevant supply buses insert power into the physical distribution network and are compensated financially. The demand buses receive power from the physical distribution network and pay for the power transaction. The outcome is that the power purchaser's buying cost is reduced, and the seller's selling price is increased resulting in market benefits for all participants.

$$P_{ij}^{l} - r_{ij}^{l} + P_{gi} - P_{dj} = \sum_{(j)} P_{jk}^{l}, \forall j \in N, l \in L \tag{12.33}$$

$$Q_{ij}^{l} - x_{ij}^{l} + Q_{gj} - Q_{dj} = \sum_{(j)} Q_{jk}^{l}, \forall j \in N, l \in L \tag{12.34}$$

$$V_j = V_i - 2\left(P_{ij}^{l} r_{ij}^{l} + Q_{ij}^{l} x_{ij}^{l}\right) + \left(z_{ij}^{l}\right)^2 I_{ij}^{l}, \forall l \in L \tag{12.35}$$

$$V_i I_{ij}^{l} = \left(P_{ij}^{l}\right)^2 + \left(Q_{ij}^{l}\right)^2, \forall l \in L \tag{12.36}$$

Equations (12.33)–(12.36) are the fundamentals applied in the model. Where N is the set of prosumers connected to a bus; L is the distribution line set; the distribution network between initial bus i and network line l is abbreviated ij; the line impedance is z_{ij}^{l}; V and I represent voltage and current, respectively; and P and Q are the active and reactive power, respectively. The model assumes functional energy sharing infrastructure is already in place on the distribution network. It is assumed that the energy-sharing scheme tariffs for buying or selling power are always lower than those of the conventional power utilities, but this could vary in different markets or circumstances [58]. Maisch [59] wrote of the Australian government's pilot program RENeW Nexus Project, which was executed from December 2018 until January 2020. The author showed how peer-to-peer energy trading incorporated with a price signal tool to provide localized energy markets had potential to contribute to lower-cost grid stability, which was beneficial to network or grid operators and to Australian citizens. The virtual power plant control strategy was used with Blockchain for the project [59]. In Sydney, New South Wales, Australia a model that used network peak load as being the electrical network capacity was promoted in [60], which may limit the power trading potential of the infrastructure unnecessarily. Market participants' broadcasted information and observations. These were used as input to the market participant prediction and action probability calculations. This step was utilized to learn the market policy. Considering the environmental conditions, electrical distance, and the participant input, the prosumers were grouped into clusters. Double auction strategy committed orders to supply and determined the associated electricity bills. Calculations are implemented on an hourly basis [60]. Distribution networks were designed to accommodate system peak loads, that are temporary, leaving the networks underutilized throughout the most part of the year. A Share Network Access business model allowed the Distribution System Operator to utilize and benefit from the spare capacity that is available for most parts of the year. The result was that the distribution network operator benefited from the revenue generated by renting out the unused network capacity during periods of high system integrity or off-peak loading. Further research is required for the rent revenue and rental agreement strategy, which would enable distribution system operators to receive incentives for improved network utilization efficiency and not network investment. The model is dependent on a fixed reliability factor of power supply from the grid, which is a variable factor in markets such South Africa [61].

The Federal Energy Regulatory Commission [62] stipulates that in the United States, electricity network interconnectivity was initially implemented to attain electricity supply reliability at the cheapest cost and to avoid the financial implications of generation infrastructure for distribution system firmness. Interconnectivity is now used for energy-sharing. The energy-sharing transactions are facilitated by direct contact between the purchaser and procurer, or through an agent, or via digital trading platforms. The identities of the purchasers and procurers are seen by all power-sharing scheme participants. The physical transportation of electricity between suppliers and consumers necessitated the stakeholders to reserve grid capacity to accommodate the movement of power. The transmission grid utility is required to disclose available transmission capacity on an open access same-time information system online platform, from which traders can reserve capacity via a power contract [62]. Tushar et al. [63] utilized an auction-type Stackelberg Game-based tactic to apply peer-to-peer energy sharing and to ensure social welfare within the market. Clusters of prosumers were incorporated into the distribution network by singular shared facility controllers. The approach resulted in market participants achieving a 60% reduction on their electricity utility bill. The model, being an adaptation of Blockchain technology, focuses on financial benefits [63]. The impact of market sharing infrastructure capital, operating and maintenance costs are not reflected in the model proposed by Vinay et al. [64]. These factors could significantly reduce the projected savings and profits. Participants sell excess power either to the grid or to the local power-sharing scheme. Power cost savings are achieved when at least one (ideally more) market participant has excess energy stored and is selling to the power-sharing system. Power from the power-sharing scheme must be exhausted before power can be purchased from the grid. Savings are achieved through competitive selling and buying pricing of the power-sharing scheme compared to the time of use tariff of the grid. Cost allocation is achieved through the cooperative game [64].

A model that aims to optimize the cost and well-being of the power trade participants by solving objective function Equation (12.37) was explored in Singapore by Foo et al. [65]:

$$F = \min_{e^S, e^B, E^S, E^B} \vartheta^{P2P} + \min_{p^G, q^G, v, P, Q} \vartheta^{DSO} \tag{12.37}$$

$$\min_{e^S, e^B, E^S, E^B} \vartheta^{P2P} := \sum_{i \in P} G_i\left(e_i^S\right) - \sum_{j \in B} U_j\left(e_j^B\right) \tag{12.38}$$

$$\min_{p^G, q^G, v, P, Q} \vartheta^{DSO} := \sum_{g \in NG} \left[C_g^P\left(P_g^G\right) + C_g^Q\right] + \left(v - \tilde{V}\right)^T W \left(v - \tilde{V}\right) \tag{12.39}$$

where ϑ^{P2P} was the peer to peer power-sharing market objective function, ϑ^{DSO} was the distribution system operator's objective function, G_i was the convex cost function of distributed generation i ϵ P, U_j was the concave welfare function of power consumer j ϵ B, C_g^P was the convex active power cost function, C_g^Q was the convex reactive power cost function, $\tilde{V}$ was the required magnitude of voltage, W was a transverse matrix, $v \in R^{|N|}$ was a voltage maginutde vector, E^B equated to the power purchased by consumer, e^B equated to the consumption vector of consumers, E^S represented power sold by producer, e^S was the production vector of the distributed generation, P was active power flow, Q represented reactive power flow, p^G was the active power generation dispatch vector of the distribution system operator, and q^G was the reactive power generation dispatch vector of the distribution system operator. Equations (12.38) and (12.39) were used to originate Equation (12.37). In the model discussed by Gupta et al. [66], a control module is placed between the distributed generation sources' busbar and the point of common coupling, connecting the peer-to-peer power-sharing system to the grid. Being used in Bhopal, India, on the off-grid side of the system, continuous current and power regulation are applied to grid connected inverters. On the grid-connected side, PQ regulation is used on inverter interfaced power sources. The model considers the power available from distributed generation sources. This is summated with the system load/demand to determine the power requirements. If the power reserve is greater than the load, the excess power is exported to the main grid for power-sharing use. If the demand is equal to the instantaneous distributed generation capacity, all locally generated power is distributed to the load and power from the main grid is used to charge the battery storage schemes (where available) that are in the uncharged state. Should the load power requirements be greater than the distributed generation sources available, then power from the grid is used to supplement the power supply shortage. Active power control is achieved using an automated controller. In the event of a sudden change in load, the PI controller uses fuel cell energy to supplement the change and maintains a constant voltage regardless of the variation in load. In the event of a phase fault, the controller detects the imbalance and uses a reference current to manage power and voltage under abnormal conditions. The controller ensures that the reference voltage is maintained under various network conditions and provides a smooth power transfer between the sources and loads connected to the system [66]. The physical implementation of the system is not discussed or prescribed. The communication channels and infrastructure between the controller and the grid operator are not discussed. The system is only adaptable to step change and load imbalance variations in the operating environment. For India [67] suggested a model that is based on a direct current network power-sharing arrangement between prosumers. Control of the power-sharing scheme in the publication's linear scheme strategy is achieved through virtual resistance value determination considering a self-adjusting variable gain k measured in ohms per amp. The virtual resistance value for incorporation into the droop equation via the virtual resistance voltage drop is calculated by subtracting the product of k and the distributed energy resource output current, from the nominal value of the virtual resistance. In the non-linear scheme, the virtual resistance value is the difference between the nominal value of the virtual resistance and the product of k and the virtual resistance power coefficient. For both control mechanisms, the virtual resistance voltage drop is subtracted from the nominal voltage. The result is input to a proportional-integral controller that regulates the pulses sent to a DC-to-DC boost converter. In this way, the virtual resistance value is varied based on the system output current. Mathematically this is offered as Equations (12.40), (12.41), and (12.42):

$$V_{ref} = V^{*} - v_{drop} \tag{12.40}$$

$$v_{drop} = i_0 r_v \tag{12.41}$$

$$r_v = R_v - k i_0^{\mathrm{Y}} \tag{12.42}$$

where V_{ref} is the power-sharing scheme voltage, V^{*} is the input voltage to the distributed generation source, and v_{drop} is the voltage drop attributed to the virtual resistance. The figure i_0 represents the distributed generation output current, r_v is the total virtual resistance, R_v is the nominal virtual resistance, k is the self-adjusting droop variation gain, and Y is the power coefficient. The model is favorable in that it does not require additional communication infrastructure that has potential to increase capital and operational costs. It is heavily reliant on voltage measurements and flexibility with regards to other power system parameters is not thoroughly regarded [67]. A virtual impedance methodology was explored in [68] where the power equations at the point of common coupling are represented by Equations (12.43) and (12.44).

$$P_i = \frac{E_i}{R_i^2 + X_i^2}\left[R_i\left(E_i - V\cos\delta_i\right) + VX_i\sin\delta_i\right] \tag{12.43}$$

$$Q_i = \frac{E_i}{R_i^2 + X_i^2}\left[X_i\left(E_i - V\cos\delta_i\right) + VR_i\sin\delta_i\right] \tag{12.44}$$

where E represents the nominal voltage, V represents the voltage at the point of common coupling, X represents the network impedance, R represents the network resistance, and δ is the power angle between the point of common coupling and a distributed generation source connected to the power system. The assumptions that $\cos\delta \rightarrow 1$ and $\sin\delta \rightarrow \delta$ were applied for functionality of the model. A vitual impedance X_v was factored into the reactance at each point of common coupling. The inclusion of the virtual impedance facilitates improved reactive power-sharing within the power system. The model is dependent on numerous assumptions such as high feeder impedance and can be ineffective on power systems where the assumptions were not valid [68]. Ashraf et al. [69] used a cost objective function that takes into account the product of the grid energy purchasing price and the prosumer price to sell electricity subtracted from the product of the grid energy selling price, and the prosumer grid energy purchased cost added to the cumulative sum of the product of the peer-to-peer scheme power price, the electrical distance between the power-sharing entities, and the energy purchased from the grid by the seller in the transaction. Equation (12.45) is utilized to optimize the function.

$$Cost = C^{-}_{P2G}(t)g^{-}_{P2G}(t,p) - C^{+}_{P2G}(t)g^{+}_{P2G}(p) + \sum_{\substack{f \in P \\ f \neq P}} C_{P2P}(f)g^{-}_{P2P}(f,p)d_{pf}(f,p) - \sum_{\substack{f \in P \\ f \neq P}} C_{P2P}(f)g^{+}_{P2P}(f,p) \quad (12.45)$$

Where C^{+}_{P2G} is the cost of buying electricity from a prosumer, C^{-}_{P2G} is the monetary value of buying power from a prosumer, C_{P2P} is the cost of energy shared within the power-sharing scheme, g^{+} is the electricity sold to the distribution system operator grid by prosumer p and g^{-} is the electricity bought from the grid, and d_{pf} represents the electrical distance between prosumer p and consumer f. Common voltage limit values are prescribed for the network nodes. Power demand and supply is sustained within the system considering the main grid supply of reactive and active power. Power flow is monitored at network node level. Physical network reconfiguration is achieved by switching between distribution lines in the network to realize the desired results and quality of supply. This has an impact on the electrical distance between consumers and prosumers. The method was assessed for networks in Kerman Iran using the IEEE-33 bus. Power is only shared between entities that have a physical electricity path between them [69]. An internet of energy approach to power-sharing is used in Malasia by Hamdan et al. [70]. A centralized platform collects data from the distribution grid, power-sharing market participants and computes the market prices as well as supplier/consumer matching. The results are prescribed to the market participants. From the study, the following challenges were identified with peer-to-peer power-sharing: market policy and regulation, ambiguous market end state, market contract design, complicated end user interface, grid impact and security of market data [70].

Tushar et al. [71] provided a detailed dialogue of the key features of peer-to-peer power trading networks, associated markets, challenges, and the technical methods to achieving power trade. The peer-to-peer power-sharing was segmented into the following primary elements:

1. Virtual layer components – information systems, market operation, pricing mechanism, energy management system.
2. Physical layer components – connection to the grid, metering, and communication infrastructure. The additional influential factors were the market participants and market regulation. The power-sharing markets were differentiated into either a fully decentralized, community-based, or composite market. The challenges identified in the virtual layer were energy cost reduction, equilibrium between generation and demand, compensating prosumers, engaging participants, charges development, uncertainty, and transaction security. Physical layer challenges were voltage and infrastructure constraints, power losses, and system strength [71]. The physical and virtual layers were also advocated for by Borghetti et al. [72] as the fundamentals of power-sharing. The application of Game Theory solved using

the Generalized Nash Equilibrium was recommended for transactional decision-making. To satisfy a market focused on individual rationality and compensation the double action strategy was prescribed. Power flow constrained systems' solution was resolved through constrained optimization strategies. The constrained optimization approaches discussed included the Alternating Direction Method of Multipliers, linear programming, mixed integer programming, and nonlinear programming. The incorporation of stochastic optimization into the power-sharing strategy was recommended for the day-ahead forecast arrangements. Blockchain utilization was discussed and endorsed for decentralized network systems [72]. The following mandatory elements of a power-sharing scheme were prescribed in [73]: individual elements, technology, resources, and environment. The individual element consisted of prosumers and consumers; the resource element comprised the distributed generation resources infrastructure. The communications infrastructure and methods, energy conversion machinery such as power-to-gas, and gas-to-power conversion apparatus, and software developments like Blockchain were categorized in the technological element of power-sharing schemes. The environmental division was made of policies, laws, power-sharing market regulation, and communal ethos. A market aggregator-centered objective function Equation (12.46)–(12.49) are recommended to achieve market participant welfare.

$$\max_{X,\pi} f^{NB} = \left(-r^{A,0} + r^{A} - \sum_{i=1}^{N} \pi_i\right)\prod_{i=1}^{N}\left(c_i^{U,0} - c_i^{U} + \pi_i\right)^{\alpha_i} \tag{12.46}$$

$$X \epsilon \varphi \cap \varphi^{S} \tag{12.47}$$

$$-r^{A,0} + r^{A} - \sum_{i=1}^{N} \pi_i \geq 0 \tag{12.48}$$

$$c_i^{U,0} - c_i^{U} + \pi_i \geq 0 \tag{12.49}$$

The denotations X and π are input decision variables based on operation and market settlement, φ and φ^S are the operation and aggregation feasible regions respectively in applying Nash equilibrium bargaining, the market operating cost before and after power-sharing are represented by $c_i^{U,0}$ and c_i^{U}, the anticipated monetary income before and after electrical energy sharing is represented by $r^{A,0}$ and r^{A}, α_i is the available power and i ϵ N. The platform-based power-sharing strategy is suggested as an alternative to the aggregator-based market. The platform-based strategy is three-layered, namely platform trading scheme design (transaction and pricing formula), decision-making process (cloud-based platform where market participants can submit their information), and market clearing (operating constraints and market participant conditions are satisfied) [73]. The model by Lehtonen et al. [74] consists of the distribution system operator, downstream market aggregators which are regarded as a cluster of prosumers linked to the distribution via one bus and aggregated by a single entity, upstream market aggregators that can buy, and sell power to the distribution grid as well as directly from downstream market aggregators. At a predetermined transaction interval, market participants declare the role they are fulfilling for that period. The distribution system operator disconnects downstream market aggregators that are not buying or selling power to the system. Historical data of the remaining downstream and upstream aggregators is used to evaluate the reliability of each and their capacity to meet supply-to-demand obligations. Equations (12.50)–(12.51) are utilized to exercise the evaluations.

$$P_i^{cur} = \frac{N_i^{cur}}{T_{year}/\Delta t};\forall \in S \tag{12.50}$$

$$P_{uma}^{cur} = \frac{N_{uma}^{cur}}{T_{year}/\Delta t} \tag{12.51}$$

where P_i^{cur} is the downstream market aggregator i's probability of load curtailment, P_{uma}^{cur} is the upstream market aggregator's probability of load curtailment, N_i^{cur} is the number of curtailments by seller i over the last 12 months, N_{uma}^{cur} is the number of curtailments by the upstream market aggregator over the previous 12 months, T_{year} is the time duration of a year and Δt is the time interval length (is very similar to T_{year}). Each downstream market aggregator and upstream market aggregator have their own internal optimization function to determine market participation preferences. The distribution system operator publishes the electricity selling price, capacity, and reliability of each market participant. Market transactions and facilitation of power-sharing are conducted by the distribution system operator. In the event of a fault within the power system, the distribution system operator isolates the parts of the network that are impacted by the fault and publishes the amount of time that will be required by the repair team to normalize the network [74]. The model improves the efficiency of peer-to-peer power-sharing however it is very dependent on the distribution network infrastructure having smart technological operating systems capable of self-detection of fault location. In typical distribution networks across South Africa, human resources are dispatched to investigate and determine the location of the fault. The repair teams are often unable to accurately determine the extent of the fault and the time required to fix it until for example the faulty cable is exposed and assessed, and the conditions that caused the fault are determined.

A power-sharing scheme that aims to facilitate power-sharing while protecting the privacy of all market participants is proposed in [75]. The model utilized the main distribution grid operator as the aggregator and smart transactive metering infrastructure to implement the power-sharing. Market participants submit their requests as energy bids to the aggregator based on their preference to buy or sell electricity. The distribution grid operator utilizes the general Nash Equilibrium strategy to determine the optimal energy sharing price and quantities based on the market participants' request and the distribution network constraints. Unfulfilled requests are quantified and communicated to the users' smart meter. Where feasible the smart meter makes modifications to the rejected bid and resubmits to the distribution system operator. The bid is reassessed and evaluated. Should an equilibrium be reached then the energy sharing is enabled by the aggregator. By using the distribution system operator as a market aggregator, the model enhanced consumer and prosumer information privacy. Communication requirements are reduced to only the bid-related information. The model does not protect the market including the aggregator from irrational or unethical bidding strategies by market participants [75]. Jogunola et al. [76] used the Physarum Polycephalum method to match electricity supply to demand within a constrained power system. This method is based on tubular networks with fluctuating capacities with network segments that are not used, declining and eventually vanishing [76]. The presence of any power node cannot be ignored within an active distribution-level power system due to power quality requirements that exist within markets.

The outcomes of peer-to-peer sharing projects in the United States, Germany, the United Kingdom, Australia, Belgium, Japan, Singapore, Switzerland, and others were examined by Christen et al. [77]. The projects revealed that Ethereum is the most effective and utilized distribution ledger technology, possibly due to it having entered the market ahead of its competitors. Fifty percent of the successful projects utilized a proof of work blockchain algorithm, while the use of the alternative strategies is distributed over the remaining 50%. The peer-to-peer power trading process is described as consisting of the device level, smart metering level, transmission level, communication level, and the management level. The system control hierarchy is arranged such that the management layer consisting of grid stability parameters, tariffs, consumption and generation forecasts, and power flow, sits at the top. Below this is the communication level where the blockchain, smart contracts, side chain, and other transactional algorithms are implemented. This level is proceeded by the transmission level, composed of physical network communication methods. The smart meter and device level (load and electrical infrastructure, respectively) are the lowest layers of the system control hierarchy. The peer-to-peer power trading system reporting hierarchy is structured in the opposite manner with the device level placed on top followed by smart metering, transmission, communication, and management level at the bottom [77].

A power-sharing model where only entities with electricity generation capability may participate is discussed in [78]. Social welfare is increased through market participants being able to borrow or use energy from other market members at a possibly cheaper generation cost. Equation (12.52) is the social welfare modeled mathematically.

$$\begin{aligned} SW := max\sum_{i=1}^{N}\Big[& a_i D_i - 0.5 b_i D_i^2 - \alpha_i S_i - 0.5\beta_i S_i^2 + c_{ch,i} Q_{ch,i} - 0.5 d_{ch,i} Q_{ch,i}^2 \\ & - c_{dis,i} Q_{dis,i} - 0.5 d_{dis,i} Q_{dis,i}^2 - \gamma_i \left(Q_{ch,i} + Q_{dis,i} \right) \Big] \end{aligned} \tag{12.52}$$

where D_i is the actual consumption by prosumer i, a_i, and b_i are the utility function coefficients of prosumer i, α_i and β_i are the cost of generation function fitting coefficients for prosumer i. In Equation (12.52) the prosumer power consumption utility function is accounted for using $a_i D_i - 0.5 b_i D_i^2$ and $\alpha_i S_i - 0.5\beta_i S_i^2$ accounted for the prosumer's cost of generation. The expressions $c_{ch,i} Q_{ch,i} - 0.5 d_{ch,i} Q_{ch,i}^2$ and

$c_{dis,i}Q_{dis,i} - 0.5d_{dis,i}Q_{dis,i}^2$ are not applicable if the power-sharing scheme does not consist of energy storage infrastructure, γ_i represents the energy loss during charging and discharging of energy storage systems therefore the expression $\gamma_i\left(Q_{ch,i} + Q_{dis,i}\right)$ is also not applicable in the absence of energy storage. S_i is the electricity generated by the prosumer. Social welfare is further increased with a rise in market partaker numbers in the power-sharing pool. Entities without the capital required for power generation and market participation infrastructure are unable to benefit from this model [78]. On the contrary, the model developed in [79] required more iterations thus potentially leading to system-related delays, with an increase in market participants, a negative effect. The power-sharing platform is provided and managed by the grid distribution system operator. Market participants only have access to their own credit or debit transaction information. The model was designed such that prosumers with excess energy benefit from selling the energy before market clearing takes place. Likewise, those market participants who only consume are penalized through obligation to purchase energy at elevated prices when market clearing takes place [79]. The peer-to-peer energy-sharing model by Pindoriya et al. [80] is segmented market participants into three clusters, namely Type 1 (do not participate in energy sharing and buy power from the grid), Type 2 (prosumers who sell excess energy but do not modify their load profile), Type 3 (prosumer with flexible loads). The k means learning algorithm is applied to place the market participants in clusters. Voltage is sampled at each point of connection to the power system and used as one of the parameters for centroid selection. One of the highlights of the model is its use of load flattening techniques. Equations (12.53)–(12.55) are applied to enable the load flattening exercise.

$$f_2 = \sum_t^{\tau}\left(P_t^{pcc} - \bar{P}\right)^2 \tag{12.53}$$

$$P_t^{pcc} = \sum_i^{V}\left(P_{t,i}^{d} + P_{t,i}^{loss}\right) \tag{12.54}$$

$$\bar{P} = \frac{\sum_t^{\tau} P_t^{pcc}}{\tau} \tag{12.55}$$

where f_2 is the load flattening function, $\bar{P}$ is the daily (24 hours) average power at the point of common coupling, P_t^{pcc} is the active power at the point of common coupling at time t, $P_{t,i}^{d}$ is the ith aggregator's total power demand at time t, $P_{t,i}^{loss}$ is the ith aggregator's total power loss at time t and τ is the time period within t. The performance of the load flattening is evaluated using the peak to average ratio at the point of common coupling is determined using Equation (12.56):

$$PAR = {max\left\{P_t^{pcc}\right\}}\Big/{\bar{P}} \tag{12.56}$$

The model demonstrated that the application of real-time pricing to energy sharing schemes increases the motivation for participation within the market. However, the study failed to account for the distribution network constraints and further relied on constant reactive power within the system despite variations in the power system operating conditions [80].

12.3 MAJOR POWER TRADING FRAMEWORKS

The recent success of the energy sharing economy has resulted in the methodology being regarded as an auspicious concept for solving challenges in peak demand reduction, power supply vs demand shortages and dealing with power system [81]. The most common procedures for implementation of power-sharing include blockchain, constrained optimization, auction theory, and game theories. These strategies can further be differentiated into subsection thus leading to the uniqueness of each technical approach proposed in the literature and exercised in the industry. Prevalent methods of solving constrained optimization problems include linear or nonlinear programming, alternating direction method of multipliers, and mixed integer linear programming. The blockchain approach is a dispersed database scheme that is cryptographic. The data is linked and organized in blocks making a decentralized multi-point maintenance algorithm. It enables information traceability. It is very difficult to tamper with and achieves secure transparent data storage. These attributes are very imperative in the functional success of a trading platform. The technologies include hyper-ledger, Ethereum, and smart contracts [82, 83].

Mathematically constrained optimization can be defined as the process of augmenting an objective function in reference to established choice variables that are subjected to constraints [84]. Constrained optimization can also be viewed as the establishment of a feasible region within a predetermined set of boundaries or constraints [85]. Constrained optimization problems can be solved using various methods dependent on the nature of the constraints. These methods are not limited to but include:

1. Linear programming, a linear algebra instrument for optimizing results through mathematical modeling [86]. Linear programming is often utilized in the modeling of real-world conditions [87].
2. Mixed Integer Linear Programming is an optimization problem containing vector variables [88].
3. Alternating Direction Method of Multipliers (ADMM), a procedure that resolves convex optimization problems through segregating the problem into smaller components that are less complex to solve [89]. ADMM produces precise solutions using minimal iterations however the number of iterations utilized is directly proportional to the accuracy level achieved in the solution [90].
4. Nonlinear Programming is an objective function that is nonlinear and/or has a feasible region obtained through nonlinear boundaries [91].

Auction Theory is one of strategies frequently used in energy trading to derive market pricing. Auction Theory is among the oldest known practices for establishing prices within a market. Based on the reason for utilizing auctions instead of the conventional buy and methods, the appropriate auction form may be selected [92]. There are different forms of auction theory that are available, these include Vickery auctions, Scaled Bid auctions, Revenue Equivalence auction, Double auction, and Common Value auctions [93]. Game theory is a mechanism commonly utilized to establish social welfare within markets. It is an evaluation of why and how entities or individuals (commonly referred to as players) make choices regarding their circumstances. The Game Theory mechanism is utilized in various real-world scenarios to establish different situations and the most probable outcome [94]. It is a mathematical method that scrutinizes tactics in competitive situations, where the results of a player's decision are reliant on the actions taken by other players. Game Theory consists of two sections namely, noncooperative game theory and cooperative game theory. These sections are further segmented based on desired outcomes within the applicable domain [78]. Blockchain is a connected record of blocks of data that creates a log of updates the data. Blockchains are extremely resistant to interference, changes to data can only be done by adding digitally signed annals to the blockchain. They are ideal for deployment in a trustless environment. In power trading, the blockchain strategy is often utilized in support of other strategies [95].

12.4 STATE-OF-THE-ART RELATING TO CURRENTLY APPLIED MODELS AND ALGORITHMS

The models discussed in literature use various combinations of strategies predominantly constrained optimization in the physical layer and smart contracts and action theory in the virtual layer. Figure 12.1 provides a summary of the power trading

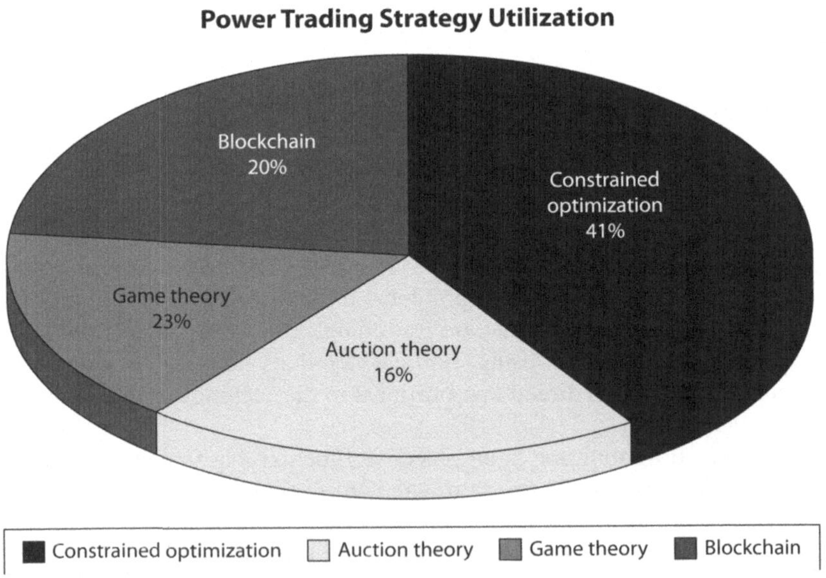

FIGURE 12.1 Power trading models utilization overview.

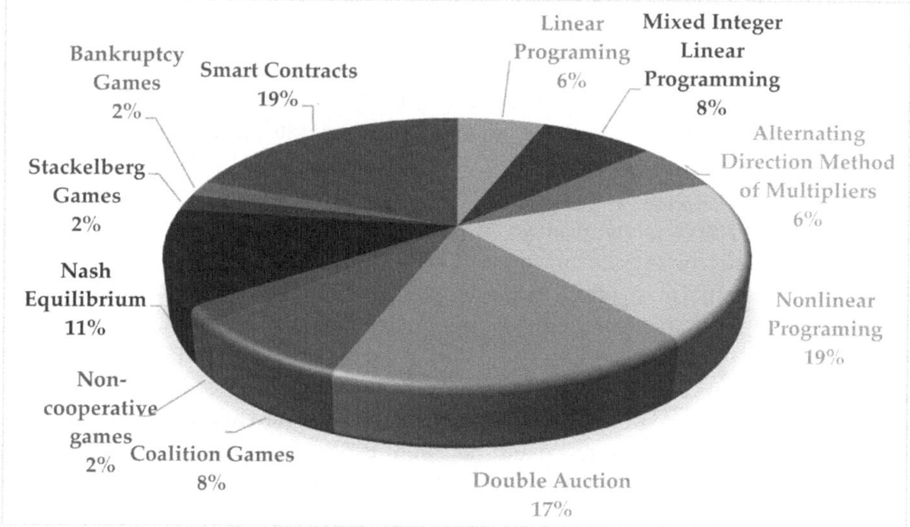

FIGURE 12.2 Power trading models strategy overview.

strategy utilization across the appraised literature. Constrained optimization is the most widely utilized strategy in achieving effective peer to peer mechanism. Optimization is mostly used in models that hold paramount the physical attributes of peer-to-peer power trading. Game Theories display superiority in market welfare establishment and are notable solved using Nash Equilibrium processes.

The methods described in Figure 12.1 are broad and can further be segmented into subdivisions. Figure 12.2 provides a summary of the variant of each strategy that was used in the considered literature. The models discussed in the literature use various combinations of strategies predominantly constrained optimization in the physical layer and smart contracts and action theory in the virtual layer.

An in-depth breakdown of each mechanism was provided in [71]. This description incorporated all the components included in literature although in most cases not explicitly expressed but were essentially present in every model. The model in [32] required soft open points to be embedded within the distribution network. This created a challenge as the majority of distribution network infrastructure is aged and the modifications required to implement the model would be a capital investment intensive exercise, more specifically within the South African electricity distribution industry, as per indicated by Davidson et al. [11]. An audit by the national energy regulator revealed that the infrastructure age was frequently over 30 years of age [96]. The power trading model proposed in [48] for peer-to-peer power trading among building structures is of interest given that the South African Electricity Regulation Act of 2006 only permits onsite consumption of distributed generation with excess generation allowed to be stored or dumped [40]. However, this restriction is being re-evaluated by the national energy regulator and the South African electricity distribution industry. This is evident by the number of distribution utilities that have begun incorporating a wheeling tariff into their tariff structures. There is great potential for the power trading structures proposed for buildings to be expanded microgrids and even distribution utility level. For a successful transition implementation, it is essential that there is buy-in from all stakeholders involved. The model prescribed by [34] retains the main grid operator as the market aggregator thus allowing involvement and a great degree of control in the power-sharing scheme by the main grid operator.

A non-cooperative game strategy utilizing linear integers to implement social welfare within the power-sharing market is suggested by [58]. This structure was based on market participants not colluding to manipulate the market and on the idea that those participants were providing honest and true information to the market aggregator. With changing world market conditions and inflation rates, this expectation of market participants may have been too ambitious. Nonetheless non-cooperative game strategies are a popular method of power-sharing market orientation and regulation. The literature and research results have shown that the method worked, however the challenge to encourage prosumer participation in the market and to develop a pricing mechanism that was suitable to all market participants remained a challenge in the industry [57]. The authors in [19] have concluded that the centralized approach was more secure for prosumers unless they refused to share data with a central unit. While there were different models of trading, the most promising model was the one that considered the third party (aggregator) because of the management simplicity of energy sharing and communication with a distribution system operator. This was due to the system operator and aggregator having a broader picture of the market, being able to take accountability and able to produce more effective prosumer to consumer matching [55]. The authors of [46] and [66] pointed out that existing studies mainly focused on the theoretical design of demand response schemes and did not verify the proposed schemes through implementation. The development in communication and solar energy production infrastructure has encouraged more passive electricity consumers to become active prosumers. Given this context, feed-in-tariffs were developed to enable prosumers to participate in energy trading as energy providers and it demonstrated great potential in reducing the main grid's pressure. The challenge in peer-to-peer power trading is that it does not enforce benefits in the sharing economy, for example, pure consumers were often able to benefit from the power-sharing without making any energy contributions within the system [79]. Peer-to-peer power-sharing from geographically proximity electricity distribution/transmission entities reduces electrical energy transmission distances and corresponding energy losses. The localized load supply can be expected to reduce utility network and transformer loading thus minimizing network congestion and imbalances. Power-sharing relieves the pressure on traditional generators and networks of electricity thus also reducing infrastructure upgrades aimed at accommodating increased loads. The presence of distributed generation and power-sharing imposes additional technical constraints and power flow on distribution networks. Factors such as system protection, fault recovery, reliability, and system voltage are impacted. A comprehensive assessment of the technical impacts on the grid is required to provide assurance that peer-to-peer power trading will not result in negative contributions to power system reliability and performance [97].

12.5 DISCUSSION

The literature reviewed indicates a reduction in dependence on the main power grid being one of the benefits derived from peer-to-peer power trading [24, 32, 40, 44, 47, 51, 60]. The physical network facilitation of the power trading transactions introduces additional power losses to the power system. The proposed power loss mitigation strategy proposed by [34] may lead to the undesirable effect of increased electricity tariffs. Expensive electricity prices may lead to the commodity being unaffordable and therefore inaccessible to some segments of society. The model is therefore partially not aligned with the United Nations and South Africa's universal access endeavors. The model does however promote the use of green renewable energy in alignment with the carbon reduction goals targeted at combating the global warming challenges faced by modern society. Other models that promote renewable energy consumption are [44, 49, 60]. The results from [49] revealed an appetite for hydrogen generated power consumption within power-sharing schemes. Power loss reduction was also achieved by [42] and [43] by mitigating the effects of reactive power-sharing. Reactive power compensation and optimization is used by [37] to improve the power-sharing system voltage profile for compliance to system regulatory requirements. Peer-to-peer power trading was demonstrated by [36] to reduce power outages that are a result of insufficient power generation capacity. Insufficient capacity is according to [1] the challenge faced by South Africa, necessitating the implementation of rotational power cuts. Again, this highlights the positive attributes of peer-to-peer power trading as a viable solution to the South African electricity distribution market. The potential for electricity cost reduction is proven by [43] and [59], with [63] exhibiting results as far as a 60% reduction in the cost of electricity because of power-sharing between power systems. Having successfully implemented peer-to-peer power-sharing within their market [53] describes the low market share of 10% held by German municipal electricity utilities in the power-sharing unregulated power system scheme. The challenge of unwillingness by large industrial power users to surrender control of their power systems to municipal utilities was expressed in [53]. Other challenges faced by power-sharing between power systems include data security, suitable user interfaces, the impact on the main power grid, market agreement designs, policy, and market regulation. Given the heavy dependence on electricity sales by South African municipal utilities, further research from this work can be undertaken that pivots on power-sharing scheme financial and operational welfare between the market participants.

12.6 CONCLUSIONS

The commonality in the literature is that peer-to-peer power trading has the potential to alleviate the supply vs demand imbalance and reliance on the main power grid by the distribution power systems. This is an imperative trait that could be employed to assist power systems with generation or transmission capacity shortages and constraints. Constrained optimization is the most utilized mechanism to enable peer-to-peer electricity trading. This can be attributed to the method's flexibility to enabling adherence to a wide variety of multidimensional parameters. Electricity distribution markets such as South Africa could benefit from peer-to-peer power trading strategies, more so given that some electricity distribution utilizes are located geographically in proximity. With numerous distributed generation initiatives currently enrolled, the potential benefit from funding of power-sharing initiatives is increased. From a technical and physical perspective, it can be concluded that power sharing between power systems has potential for positive social and economic benefits.

REFERENCES

1. Akpeji K.O., Awodele K.O., Folly K.A., Gaunt C.T., Olasoji O.A., Oyedokun D.T.O. (2020). Economic impact of electricity supply interruptions in South Africa.
2. South African Government (2021). The Integrated National Electrification Programme.
3. Meyer E.L., Overen O.K., (2021). Towards a sustainable rural electrification scheme in South Africa: analysis of the Status quo.
4. CSIR (2021). South Africa load shedding statistics
5. Eskom (2023). Eskom integrated report 2023.
6. Tuunanen, J., Honkapuro, S., & Partanen, J. (2016). Power-based distribution tariff structure: DSO's perspective.
7. Deutsche Gesellschaft für Internationale Zusammenarbeit (GIZ) GmbH (2017). The electricity distribution industry in Germany and South Africa - a review of policy and regulation.
8. City of Ekurhuleni (2021). Supply of electricity tariffs for the 2021/2022 financial year.
9. City of Tshwane (2021). Tariff Policies 2021/22 (Policy Number 21144c).
10. Ndlela N. W., Davidson I. E. (2023). Progress in the development of the southern african regional grid.
11. Mbanda N. L., Simiyu P., Wu Z., Davidson I. E. (2021). Tracking the progress of universal access to electricity in Africa: status quo, challenges and opportunities.
12. Hlongwane N. W., Daw O. D. (2022). Electricity consumption and population growth in South Africa: a panel approach.
13. Madoda N., Telukdarie A. (2021). Life extension of systems for coal-fired power plants.
14. Stemmet W.C., Atkinson-Hope G. (2010). HVAC/HVDC strategy for solving power delivery shortages to a localised area of a national grid.
15. Mzimba P. T., Langerman K. E., Calitz J. (2023). The cost implications of operating a coal-fired power plant in a cycling mode.
16. Buraimoh E., Adebiyi A. A., Ayamolowo O. J., Davidson I. E. (2020). South Africa electricity supply system: the past, present and the future.
17. Mulongo N. Y., Mushavhanamadi K. (2023). Emerging role of artificial intelligence in addressing the electricity crisis.
18. International Trade Administration (2021). This is a best prospect industry sector for this country includes a market overview and trade Data.
19. National Energy Regulator of South Africa (2019). The South African Grid Code: Distribution Tariff Code. Retrieved 3 24, 2021, from https://www.nersa.org.za/electricity-overview/electricity-grid-code/.
20. International Trade Administration (2022). This is a best prospect industry sector for this country.
21. StatSA (2017). Electricity: big business for municipalities.
22. South African Government (2011). National Development Plan 2030 our future-make it work.
23. South African Government (2019). Medium term strategic framework.
24. Wu Z., Zhou M., Wang J., Du E., Zhang N., Li G. (2020). Profit-sharing mechanism for aggregation of wind farms and concentrating solar power, in: IEEE Transactions on Sustainable Energy, Volume 11 (4) pp. 2606–2616.

25. Zhou Y., Wu J., Gan W. (2023). P2P energy trading via public power networks: practical challenges, emerging solutions, and the way Forward.
26. Valickova P., Elms N. (2021). Potential gains from regional integration to reduce costs of electricity supply and access in Southern Africa
27. Xie M., Lui S., Cai H., Jai Y. (2023). Optimization method for medium- and long-term electricity market trading of clean energy power plants.
28. Wang H, Xu Y. (2024). Green power trading strategies for power sales retailers considering demand side response in a multi-market environment.
29. Sambasivam B. Balachandra P. (2023). Demand response an effective solution for dynamic balancing of variable supply with variable demand: evidence from Indian electricity system.
30. Davidson I. E., Reddy R. (2019). Performance evaluation of solar roof-top PV on Eskom's LV electric power distribution networks.
31. Khan M. T., Thopil G. A., Lalk J. (2016). Review of proposals for practical power sector restructuring and reforms in a dynamic electricity supply industry.
32. Ji H., Jian J., Yu H., Ji J., Wei M., Zhang X., Li P., Yan J., Wang C. (2021). Peer-to-peer electricity trading of interconnected flexible distribution networks based on distributed ledger.
33. Liu X., Yan L., Shang W., Dang Q. (2021). Power business data sharing system based on blockchain and cryptography technology.
34. Chen L., Lui N., Li C., Zhang S., Yan X. (2021). Peer-to-peer energy sharing with dynamic network structures.
35. Dugan R.C., McGranaghan M.F., Santoso S., Beaty H.W. (2014). Electrical power systems quality.
36. Lui K., Yamada H., Iwatsuki K., Otsuji T., (2022). Study on the impact of power sharing between microgrids on the usage of renewable energy and system stability.
37. Lui N., Tan L., Sun., Zhou Z., Guo B. (2021). Bilevel heat–electricity energy sharing for integrated energy systems with energy hubs and prosumers.
38. Chen Y., Zhu Z., Lin H., Xu C., Sun Y., Wang Y. (2023). Shared energy storage optimization considering electricity price and PV output uncertainty.
39. Khan A. A., Niloy R. S., Mohammad N. (2024). Role of power system operator in initiating peer-to-peer energy trading in the distribution networks.
40. Kusakana K. (2019). Modelling an optimal peer-to-peer energy sharing between prosumers in a South African context.
41. Shukla K., Suryawanshi H.M., Talapur G.G. (2020). A modified load sharing technique for area electric power system.
42. Han Y., Li H., Shen P., Coelho E. A. A., Guerrero J.M. (2017). Review of active and reactive power sharing strategies in hierarchical controlled microgrids.
43. Wei J., Ye S., Han Y., Lui X., Li., Zhao D., Wang S., Ren W., Li C., Gao H., (2020). Optimal energy sharing model for multi-microgrids integrated into the distribution network.
44. Wang D., Lui B., Jia H., Zhang Z., Chen J., Huang D., (2020). Peer-to-peer electricity transaction decisions of the user-side smart energy system based on the SARSA reinforcement learning.
45. Scherer M., Anderson G., (2018). An outline of an active power reserve sharing process across continental Europe.
46. Piclo Energy (2023). Piclo mission statement.
47. Oikonomakou M., Antonopoulos A., Alonso L., Verikoukis C., (2019). Energy sharing and trading in multi-operator heterogeneous network deployments.
48. Di Lorenzoa G., Rotondoa S., Araneoa R., Petroneb G., Martiranoa L. (2021). Innovative power-sharing model for buildings and energy communities.

49. Mujeeb A., Hong X., Wang P. (2019). Analysis of peer-to-peer (P2P) electricity market and piclo's local matching trading platform in UK.
50. Gomes L., Morais H., Gonçalves C., Gomes E., Pereira L., Vale Z. (2022). Impact of forecasting models errors in a peer-to-peer energy sharing market.
51. Saatloo A. M., Mirzaei M. A., Mohammadi-Ivatloo B. (2023). A robust decentralized peer-to-peer energy trading in community of flexible microgrids.
52. Sonnen Community (2024). Sonnen - empowering your clean energy future, together.
53. Deutsche Gesellschaft für Internationale Zusammenarbeit (GIZ) GmbH (2018). New business models for municipalities in the electricity and energy sector German approaches.
54. Oliva F., Moretti M., Gallo E., Faranda R. (2023). Energy and economic comparison of virtual and physical power sharing.
55. Petri I., Alzahrani A., Reynolds J., Rezgui Y. (2021). Federating smart cluster energy grids for peer-to-peer energy sharing and trading.
56. Zahraoui Y., Korõtko T., Rosin A., Zidane T. E. K., Agabu H., (2024). A competitive framework for the participation of multi-microgrids in the community energy trading market: a case study.
57. Moncecchi M., Meneghello S., Merlo M. (2020). Energy sharing in renewable energy communities: the Italian case
58. Zheng B., Fan Y., Xu Y., Huang S., Xu Y., Mei M. (2021). Distribution optimal power flow with energy sharing via a peer-to-peer transactive market.
59. Maisch M. (2020). Blockchain P2P solar energy trading proves feasible and popular in world-first trial.
60. Sanayha M., Vateekul P. (2022). Model-based approach on multi-agent deep reinforcement learning with multiple clusters for peer-to-peer energy trading.
61. Li R., Li F., Zhang Z., Ahokangas P., (2018). A shared network access business model for distribution networks.
62. Federal Energy Regulatory Commission (2020). Energy Primer: A Hand Book to Energy Market Basics.
63. Tushar W., Yuen C., Mohsenian-Rad H., Saha T., Poor H. V., Wood L. K. (2018). Transforming energy networks via peer-to-peer energy trading: the potential of game-theoretic approaches.
64. Babu K. V. S. M., Vinay K. S. S., Chakraborty P. (2023). Peer-to-peer sharing of energy storage systems under net metering and time-of-use pricing.
65. Sampath L. P. M. I., Paudel A., Foo E. Y. S., Nguyen H. D., Gooi H. B. (2021). Peer-to-peer power trading with voltage and congestion management for distribution grids.
66. Gupta P., Swarnkar P. (2019). Intelligent power sharing control for hybrid system.
67. Vijay S., Umarikar A. C. (2023). Output impedance shaping schemes for enhanced power sharing amongst distributed energy resources in DC microgrids.
68. Jin Z., Wang., Lui F., Xin K., Lui Y. (2020). A voltage-droop-free reactive power sharing solution in microgrids.
69. Ashrafi M., Abbaspour A., Fotuhi-Firuzabad M. (2023). Topology-based peer-to-peer energy sharing or trading for active distribution network.
70. Esa M. N., Mohamad M. M., Mohtar M. S., Daud M. A., Ridwan M. I., Hamdan M. F. (2022). Towards realising value of internet of energy (IoE) application for peer-to-peer (P2P) energy trading: a PETRONAS case study.
71. Tushar W., Saha T.K., Yuen C., Smith D., Poor H.V. (2020). Peer-to-peer trading in electricity networks: an overview.
72. Zhu J., Li S., Borghetti A., Lan J., Li H., Guo T. (2023). Review of demand-side energy sharing and collective self-consumption schemes in future power systems.
73. Wu Z., Wang J., Zhong H., Gao F., Pu T., Tan C., Chen X., Li G., Zhao H., Xia Q. (2023). Sharing economy in local energy markets.

74. Tofighi-Milani M., Fattaheian-Dehkordi S., Saber H., Rehtanz C., Lehtonen M. (2024). A New Management method for reliable peer-to-peer energy sharing in power distribution systems.
75. Chen Y., Wei W., Li M., Chen L., Catalão J. P. S. (2021). Flexibility requirement when tracking renewable power fluctuation with peer-to-peer energy sharing.
76. Jogunola O., Wang W., Adebisi B. (2020). Prosumers matching and least-cost energy path optimisation for peer-to-peer energy trading
77. Mazzola L., Denzler A., Christen R. (2020). Towards a peer-to-peer energy market: an overview.
78. Yang S., Hu X., Wang H., Ligao J., Meng L., Zhou W., Zhou H. (2022). A prosumer-based energy sharing mechanism of active distribution network considering household energy storage.
79. Lui G., Lai S., Qui J., Tao Y., Zhao J. (2023). Credit-based Peer-to-Peer energy sharing mechanism under the distributed negotiation framework.
80. Tiwari A., Bablesh K. J., Pindoriya N. M. (2024). Multi-objective optimization based demand response program with network aware peer-to-peer energy sharing.
81. Chen Y., Zhao C., Low S.H., Wierman A. (2022). An energy sharing mechanism considering network constraints and market power limitation.
82. Cheng Y., Shaoqin H. (2021). Research on blockchain technology in cryptographic exploration.
83. Li D., Guo Q., Bai D., Zhang W. (2022). Research and implementation on the operation and transaction system based on blockchain technology for virtual power plant.
84. Luu Q. T. (2024). What is constrained optimization?
85. He X., Liang X., Wang H. (2021). Distributed neurodynamic algorithm for multi-objective problem optimization and its applications to isolated micro-grid energy management.
86. Kunwar R., Sapkota H. S. (2022). Linear programming, a linear algebra instrument for optimizing results through mathematical modeling.
87. Miller J. M., College W. (2007). An introduction to linear programming.
88. Módos I. (2023). Mixed Integer Linear Programming: formal definition and solution space.
89. Parikh N., Boyd S. (2014). Proximal algorithms.
90. Tibshirani R. (2017). Alternating direction method of multipliers.
91. Fourer R., Gay D. M., Kernigham B. W. (2003). Nonlinear programs.
92. Das S. R., Sundaram R. K. (1997). Auction Theory: a summary with Applications to Treasury Markets.
93. Levin J. (2004). Auction theory.
94. Hayes A. (2024). Game Theory: a comprehensive guide.
95. Mohanty D., Anand D., Aljahdali H. M., Villar S. G. (2022). Blockchain interoperability: towards a sustainable payment system.
96. National Energy Regulator of South Africa (2020). Consolidated report on the findings of the electricity distribution industry compliance audits conducted in 2019.
97. Cuenca J. J., Jamil E., Hayes B. (2021). State of the art in energy communities and sharing economy concepts in the electricity sector.

13 Modeling and Simulation of Three-Phase Induction Motor Under Faulty Conditions

Puja Pohakar, Ravi Gandhi, and Biswajeet Champaty

13.1 INTRODUCTION

Induction motors play an integral part in modern manufacturing industries. More than 90% of the machines in almost all industries use an induction motor as a prime mover, an essential driving unit of the machine. With around-the-clock operation, electrical machines are subjected to various stresses, leading to faults and sudden breakdowns. Every minute of production is critical, and every minute of downtime avoided is valuable to prevent supply chain disruption and maximize profitability. The need for intelligent functions to anticipate faults before they occur or cause significant harm to induction motors stems from the industry's increased focus on lowering downtime costs and improving production output. Recently, research has been carried out to develop accurate and reliable fault prediction methods for induction motors. In recent years, research has been carried out to develop accurate and reliable fault prediction methods for induction motors.

Industry uses induction motors frequently, and choosing the right model can save control and operating expenses. This chapter presents the procedure for determining the the self and mutual inductances of the stator and rotor, as well as the mutual inductances between them, as a function of the rotor rotation angle [3]. Hamidreza Heidari suggests modifying the linear induction motor's dynamic equivalent circuit model while taking the transverse edge effect and longitudinal end effect into account [4]. Over the years, motors have revolutionized the mining and automation industry. Processes such as hoisting conveyor belt systems for moving minerals, e.g., gold, coal, diamonds, etc., from underground and opencast mines depend largely on the utilization of induction motors [5]. Thus, for the reliable operation of these machines, proper protection needs to be implemented for safe operation under load conditions. Any malfunction of an induction motor can be described as an electrical fault, environmental factor, or mechanical breakdown. Rotor bearings could result in overheating, wear, and tear due to mechanical stresses [6]. Drawing enormous magnitudes of currents ensures high temperatures. Modeling an induction motor is somewhat complex, stemming from its non-linear behavior Triggered by electromagnetic exhaustion and the significant temperature influence from the synchronous motor

DOI: 10.1201/9781003537823-17

settings [7]. Furthermore, the shaft time constant of an induction motor can change due to rotor heat. Such characteristics render the mathematical modeling of induction motors somewhat insurmountable. The majority of researchers employ simple models that ignore the factors mentioned above. Production of these machines is imperative and requires urgency in reproduction during their idle state [8]. Chitra and Prabhakar [9] presented a simulation of an induction machine by utilizing the fuzzy logic approach. The authors applied the approach in their study to regulate the velocity of an induction machine to obtain the optimum torque with the least amount of loss. They used the field-oriented control approach to create a fuzzy logic controller that enables improved control of motor torque with greatly variable performance. Their simulated design was evaluated by utilizing multiple MATLAB toolboxes. They observed that the induction motor's efficiency increased in stable conditions. The results show that the suggested speed regulator was efficient and reliable. Elnaghi et al. [10] proposed using a genetic algorithm (GA) to process experimental loads on an inductive machine. The principle of predicting motor parameters from testing data was demonstrated using a genetic algorithm-based technique. The specifications were determined using typical no-load and blocked rotor experiments. The cost equation—the graded sum of the stator currents and rotor velocity—was studied and improved for various motor parameter values. The impact of differential equations on the estimates was also shown. The estimated speed and torque parameters from the mathematical equation were compared to the experimental findings, and both exhibited a strong connection, proving the validity of the mathematical equation and the genetic algorithm method for improvement. Sadasivan and Mammen [11] applied the same algorithm to obtain parameters that linked the proposed technique and the loading of the electric motor using the evaluation function. They used the genetic algorithm on three separate situations of simulated loading and found that the outcomes were superior in terms of the overall losses induced by the motor. The authors' technique proved to be effective in terms of parameter estimation. Jirdehi and Rezaei [12] presented a simulation of an induction motor by utilising an artificial neural network (ANN) and an adaptive neuro-fuzzy inference system (ANFIS) to investigate variables that are often difficult to obtain. They used both methods to test 20 induction motors of varying power outputs. The experimental results consisted of the starting torque, current, maximum torque, full-load slip, efficiency, rated active power, and reactive power. The authors compared the findings produced by the proposed ANN and ANFIS models and the practical results. They discovered a good relationship between the projected values and the practical data. However, the proposed ANFIS model was more precise than the proposed ANN model. Keerthipala et al. [13] explained the ANN algorithm and how it may be used to monitor an induction motor's torque and speed regulation using linear and non-linear models. The Energies 2022, 15, 3564 3 of 21 authors reported that the linear observer approach is simple to apply in realtime; however, it does not accurately estimate the rotor and vector angle since the induction machine generally works in the saturated region. The non-linear observer approach considers the impact of the magnetic saturation of the induction machine; however, it cannot be practically applied using conventional techniques because estimating the angle requires too much time. Their suggested technique compensates for the effect of saturation and estimates the angle in a few milliseconds, which is well within the real-time limit. To account for the longitudinal end effect and the transverse edge effect, several coefficients were added to the secondary resistance and magnetic inductance in [14].

Duncan's model is widely used for the design of variable-speed drive systems because it considers both the dynamic and steady-state performance of LIMs. In this regard, the d-q equivalent circuit model and space-vector model were presented in [15–17] respectively. A modified steady-state Duncan's model was developed in [18] to cover special phenomena such as the transverse edge effect. All aforementioned DAMs only take into account the dynamic end effect, which is related to linear speed. The dynamic d-q and steady-state equivalent circuit models with consideration of both dynamic and static end effect (or speed-independent end effect) have been investigated in [18–20] respectively.

Modeling of motors is crucial for many objectives, including life-cycle analysis, performance analysis, and more importantly, control purposes [21–23]. Thus, there have been several recent literature surveys on the dynamic modelling of induction motors that highlight the importance of accurate mathematical models for motor design, control, and maintenance. These models can consider various parameters that affect the dynamic behavior of the motor and can be used to develop effective control strategies and predictive maintenance techniques.

The design, analysis, and implementation of wireless shaded-pole induction motors were covered in [24]. Modeling and simulation of a three-phase electric traction induction motor using MATLAB/Simulink were presented in [25]. Currents in bearings of induction motors of electric drives with semiconductor converters were investigated in [26]. An improvement in the accuracy of determining the parameters of induction motors in transient starting modes was proposed in [27]. Research on intelligent power supply control based on sensor-less temperature identification of electric submersible motors was presented [28]. A fuzzy logic-based speed control system for three-phase induction motors was proposed in [29]. Load testing of induction motors based on the PWM technique using genetic algorithms was described in [30]. Enhancing grid reliability and renewable integration through AI-based predictive maintenance was discussed in [31]. Induction motor bearing fault classification using extreme learning machines based on power features was presented in [32]. An integrated approach of adaptive neuro-fuzzy inference system and dimension theory for the diagnosis of rolling element bearings was described in [33]. Fault diagnosis in industrial rotating equipment based on permutation entropy, signal processing, and a multi-output neuro-fuzzy classifier was explored in [34]. The application of an adaptive neuro-fuzzy controller for switched reluctance motors (SRMs) was studied in [35]. Condition monitoring using machine learning, including a review of theory, applications, and recent advances, was covered in [36]. A Takagi-Sugeno fuzzy inference system for developing a sustainability index of biomass was proposed in [37].

In this chapter, mathematical modeling of induction motors is incorporated subjected to healthy and faulty conditions. Section 2 highlights the problem statement and objective. The different types of methodologies have been elaborated in section 3. The equivalent circuit of three-phase induction motor along with the equation for voltage, flux linkage, current, torque, and speed, is elaborated in section 4. Section 4 also represents the mathematical model along with the parameters under consideration. Section 5 gives an overview of the performance characteristics of the phase Induction motor under healthy conditions. It reflects different fault conditions and the performance of Induction motors under dynamic conditions. Finally, the conclusion is presented in section 6.

13.2 PROBLEM STATEMENT AND OBJECTIVE

This chapter aims to develop models for a three-phase induction machine to meet demanding performance standards and rectify potential errors. The differential equations for the dynamic model are derived using the dq-0 direct axis method and executed within the MATLAB/Simulink environment. The behavior of these models is assessed in steady-state conditions. Static and dynamic motor models are represented mathematically to eliminate human error, ensure high performance, and create efficient induction motors [1, 2]. To overcome the problems mentioned above, this work aims at the following objectives:

- To comprehensively implement modeling of the In3-Φ motor, which could be used to determine the starting and working characteristics of the motor to observe the performance of the 3-Φ induction motor along with its characteristics.
- To simulate a flexible model of a 3-Φ induction motor in MATLAB/Simulink, incorporating dq-0 axis frame as a reference modeling methodology for the machine.
- The differential equations for the dynamic model were initially generated and put into practice in MATLAB/Simulink using the direct dq-0 direct axis approach to evaluate the performance characteristics in the steady state.

13.3 MODEL-BASED FAULT DETECTION

Fault detection in induction motors can be approached using various methods, each with its advantages and applications. Different methodologies that can be implemented are model-based, soft computing, and signal processing techniques (Figure 13.1). The model-based fault prediction technique is accurate and well-suited

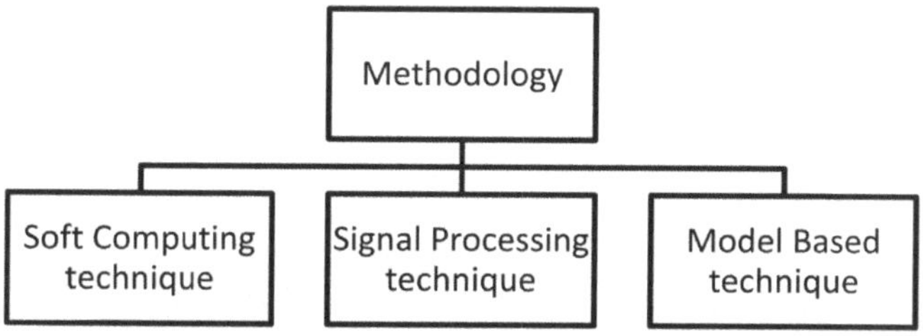

FIGURE 13.1 Different methodology.

for complex systems where the behavior can be accurately modeled. Moreover, it has limited applicability in cases where the system dynamics are not well understood or difficult to model accurately [2]. Signal processing methods analyze the electrical or mechanical signals from the motor to identify patterns associated with faults. Techniques like Fourier analysis, motor current signature analysis, sequence components, parks transform, wavelet transform, and envelope analysis are commonly used. These techniques can be applied directly to sensor data without the need for complex mathematical models and are effective in identifying specific frequency components associated with faults. These are limited to detecting faults that manifest as distinct patterns in the acquired signals. Performance can be affected by noise and variations in operating conditions. Soft computing techniques, including ANN, FIS, and GA, use computational models inspired by biological processes to learn and make decisions based on input data. These techniques can handle complex and nonlinear relationships in the data. Adaptability and self-learning capabilities make them suitable for real-time applications.

Modeling of motors plays an important role in motor design and control, as it provides parameters and characteristics under different operating conditions. The model, particularly the dynamic induction motor, accurately predicts the motor's behavior under varied operations. Modeling of 3-Φ motors is useful for enhancing comprehension of motor behavior, optimizing performance, and developing more efficient systems. As a result, it holds significant importance across various industries such as manufacturing, transportation, and energy. The performance of induction motors under different conditions is analyzed, simulated, and predicted using these mathematical models. Assessments of motor function, efficiency, and reaction to various control schemes can be aided by simulations. MATLAB/Simulink is used to carry out digital simulation, which involves creating models for both ideal and imperfect operating circumstances.

A dependable failure prediction technique can only be achieved using accurate system models. On the other hand, real-time monitoring is possible with models, enabling ongoing evaluation of motor condition. When it is developed, mathematical modeling can be utilized across any motors in an industrial setting, providing a solution for fault prediction. Organizations can benefit from the historical study of the data produced by mathematical model training, which can be used to identify patterns and trends in motor behavior and malfunctions. All things considered, developing a mathematical model to anticipate induction motor defects might be a wise investment for sectors that significantly rely on motor-driven machinery, providing increased operational reliability, lower costs, and increased efficiency. Indirectly, simulating and modeling the performance of electrical machines in both ideal and faulty scenarios would be more helpful.

The dynamic model of a motor provides a comprehensive insight into its behavior, encompassing both transient and steady-state analysis. The main aim of this research is to establish the mathematical relationship between the parameters of stator and rotor variable models, facilitating a deeper understanding of the machine's operation. These parameters also streamline calculations and representations. The chapter offers a generalized modeling approach for induction machines using MATLAB/Simulink [1].

Initially, we convert 3-phase (3-Φ) vectors to 2-phase (2-Φ) vectors in the stationary reference frame (α-β). Subsequently, these stationary 2-phase vectors (α-β) are transformed into rotating 2-phase vectors (d-q-0). Two windings are located at 90 degrees. We thus get flux and torque, which are coupled. With the change in the current id, the flux changes. Similarly, with a change in the current iq, torque changes and not the flux. The winding has resistance and inductance, as shown in the circuit equivalent to the induction motor [1].

The stator dynamics are represented by equation (13.1)

The equivalent circuit for a 3-Φ motor in the d and q-variables is shown in Figure 13.2. This chapter systematically examines the various components and parameters that constitute the equivalent circuit, encompassing the stator and rotor windings, core losses, and magnetizing inductance, among others. Through meticulous analysis and experimentation, we elucidate the intricacies of the equivalent circuit and its ability to accurately represent the induction motor's dynamic behavior. Furthermore, our research explores the practical implications of this modeling approach, showcasing its utility in predicting motor performance, assessing efficiency, and facilitating the design of effective control strategies [1].

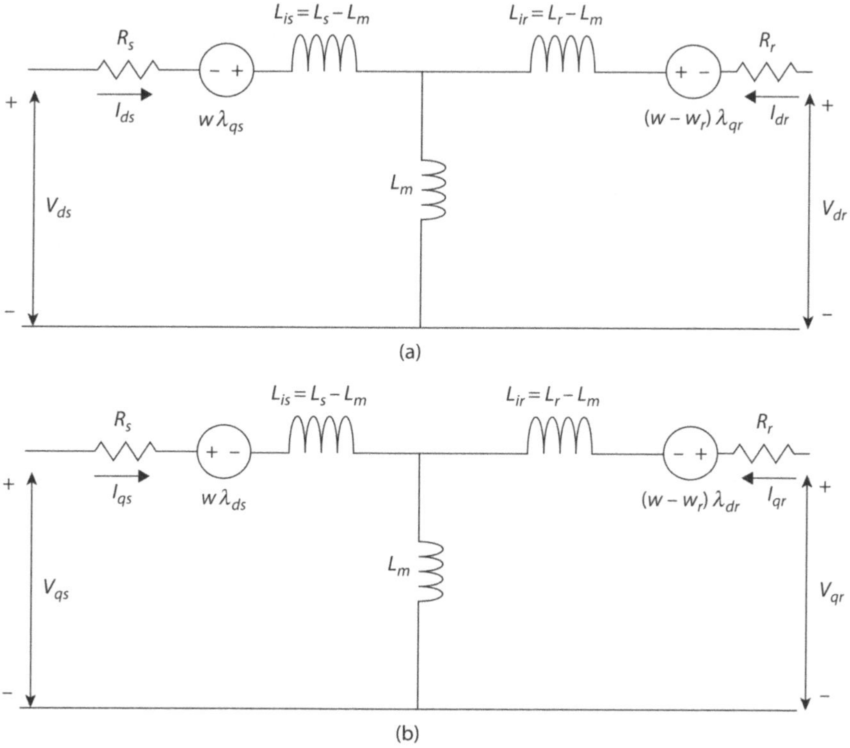

FIGURE 13.2 (a) d-Axis equivalent circuit of induction motor (b) q-axis equivalent circuit of induction.

In the following equations, the subscript "s" represents variables and parameters linked with the stator circuits, while the subscript "r" denotes variables and parameters associated with the rotor circuits [31].

$$v_{qs} = r_s i_{qs} + \omega\lambda_{ds} + \frac{\partial}{\partial t}\lambda_{qs} \tag{13.1}$$

$$v_{ds} = r_s i_{ds} + \omega\lambda_{qs} + \frac{\partial}{\partial t}\lambda_{ds} \tag{13.2}$$

$$v_{os} = r_s i_{os} + \frac{\partial}{\partial t}\lambda_{os} \tag{13.3}$$

$$v_{qr} = r_r i_{qr} + (\omega - \omega_r)\lambda_{dr} + \frac{\partial}{\partial t}\lambda_{qr} \tag{13.4}$$

$$v_{dr} = r_r i_{dr} + (\omega - \omega_r)\lambda_{qr} + \frac{\partial}{\partial t}\lambda_{dr} \tag{13.5}$$

$$v_{or} = r_r i_{or} + \frac{\partial}{\partial t}\lambda_{or} \tag{13.6}$$

where

V_{qs}, V_{ds}	stator voltage(s) for the q and d axes of the
V_{qr}, V_{dr}	q and d axes of the rotor voltages
λ_{mq}, λmd	q and d axes magnetizing the flux linkages
R_S	stator resistance
R_r	rotor resistance
X_{ls}	stator leakage reactance
X_{lr}	rotor leakage reactance
P	number of poles
T_{em}	magnetic torque
Ω	arbitrary speed
λ_{qs}	flux linkage of q-axis
$(\omega - \omega_r)$	relative Speed
$\omega\lambda_{ds}$	sped emf due to rotation of the axis
λ_{qs}	stator winding flux linkage

Here,

$$\lambda_{qs} = \left(i_{qs}L_s + L_{m\,i_{qr}}\right) \tag{13.7}$$

$$\lambda_{ds} = \left(i_{ds}L_s + L_{m\,i_{dr}}\right) \tag{13.8}$$

$$\lambda_{qr} = \left(i_{qr} L_r + L_{m\, i_{qs}}\right) \tag{13.9}$$

$$\lambda_{dr} = \left(i_{dr} L_r + L_{m\, i_{ds}}\right) \tag{13.10}$$

Using equations (13.7) and (13.8) in (13.1), we can derive equations from equations (13.11)–(13.14)

$$i_{qs} = \int \frac{1}{L_s}\left(v_{qs} - r_s i_{qs} - L_m \frac{\partial}{\partial t}\left(i_{qr}\right) - \omega L_s\, i_{ds} - \omega L_m\, i_{dr}\right) \tag{13.11}$$

$$i_{ds} = \int \frac{1}{L_s}\left(v_{ds} - r_s i_{ds} - L_m \frac{\partial}{\partial t}\left(i_{dr}\right) - \omega L_s\, i_{qs} - \omega L_m\, i_{qr}\right) \tag{13.12}$$

$$i_{qr} = \int \frac{1}{L_r}\left(v_{qr} - r_r i_{qr} - L_m \frac{\partial}{\partial t}\left(i_{qs}\right) - \omega L_r\, i_{dr} - \omega L_m\, ids\right) \tag{13.13}$$

$$i_{dr} = \int \frac{1}{L_r}\left(v_{qr} - r_r i_{dr} - L_m \frac{\partial}{\partial t}\left(i_{ds}\right) - \omega L_r\, i_{qr} - \omega L_m\, iqs\right) \tag{13.14}$$

Torque and speed equations are as shown below:

$$T = \frac{3}{2}(P/2)\left(L_m\left(i_s i_{dr} - i_{ds} i_{gr}\right)\right) \tag{13.15}$$

$$\omega_r = \int \left(T_e - T_l\right)(P/2) \tag{13.16}$$

13.4 SIMULATION RESULTS

This section represents the implementation of the dynamic model of the induction motor in MATLAB/Simulink

The dynamic circuit of a 3-Φ induction motor describes its behavior over time, taking into account various electrical and mechanical factors. Figure 13.3 illustrates the dynamic model implementation for a 3-Φ induction motor. The dynamics of stator like supply voltage, current, torque, and speed are represented by Figure 13.3a, b, c, and d, respectively. the torque produced by the motor, which depends on factors like the current flowing through the stator windings, the rotor speed, and the motor's parameters such as the rotor resistance and reactance. The relationship between the motor's speed and the torque produced is represented by Figure 13.3c and d, respectively. It typically incorporates factors like the motor's mechanical load and inertia, as well as the electromagnetic torque generated within the motor.

By analyzing these equations and waveforms, we can understand how the motor responds to changes in voltage, current, load, and other operating conditions. This

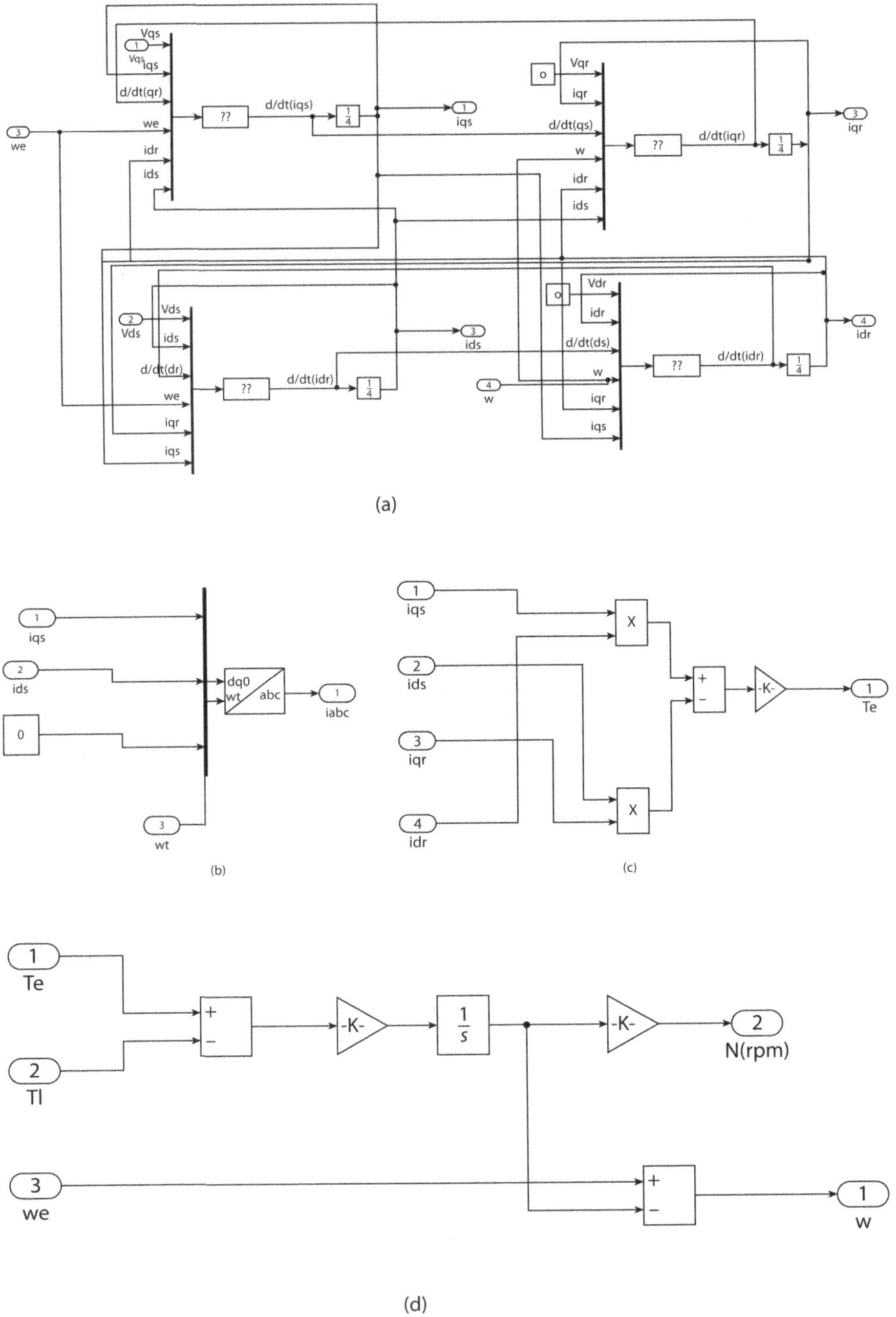

FIGURE 13.3 Mathematical model represented in MATLAB/Simulink of induction motor. (a) Sub-system for motor, (b) current equation, (c) torque equation, (d) speed equation of induction motor.

TABLE 13.1
Induction Motor Parameter [1]

Sr. No	Parameter	Value	Unit
1	P_{rated}	7.5	KW
2	f	60	HZ
3	V_m	220	V
4	J	0.4	kgm^2
5	r_s	288e–3	Ω
6	r_r	158e–3	Ω
7	L_s	42.5e–3	H
8	L_r	41.8e–3	H
9	L_m	41.2e–3	H
10	T_{rated}	40	Nm
11	P	4	-

understanding is crucial for designing motor control systems, optimizing motor performance, and predicting motor behavior in different scenarios.

The simulation parameters used for the 3-Φ induction motor are presented in Table 13.1.

13.4.1 Case I: No Load Condition

The simulation was carried out to understand whether the induction motor operational equations can be utilized for fault detection. When an induction motor is operated under no-load conditions, it exhibits certain performance characteristics. Figure 13.4a indicates the voltage supply under no load, and Figure 13.4b represents three phase currents, which is very high at the start.

After the transient condition, the current drawn is a very small amount of no-load current to generate the required magnetic field. The motor develops a small amount of no-load torque, as shown in Figure 13.4c, which is necessary to counteract friction and windage losses to initiate the rotation of the motor. Speed under no-load: To meet the increased demand for torque output, the induction motor generates higher torque, but there is a temporary dip in speed.

13.4.2 Case II: Half Load Condition

When a 3-Φ motor is connected to a load that requires half of its rated power (20 N.m), it exhibits certain performance characteristics [31].

Figure 13.5a indicates the current drawn by the motor. It can be observed that the current drawn at 50% loading is increased compared to the no-load condition. The motor develops a significant amount of torque to drive the load. The motor runs at a speed slightly below the synchronous speed, as depicted in Figure 13.5b.

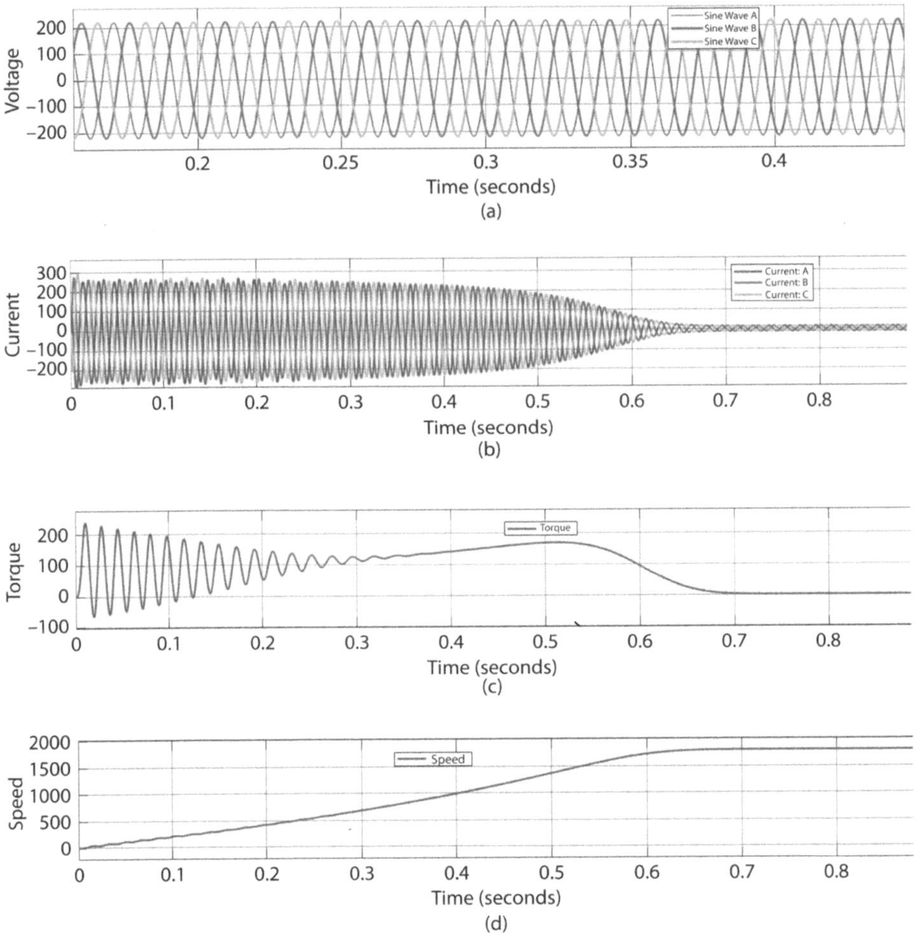

FIGURE 13.4 (a) Voltage supply under no-load, (b) line current under no-load, (c) electromagnetic torque under no-load, (d) speed under no-load.

13.4.3 CASE III: Full-Load Condition

The performance characteristics at 100% loading (40 N.m) are as follows.

The induction motor draws its rated current. The torque and speed characteristics are shown in Figure 13.6b and c. [31]

The induction motor is subjected to varied operating conditions like variation in the mechanical load, instability in the supply voltage, phase shift, and grounding of phase. The induction motor gets subjected to varied operating conditions like variation in the mechanical load, instability in the supply voltage, phase shift, and grounding of phase.

Table 13.2 displays the operational characteristics of the induction motor across different load scenarios.

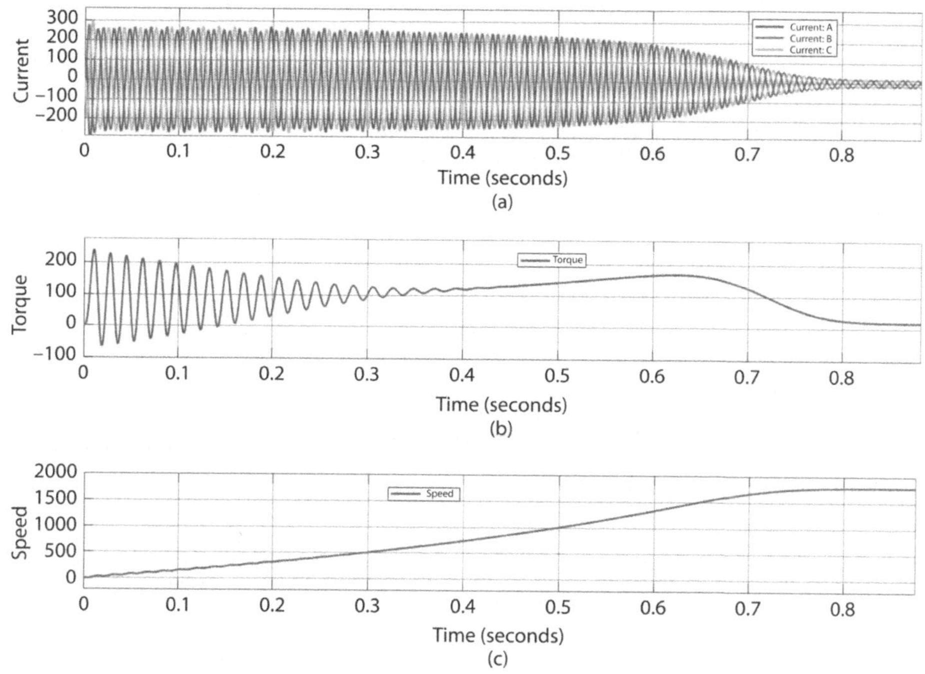

FIGURE 13.5 (a) Current at 20 N.m, (b) torque at 20 Nm, (c) speed at 20 N.m.

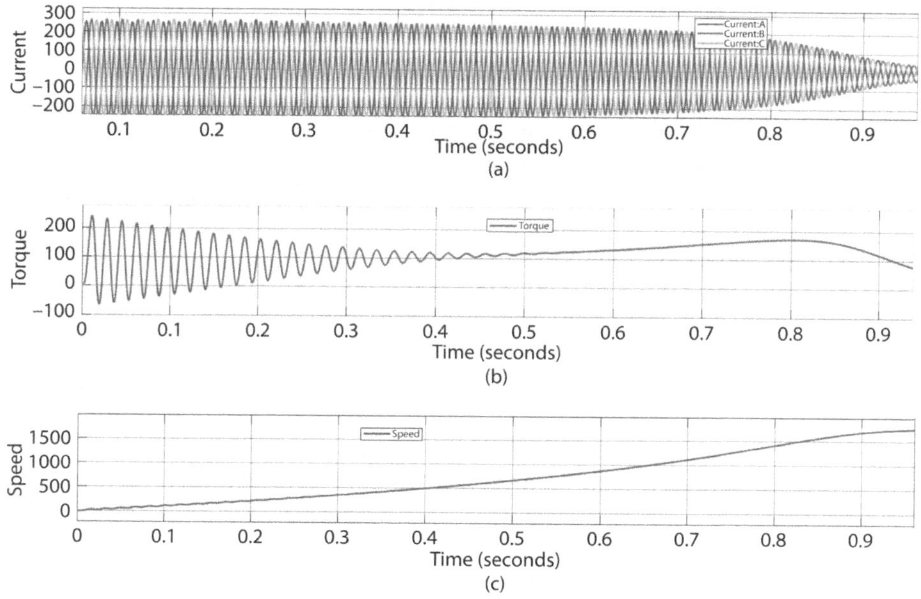

FIGURE 13.6 (a) Current at 40 N.m, (b) torque at 40 N.m, (c) speed at 40 N.m.

TABLE 13.2
Performance Parameter of Induction Motor

Torque Load (%)	Current (A)	P_{in} (Watt)	P_{out} (Watt)	Q_{in} (Var)	Efficiency (%)	Speed (rpm)	T_{em} (N-m)
0	9.705	81.42	1.14×10^{-7}	4530	0	1800	6.05×10^{-10}
10	9.826	837.4	752.7	4509	89.88	1797	4
20	10.23	1598	1503	4499	94.01	1794	8
30	10.9	2365	2250	4502	94.11	1790	12
40	11.78	3136	2994	4517	94.52	1787	16
50	12.85	3913	3736	4544	94.97	1784	20
60	14.06	4695	4475	4584	95.21	1780	24
70	15.39	5482	5210	4638	95.32	1777	28
80	16.81	6276	5943	4705	95.49	1773	32
90	18.3	7075	6672	4706	95.49	1770	36
100	19.86	7881	7399	4881	95.31	1766	40
110	21.48	8692	8122	4990	95.04	1763	44
120	23.14	9510	8841	5115	94.69	1759	48
125	23.99	9922	9199	5184	94.54	1757	50

The behavior or operational aspects of an induction motor can be described by various parameters under various dynamic conditions. The loading condition is often expressed as a percentage of the motor's rated load which is 40 N.m.

Here's an overview of how an induction motor typically behaves at different loading conditions:

- Under no-load conditions, the motor's speed approaches the synchronous speed, and the current aligns with the no-load current. At light load (0%–25% loading), speed is slightly less than synchronous speed, and current is around 50% less than the rated current.
- At half load (50% loading), the motor runs moderately less than the synchronous speed. Current increases and efficiency are lower than at the full load.
- At rated load (100% loading), the speed is below the synchronous speed. Rated current with reasonably good power factor. Efficiency is typically around 85%–95%.
- At overload (greater than 100%), further reduction in speed due to increased slip. The current continues to increase beyond the rated load. The power factor begins to decrease, leading to lower efficiency. The motor will start to overheat, potentially causing damage if sustained.
- At 125% loading, further decrease in speed with higher slip. The current exceeds the rated current significantly. Drop in the efficiency and motor approaches a state of thermal overloading.

13.5 CONCLUSION

It is observed from the simulation that in a healthy condition, an induction motor exhibits specific current, speed, and torque characteristics.

- Under normal operating conditions (e.g., no load), the current flowing through the stator winding of an induction motor follows an approximately sinusoidal waveform when connected to a sinusoidal power supply. The actual speed may be slightly lower than synchronous speed. The torque produced by the motor under no-load conditions is very low. It's often referred to as the "no-load torque" or "starting torque," and it's just enough to overcome the friction and windage losses in the motor and drive the unloaded rotor.
- At 50% loading condition, the motor will operate at a speed slightly below its synchronous speed (Ns) because of the load increment. The motor will produce a torque equal to 50% of its rated torque. The stator current will increase compared to the no-load condition but will be less than the full-load current.
- At 100% load torque, the operating speed will be lower than the synchronous Speed, but the motor should maintain a relatively constant speed under a steady load. The motor will produce its rated torque under these conditions, providing the maximum mechanical power output as per the design. The stator current will be at its highest level under full-load conditions due to the motor torque requirement being higher.
- Thus, the change in the operating conditions due to the increment in the % load has been verified by the model-based Simulation of the induction motor.

REFERENCES

1. Le Roux, P.F.; Ngwenyama, M.K. 2022, "Static and dynamic simulation of an induction motor using Matlab/Simulink," Energies 15, no. 10: 3564.
2. Chen, J.; Fujii, Y.; Johnson, M.W.; Farhan, A.; Severson, E.L. 2021, "Optimal design of the bearingless induction motor," IEEE Trans. Ind. Appl. 57, no. 2: 1375–1388.
3. Wcislik, M.; Suchenia, K.; Cyganik, A. 2020, "Vectorized mathematical model of a slip-ring induction motor," Energies 13, no. 15: 4015.
4. Heidari, H.; Rassõlkin, A.; Razzaghi, A.; Vaimann, T.; Kallaste, A.; Andriushchenko, E.; Belahcen, A.; Lukichev, D.V. 2021, "A modified dynamic model of single-sided linear induction motors considering longitudinal and transversal effects," Electronics 10, no. 8: 933.
5. Rahaman, N. 2014, Modelling & simulation of a three-phase electric traction induction motor using Matlab Simulink.
6. Vostrov, K.; Pyrhönen, J.; Ahola, J.. 2020. Shielding the end windings to reduce bearing currents, in: 2020 International Conference on Electrical Machines (ICEM). IEEE Press, 1431–1437.

7. Maleki, M.G.; Chabanloo, R.M.; Farrokhifar, M. 2019, "Accurate coordination method based on the dynamic model of overcurrent relay for industrial power networks taking the contribution of induction motors into account," IET Gener. Transm. Distrib 14: 645–655.
8. Abdelwanis, M.I.; Selim, F.; El-Sehiemy, R. 2015, "An efficient sensorless slip dependent thermal motor protection schemes applied to submersible pumps," Int. J. Eng. Res. Afr 14: 75–86.
9. Chitra, V.; Prabhakar, R. 2006, "Induction motor speed control using fuzzy logic controller," World Acad. Sci. Eng. Technol 23: 17–22.
10. Elnaghi, B.E.; Mohammed, R.H.; Dessouky, S.S.; Shehata, M.K. 2019, "Load test of induction motors based on PWM technique using genetic algorithm," Int. J. Eng. Manuf 9: 1.
11. Sadasivan, J.; Mammen, O. 2011, "Genetic algorithm based parameter identification of three phase induction motors," Int. J. Comput. Appl 31: 51–56.
12. Jirdehi, M.A.; Rezaei, A. 2016, "Parameters estimation of squirrel-cage induction motors using ANN and ANFIS," Alex. Eng. J 55: 357–368.
13. Keerthipala, W.; Duggal, B.; Chun, M.H. 1998 "Torque and speed control of induction motors using ANN observers, in: Int. Conf. Power Electron. Drives Energy Syst. Ind. Growth Proc., 1, 282–288.
14. Duncan, J.1983, "Linear induction motor-equivalent-circuit model," IEE Proc. B Electr. Power Appl 130: 51–57.
15. Kang, G.; Nam, K. 2005, "Field-oriented control scheme for linear induction motor with the end effect," IEE Proc. Electr. Power Appl, 152: 1565–1572.
16. Mirsalim, M.; Doroudi, A.; Moghani, J. 2002, "Obtaining the operating characteristics of linear induction motors: a new approach," IEEE Trans. Magn 38: 1365–1370.
17. Kim, D.K.; Kwon, B.I. 2006, "A novel equivalent circuit model of linear induction motor based on finite element analysis and its coupling with external circuits," IEEE Trans. Magn 42: 3407–3409.
18. Amiri, E.; Mendrela, E. 2014, "A novel equivalent circuit model of linear induction motors considering static and dynamic end effects," IEEE Trans. Magn 50: 120–128.
19. Heidari, H.; Rassõlkin, A.; Kallaste, A.; Vaimann, T.; Andriushchenko, E.; Belahcen, A.; Lukichev, D.V. 2021, "A review of synchronous reluctance motor-drive advancements," Sustainability 13: 729.
20. Heidari, H.; Rassõlkin, A.; Vaimann, T.; Kallaste, A.; Taheri, A.; Holakooie, M.H.; Belahcen, A. 2019, "A novel vector control strategy for a six-phase induction motor with low torque ripples and harmonic currents," Energies 12: 1102.
21. Rajput, S.; Bender, E.; Averbukh, M. 2020, "Simplified algorithm for assessment equivalent circuit parameters of induction motors," IET Electr. Power Appl 14: 426–432.
22. Ahmed, M.; Vahidnia, A.; Datta, M.; Meegahapola, L. 2020, "An adaptive power oscillation damping controller for a hybrid AC/DC microgrid," IEEE Access 8: 69482–69495.
23. Iegorov, O.; Iegorova, O.; Miroshnyk, O.; Savchenko, O. 2020, "Improving the accuracy of determining the parameters of induction motors in transient starting modes," Energetika 66: 15–23.
24. Wang, H.; Chau, K.; Lee, C.H.; Cao, L.; Lam, W.H. 2020 "Design, analysis, and implementation of wireless shaded-Pole induction motors," IEEE Trans. Ind. Electron 68: 6493–6503.
25. Rahaman, N.; Govindraju, H. 2014, "Modeling and simulation of a three-phase electric traction induction motor using MATLAB/Simulink," Int. J. Electr. Electron. Comput. Syst 2: 5.

26. Kozjaruk, A.E.; Vasilev, B.U.; Shtop, S.A.; Serdukov, N.A., 2018, "Currents in bearings of induction motors of electric drives with semiconductor converter," in: Proceedings of the 17th International Ural Conference on AC Electric Drives (ACED), Ekaterinburg, Russia, 26–30; pp. 1–5.
27. Iegorov, O.; Iegorova, O.; Miroshnyk, O.; Savchenko, O., 2020, "Improving the accuracy of determining the parameters of induction motors in transient starting modes," Energetika, 66: 15–23.
28. Zhang, X.; Li, H.; Liu, J., 2015, "Research on intelligent power supply control based on sensor-less temperature identification of electric submersible motor," J. Pet. Sci. Eng., 133: 102–110.
29. Mannan, M.A.; Sulaiman, M.A.; Nordin, N.A.; Ali, M.K.; Rahman, A.; Mohamed, A., 2012, "Fuzzy logic based speed control system for three-phase induction motor," Int. J. Comput. Appl., 44, no. 16: 41–46.
30. Elnaghi, B.E.; Mohammed, R.H.; Dessouky, S.S.; Shehata, M.K., 2019 "Load test of induction motors based on PWM technique using genetic algorithm," Int. J. Eng. Manuf., 9.
31. Pohakar, P.Y.; Gandhi, R.; Champaty, B. 2024. Enhancing grid reliability and renewable integration through AI-based predictive maintenance, in: AI Approaches to Smart and Sustainable Power Systems (pp. 47–62). IGI Global.
32. Sikder, N.; Mohammad Arif, A.S.; Islam, M.M.; Nahid, A.A. 2021, "Induction motor bearing fault classification using extreme learning machine based on power features," Arab. J. Sci. Eng. 46, no. 9: 8475–8491.
33. Kumbhar, S.G.; Sudhagar P, E., 2020, "An integrated approach of adaptive neuro-fuzzy inference system and dimension theory for diagnosis of rolling element bearing", Measurement, 166.
34. Rajabi, S.; Azari, M.S.; Santini, S.; Flammini, F., 2022, "Fault diagnosis in industrial rotating equipment based on permutation entropy, signal processing and multi-output neuro-fuzzy classifier", Expert Syst. Appl., 206: 117754, ISSN 0957-4174,
35. Ali Akcayol, M., 2004, "Application of adaptive neuro-fuzzy controller for SRM, advances in engineering software", Appl. Adapt. Neuro-Fuzzy Controller SRM, 35, no. 3–4: 129–137.
36. Surucu, O.; Gadsden, S.A.; Yawney, J., (2023) "Condition monitoring using machine learning: a review of theory, applications, and recent advances, expert systems with applications", 221:119738.
37. Cavallaro, F. 2015, "A Takagi-Sugeno fuzzy inference system for developing a sustainability index of biomass," Sustainability 7, no. 9: 12359–12371.

Section V

Conclusions

14 Summary of the Contributions

The chapters in this book are the result of several years of intensive research carried out by mostly experienced national and international teams from various affiliations across the world. The sections of the book were organised based on the thematic alignment of the proposed topics. The book presents novel concepts and innovative applications in energy systems in the context of the fourth industrial revolution and the energy transition. The impact of 4IR technologies on the design of electric vehicle charging stations and on energy forecasting, which are vital for sustainable development, was a key focus of the contributions. The scientific contributions stemming from the book sections can be summarised as follows:

- **Section 1:** The knowledge produced in this section is essentially focussed on technology-enhanced energy efficiency and output voltage regulation for alternative energy systems. It also includes a case study on optimising cooperative communication by using energy harvesting tags in the context of limited energy resource availability.
- **Section 2:** highlights the latest advancements in off-grid and hybrid charging stations for electric vehicles and explores innovative design concepts of converters that serve as the foundation of these charging stations. Additionally, it proposes the steady-state and dynamic operational modelling and analysis of converters used in DC microgrids.
- **Section 3:** determines the essential role of intelligent technologies in enhancing the reliability and efficiency of energy systems. It utilises artificial intelligence and machine learning algorithms to predict the balance between supply and demand, ensuring sustainable energy consumption. This knowledge is further applied to real-time monitoring, enabling better management and performance optimisation of energy infrastructure.
- **Section 4:** proposes effective stability and control technologies of distribution systems in the context of load fluctuations and energy exchange or trade. It also includes a case study of modelling the performance of a three-phase induction motor under fault condition conditions.
- **Section 5:** provides an account of summarised contributions of the book sections.

These contributions indicate that achieving energy sustainability is not possible unless the full integration of intelligent technologies into the energy ecosystem.

DOI: 10.1201/9781003537823-19

Index

Note: **Bold** an *italic* page numbers refer to **tables** and *figures*.

K

L

M

N

O

P